Marine Phenolics

Marine Phenolics: Extraction and Purification, Identification, Characterization and Applications

Editors

Herminia Domínguez
José Ricardo Pérez-Correa

MDPI • Basel • Beijing • Wuhan • Barcelona • Belgrade • Manchester • Tokyo • Cluj • Tianjin

Editors

Herminia Domínguez
University of Vigo
Spain

José Ricardo Pérez-Correa
Pontificia Universidad Católica de Chile
Chile

Editorial Office
MDPI
St. Alban-Anlage 66
4052 Basel, Switzerland

This is a reprint of articles from the Special Issue published online in the open access journal *Marine Drugs* (ISSN 1660-3397) (available at: https://www.mdpi.com/journal/marinedrugs/special_issues/marinephenolic2020).

For citation purposes, cite each article independently as indicated on the article page online and as indicated below:

LastName, A.A.; LastName, B.B.; LastName, C.C. Article Title. *Journal Name* **Year**, *Volume Number*, Page Range.

ISBN 978-3-0365-0264-9 (Hbk)
ISBN 978-3-0365-0265-6 (PDF)

Contents

About the Editors

Herminia Domínguez (Professor). Ph.D. degree in Chemistry from the University of Santiago de Compostela. She is Professor of Chemical Engineering at University of Vigo (Spain), where she has been working on the extraction of bioactive compounds from underutilized and residual sources using green technologies, particularly pressurized solvents and membrane technology.

José Ricardo Pérez-Correa (Professor). Ph.D. degree in Chemical Engineering from Imperial College (London, UK). He is Professor and Head of the Chemical and Bioprocess Engineering Department at Pontificia Universidad Católica de Chile (Santiago, Chile). He has been working on the extraction and purification of bioactive compounds from local foods and agroindustrial discards using green technologies, particularly hot pressurized liquids and solid phase extraction.

Preface to "Marine Phenolics: Extraction and Purification, Identification, Characterization and Applications"

Phenolic compounds are a highly diverse class of ubiquitous secondary metabolites produced by various organisms playing different biological roles. They have numerous types of demonstrated bioactivities, including antioxidant, antimicrobial, anti-inflammatory, antitumoral, immunomodulator, neuroprotective, cardioprotective, and antidiabetic activities. Marine organisms produce a vast collection of unique phenolic structures that are far less studied than those from terrestrial sources. Emerging extraction and purification techniques can be applied to obtain novel bioactive marine phenolics useful for food, nutraceutical, cosmeceutical, and pharmaceutical applications. This book provides updated information from recent studies on the progress in different aspects of marine phenolics. The chemical characterization, elucidation of structures, evaluation of biological properties, and efficient extraction and purification technologies of marine phenolics are discussed here. Engineers, biotechnologists, and chemists can benefit from the exciting ideas discussed in each of the chapters of this Book, whether to develop new products or open new research lines. The first chapter surveys the main types of phenolic compounds found in marine sources and their reputed bioactive properties. The second chapter focuses on the extraction, purification, and potential applications of seaweed phenolics. Chapter 3 investigates the phenolic compounds in eight seaweeds using advanced chromatographic techniques. The potential of six common Chilean seaweeds to obtain anti-hyperglycemic polyphenol extracts is explored in Chapter 4. Emerging technologies that have been applied to extract marine phenolic are reviewed in Chapter 5. In Chapter 6, the structural diversity, extraction, and purification of phlorotannins of the edible brown seaweed *Ascophyllum nodosum* is reported. Chapter 7 describes how to apply microwave-assisted extraction to obtain *Fucus vesiculosus* phlorotannin concentrates with superior ability to inhibit the alfa-glucosidase activity, a key enzyme in starch digestion. Finally, ultrasound-assisted extraction (UAE) of phlorotannins from 11 brown seaweed species is investigated in Chapter 8.

Herminia Domínguez, José Ricardo Pérez-Correa

Editors

Review

Bioactive Properties of Marine Phenolics

Raquel Mateos [1], José Ricardo Pérez-Correa [2] and Herminia Domínguez [3,*]

[1]   Institute of Food Science, Technology and Nutrition (ICTAN-CSIC),
      Spanish National Research Council (CSIC), José Antonio Nováis 10, 28040 Madrid, Spain;
      raquel.mateos@ictan.csic.es
[2]   Department of Chemical and Bioprocess Engineering, Pontificia Universidad Católica de Chile, Macul,
      Santiago 7810000, Chile; perez@ing.puc.cl
[3]   CINBIO, Department of Chemical Engineering, Faculty of Sciences, Campus Ourense, Universidade de Vigo,
      As Lagoas, 32004 Ourense, Spain
*   Correspondence: herminia@uvigo.es; Tel.: +34-988-387082

Received: 7 August 2020; Accepted: 25 September 2020; Published: 30 September 2020

Abstract: Phenolic compounds from marine organisms are far less studied than those from terrestrial sources since their structural diversity and variability require powerful analytical tools. However, both their biological relevance and potential properties make them an attractive group deserving increasing scientific interest. The use of efficient extraction and, in some cases, purification techniques can provide novel bioactives useful for food, nutraceutical, cosmeceutical and pharmaceutical applications. The bioactivity of marine phenolics is the consequence of their enzyme inhibitory effect and antimicrobial, antiviral, anticancer, antidiabetic, antioxidant, or anti-inflammatory activities. This review presents a survey of the major types of phenolic compounds found in marine sources, as well as their reputed effect in relation to the occurrence of dietary and lifestyle-related diseases, notably type 2 diabetes mellitus, obesity, metabolic syndrome, cancer and Alzheimer's disease. In addition, the influence of marine phenolics on gut microbiota and other pathologies is also addressed.

Keywords: bromophenols; simple phenolics; flavonoids; phlorotannins; seawater; algae; seagrass; health benefits; biological activity

1. Introduction

The occurrence of dietary and lifestyle-related diseases (type 2 diabetes mellitus, obesity, metabolic syndrome, cancer or neurodegenerative diseases) has become a health pandemic in developed countries. Global epidemiological studies have shown that countries where seaweeds are consumed on a regular basis have significantly fewer instances of obesity and dietary-related diseases [1]. Among marine metabolites with biological properties, phenolic compounds have attracted great interest. However, compared to those found in terrestrial sources, their study is recent and challenging in different aspects. Some families of phenolic compounds have been reported in both terrestrial and marine organisms but others, such as bromophenols and phlorotannins, are exclusively found in marine sources. The natural production of phenolic compounds in marine organisms has been associated with external factors, particularly with environmental stressing conditions, such as desiccation, salinity, UV radiation, nutrients availability, and temperature [2–5]. Variability and dependence with species, seasonality and environmental conditions occur for macroalgae [6–8] and seagrass [9–11] and with the growing conditions on microalgae [12].

Different extraction strategies have been successfully used, from conventional solvent extraction with water or with organic solvents to alternative techniques using either greener solvents or intensification tools to enhance yields and rates [13]. Enzymatic-assisted hydrolysis provided higher extraction rate and extraction yields, with lower time and cost, but appeared less effective for polyphenols because the extraction of other fractions such as proteins and saccharides was enhanced [14,15]. Ultrasonication aided in the disruption of marine algal biomass and the enhanced extraction of components [3,16–18]; hence, it can also be applied as a pretreatment [17]. However, degradation of bioactives could occur due to sonication induced effects such as high temperatures and radical's generation. Cleaner and efficient polyphenol extraction processes using safer solvents are increasingly demanded. Supercritical CO_2 extraction complies with these requirements and offers advantages derived from the tunablity of the solvation power by modifying pressure and temperature; however, due to its apolar character it requires the addition of polar modifiers. Most studies have been reported with crude solvent extracts; therefore, the properties cannot be ascribed to a single compound, and the synergistic effects among the components should be considered. Depending on the final use, a series of fractionation stages, would be required, because more active fractions can be obtained by purification of crude extracts [19,20].

The most basic phenolics quantification relies on the colorimetric Folin–Ciocalteu assay, but modern analytical tools have contributed to the provision of information on the complex structure of marine phenolics [21,22], usually with chromatographic, IR spectroscopic and NMR methods [9,10,23,24]. Advanced and coupled techniques such as HPLC–DAD–ESI/MS and UPLC–ESI–QTOF/MS analyses [5,7,20,25,26], LC–ESI–MS/MS [27], RRLC-ESI–MS [28], UPLC [29], UPLC–MS [25], UPLC–MS/MS TIC [4], [1]D and [2]D NMR techniques (^{13}C-NMR, COSY, TOCSY, NOESY, HSQC) [24], are required to unveil the highly diverse and complex chemical structure of marine phenolics. The development of strategies for simultaneous determination and quantification of the different phenolic subclasses is needed [20]. Particularly interesting has been the identification of phlorotannins, which show an extremely large diversity and complexity, regarding the number or monomeric basic units, distribution of hydroxyl groups and structural conformations of isomers [7,25,29]. In addition, their combination with preconcentration, and hydrolysis allowed simultaneous determination of phenolics in minutes [28]. Biological resources including seaweed may contain toxic compounds, such as heavy metals, and the evaluation of toxicity is required prior to focusing on any other activity [30,31].

Abundant reviews of the bioactive properties of marine phenolics can be found [32–36]. Most of them have been focused on seaweeds, but other marine organisms deserve interest as potential worldwide distributed and ubiquitous sources of phenolic compounds. Furthermore, the extensive variety of biological activities with potential to improve human and animal health, as well as the possibility of using these compounds for the formulation of novel products, configurates the food and feed applications as an efficient route of administration to maintain health and for preventing and treating different diseases. This review presents an overview of the major phenolic compounds found in marine sources and discusses their relevant biological properties in relation to lifestyle related diseases.

2. Marine Phenolics: Sources and Phenolic Composition

2.1. Families of Phenolic Compounds Identified in Marine Sources

Marine organisms are a rich source of phenolics that include bromophenolic compounds, simple phenolic acids and flavonoids as well as phlorotannins. Figure 1 shows the basic structure of some key classes of the marine phenolics identified. Examples of each class were selected based on their biological relevance in the reported studies.

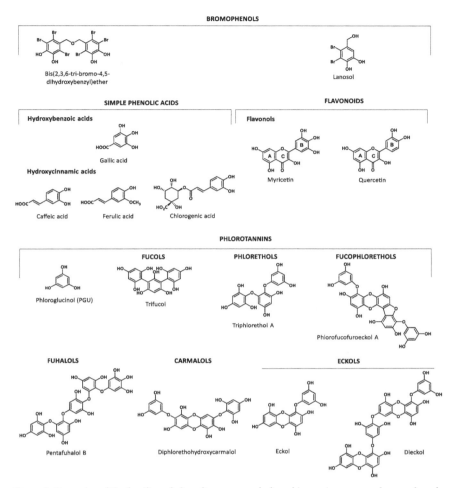

Figure 1. Examples of the families of phenolic compounds found in marine sources: bromophenols, simple phenolic acids and flavonoids, different types of phlorotannins (fucols, phlorethols, fucophlorethols, fuhalols, carmalols and eckols, as well as phloroglucinol monomeric unit).

Bromophenolic compounds have been found in several macroalgae (red, green and brown) and cyanobacteria. They can be transferred through the food chain from macroalgae to invertebrate grazers to fish. Since some of them have toxic properties similar to those of anthropogenic contaminants, their characterization is needed [37]. The lack of reports regarding the industrial production of commercially available bromophenols (hydroxylated and methoxylated bromodiphenyl ethers) suggest that they should come from natural sources and from biotransformation of natural and anthropogenic compounds [38]. Red algae are the major source of natural marine bromophenols [39], but other organisms such as fish, shrimps and crabs ingest them through the food chain. Cade et al. [38] found polybrominated diphenyl ethers (PBDEs) at higher concentration in finfish than in shellfish. Among shellfish, bivalves (clams and mussels) tended to have higher levels of hydroxylated and methoxylated PBDEs than other types of seafood. Koch and Sures [40] have compiled information on the concentrations of tribromophenols in aquatic organisms, ranging from 7 to 1600 ng/g algal ww, 0.3 to 2360 ng/g crustacean ww, 0.9 to 198 ng/g mollusks dw, 3.7 to 230 ng/g fish ww.

Phenolic acids and flavonoids have also been found in marine sources. Among phenolic acids, there are two major groups, hydroxycinnamic acids and hydroxybenzoic acids, whereas flavonols, belonging to flavonoids is the most abundant group of compounds identified in marine organisms [20,41,42]. Phlorotannins, exclusively found in brown seaweeds, are complex polymers of phloroglucinol (1,3,5-trihydroxybenzene). This structurally heterogeneous group presents a complex chemical composition, diverse linkage positions and a degree of polymerization (126 Da–650 kDa) [21] which determine its biological properties. The structural classification is based on the inter-monomeric linkages: fucols possess only aryl–aryl linkages, phlorethols aryl–ether linkages, fuhalols possess only ether linkages and additional OH groups in every third ring, fucophlorethols possess aryl–aryl and aryl–ether units, carmalols are derived from phlorethols and possess a dibenzodioxin moiety, and eckols that possess at least one three-ring moiety with a dibenzodioxin moiety substituted by a phenoxyl group at C-4 [43–45].

2.2. Sources

2.2.1. Seawater

The most abundant phenolic compounds found in seawater are sinapic acid, catechin, myricetin, kaempherol and protocatechuic acid (found at 0.8–2.8 nM/L), whereas vanillic acid, coumaric acid, ferulic acid, and rutin are below 0.5 nM/L [46]. In a recent study on the presence of free phenolic compounds in Antarctic sea water, Zangrando et al. [42] concluded that the release from phytoplankton could be the origin of phenolics in seawater, since diatoms produce exudates that contain phenolic compounds. Other possible but less plausible sources could be the intrusion of circumpolar deep water that may transport oceanic lignin; the melting of glaciers, which contain lignin that can be degraded in the snow; photooxidation in water; the photochemical and microbiological degradation of lignin contained in dissolved organic material. These authors have found vanillin, vanillic acid, acetovanillone and p-coumaric acid, both in the dissolved and particulate fractions in seawater samples, with syringic acid, syringaldehyde and homovanillic acid at residual concentrations. Bidleman et al. [37] also reported the presence of the bromophenol lanosol (2,3-dibromo-4,5-dihydroxybenzyl alcohol) in seawater.

2.2.2. Microalgae

Microalgae conform a highly ecologically diverse group of unicellular eukaryotic organisms; they are the most important primary source of biomass in aquatic ecosystems. They are able to produce a wide variety of commercially interesting compounds, such as lipids, carbohydrates, phenolics, carotenoids, sterols, vitamins, and other bioactives [47]. Microalgae offer advantages over terrestrial sources derived from their metabolic diversity and adaptive flexibility, the efficient photosynthesis and high growth rate, the possibility of large scale cultivation, simple nutritional requirements, and their ability to accumulate or secrete metabolites [48]. Microalgae can grow in different habitats such as fresh water, saltwater and marine environments. They can even grow on industrial wastewaters [49]. The valuable bioactives with pharmaceutical, food, feed, and cosmetic applications [50,51] from microalgae could be relevant regarding the higher profitability of the cultivation processes and could complement the energetic application [47]. In fact, the extraction of phenolic compounds from microalgae biomass does not interfere with already established processes such as biofuel production [27].

Microalgae produce protective antioxidant compounds in response to stress damage caused by UV radiation, temperature variation, excessive light, and others. In some cases, these are not influencing factors. Gómez et al. [52] observed that the accumulation of phenolic compounds in some microalgae was independent of the illumination condition. The production of flavonoids and polyphenols could be favored with the adequate control of selected variables of the culture process [12]. Non-natural factors, such as CuO nanoparticles, can induce the production of phenolics in *Nannochloropsis oculata* [53], lowering growth rates as well as chlorophyll and carotenoids content. Moreover, CuO nanoparticles damaged the membrane as well as increased the activity of antioxidant endogenous enzymes, such as catalase, ascorbate peroxidase, polyphenol oxidase and lactate dehydrogenase.

Some phenolics in marine microorganisms are released into the environment to form metal complexes in order to acquire micronutrients or to sequester toxic metals, and their presence can stimulate the growth of diatoms. Catechin, sinapic acid, apigenin, quercitrin, kaempferol, epicatechin, gentisic acid, syringic acid, chlorogenic acid, vanillic acid, ferulic acid, caffeic acid, protocatechuic acid, coumaric acid, rutin and gallic acid have been reported in the exudates from diatoms [42,54,55].

Data in Tables 1–3 summarize the phenolic compounds reported in marine organisms and their in vitro antioxidant characteristics, which could be used as a preliminary indication of potential bioactivities. Phenolic compounds can be efficient antioxidants acting with different mechanisms, as scavengers of singlet oxygen and free radicals, reducing agents, chelating agents, inhibiting specific oxidative enzymes or can act by mixed mechanisms. Assays to determine the reducing and antiradical properties against 2,2-diphenyl-1-picrylhydracil (DPPH), as well as 2,2′-azino-bis (3-ethylbenzothiazoline-6-sulphonic acid) (ABTS), superoxide and hydroxyl radicals, are usually preferred to screen the most active extracts from natural sources. Data in Table 1 confirm that phenolic acids, and particularly hydroxycinnamic acids, are the major families identified in microalgae.

Considering the diversity of phenolic compounds found in marine organisms, and the influence of composition on the activity, the selection of the extraction solvent is important and should be chosen with care, either individually or in mixtures [3,17]. Some examples are cited to illustrate this fact. In a comparative study, acetone provided the highest phenolic content in extracts from *Isochrysis galbana*, *Tetraselmis* sp. and *Scenedesmus* sp. The highest radical scavenging activity was observed in the acetone extract of *I. galbana*, the maximum Fe (II) chelating capacity in the hexane extract of *Scenedesmus* sp. and the in vitro inhibition of acetylcholinesterase in the water and ether extracts of both microalgae. Whereas the antiradical properties of the polar extracts can be ascribed to phenolics, in the non-polar extracts the activity could be due to fatty acids or to other lipophilic components [56]. Aqueous and methanolic extracts provided higher phenolic yield and reducing power from *Nannochloropsis gaditana* than acetone, dichloromethane or hexane; however, acetone provided the highest DPPH radical scavenging activity and cytotoxicity against human lung cancer cells (A549) [57]. Moreover, the methanolic extracts of *Chaetoceros curvisetus*, *Thalassiosira subtilis* and *Odontella aurita* were more active than those in acetone and in hexane [58].

In some cases, a linear relationship between phenolic content and antioxidant and biological properties has been established. Phenolic content is correlated with DPPH radical scavenging activity [18,59] and also with antitumoral properties [56]. Solvent extracts from *Nannochloropsis oceanica* showed reducing and antiradical properties and those from *Skeletonema costatum* and *Chroococcus turgidus* showed chelating ability; both properties are correlated with the phenolic content [59]. However, this correlation was found to be insignificant in other extracts, suggesting that these might not be major contributors to the antioxidant capacities [60]. Safafar et al. [49] reported that phenolic compounds were the major contributors to the antioxidant activity in microalgal extracts, but also carotenoids contributed to the DPPH radical scavenging activity, ferrous reduction power (FRAP), and ABTS-radical scavenging capacity activity. Maadame et al. [3] did not find correlation between the antioxidant capacities and the phenolic and carotenoids content in ethanolic extracts [57]. The low phenolic content (0.3–20 mg GAE/g DW) in microalgal extracts [48,56,58] could suggest that other compounds could be responsible for the observed activities, such as carotenoids, fatty acids, sterols, vitamins as well as other compounds such as micosporine-like aminoacids (MAAs) [61]. The TEAC (Trolox equivalent antioxidant capacity) values and antiproliferative activities of phytoplankton extracts show a strong positive correlation with the amount of the total carotenoids and micosporine-like aminoacids, but were negatively correlated with the amounts of phenolic compounds [18].

Table 1. Phenolic compounds identified in different marine organisms: microalgae, cyanobacteria, fungus, seagrasses and sponges.

Marine Organism Extraction Chemical Analysis	Phenolic Compounds Antioxidant Activity (When Provided)	Ref.
	Marine-derived Fungus	
Alternaria sp. SCSIO41014 from sponge AC, EtOAc, US HPLC-UV, HRESIMS, NMR, ECD, XRay	Perylenequinone derivatives; altenusin derivative; phenol derivatives	[62]
Arthrinium sp. MeOH, MeCN HPLC, HRESIMS, NMR	2-(2,3-Dihydroxy-5-methyl benzoyl)-6-hydroxybenzoic acid	[63]
Aspergillus sydowii, from the sponge *Stelletta* sp. EtOAc, BuOH HPLC, UV, IR, HRESIMS, NMR, OP	Diorcinolic acid; β-D-glucopyranosyl aspergillusene A; diphenylethers; chromone; xanthone first glycoside of phenolic bisabolane sesquiterpenes	[64]
Aspergillus sp. from the sponge *Xestospongia testudinaria* EtOAc, AC RP-HPLC, HRESIMS, NMR	Phenolic bisabolane sesquiterpenoid dimers (disydonols A–C), (S)-(+)-sydonol	[65]
Aspergillus sp. from the sponge *Chondrilla nucula* EtOAc HPLC-PDA, UV, HRESIMS, NMR, OP	Phenolic bisabolane sesquiterpenes; asperchondol A; asperchondol B	[66]
Aspergillus sp., from the sponge *Chondrilla nucula* EtOAc, BuOH HPLC, UV, IR, HRESIMS, NMR, OP	Phenolic bisabolane sesquiterpenes; asperchondols A and B; diphenyl ethers	[64]
Aspergillus versicolor, deep-sea fungus EtOAc, BuOH HPLC, IR, TLC, HRESIMS, NMR, OP, ECD	Aspergilols A–F; diorcinal; cordyol E; 4-carboxydiorcinal; 4-methoxycarbonyldiorcinol; 4-carbethoxydiorcinal; cordyol C; methylgerfelin; violaceol II; averythrin; averantin, 1'-O-methylaverantin; lecanoric acid; orsellic acid; orcinol; 1- methylpyrogallol; fumaric acid ABTS = 0.1–5.4 mmol Trolox/g	[67]
Cladosporium cladosporioides from *Sargassum wightii* EtOAc LC-MS	N-(2-Iodophenyl)-2-[2-oxo-5-(thiophen-2-yl)-2,3-dihydro-1,3,4-oxadiazol-3-yl] acetamide; 2-[3-chloro-4-(4-chlorophenoxy)phenyl]-1,3-dioxo-2,3-dihydro-1H-isoindole-5-carboxylic acid; 2-(2,4-dichlorophenyl)-2-oxoethyl 3,4-dihydro-2H-1,5-benzodioxepine-7-carboxylate; 4-bromo-N'-(4-fluoro-1-benzothiophene-2-carbonyl)-1H-pyrrole-2-carbohydrazide; (1R,2R,5S)-2-[3-{(2-[(2,4-dichlorophenyl)methyl]-2H-1,2,3,4-tetrazol-5-yl]methyl)-4-methyl-5 sulfanylidene-4,5-dihydro-1H-1,2,4-triazol-1-yl]-6,8-dioxabicyclo [3.2.1]octan-4-one; methyl 2-([[5-bromo-2-(4 methoxybenzamido)phenyl] (phenyl)methyl]amino) acetate; 2-[4-(2,4-dichlorophenoxy)phenyl]-5-phenyl-octahydro-1H-isoindole-1,3-dione; N-([2-[(3,4-dichlorophenyl) methoxy]naphthalen-1-yl]methyl)-2,3-dihydro-1,4-benzodioxin- 6-amine; 2-([[(2,4-dichlorophenyl) carbamoyl] methyl](propyl)amino)-N-(2,2,2-trifluoroethyl) acetamide; N-(4-bromo-2-fluorophenyl)-6-(2-tert-butylhydrazin1-yl)-5-nitropyrimidin-4-amine; N-(2-[3-[(3,4-di chlorophenyl) methyl]-2-oxo-1,3-diazina EC$_{50,DPPH}$ = 50 mg/mL; EC$_{50,DPPH}$ AS = 18 mg/mL; RP = 0.81 mg/g	[68]
Penicillium brevicompactum EtOAc RP-HPLC, UV, HRESIMS, NMR	Anthranilic acid; syringic acid; sinapic acid; acetosyringone IC$_{50, DPPH}$ = 20–30 μg/mL	[69]

Table 1. *Cont.*

Marine Organism Extraction Chemical Analysis	Phenolic Compounds Antioxidant Activity (When Provided)	Ref.
Penicillium ianthinellum EtOAc, MeOH HPLC, UV, IR, HRESIMS, GC-MS, NMR	6-(2-Acetyl-3,5-dihydroxybenzyl)-4-hydroxy-3-methyl-2H-pyran-2-one; 7-hydroxy-2-(hydroxymethyl)-5-methyl-4H-chromen4-one; 3,5-dihydroxy-2-(2-(2-hydroxy-6-methylphenyl)-2-oxoethyl)-4-methylbenzaldehyde; 3-hydroxy-5-methylphenyl 2,4-dihydroxy-6-methylbenzoate; lecanoric acid; orsellinic acid; orcinol	[70]
Penicillium griseofulzum EtOAc, BuOH HPLC, UV, IR, HRESIMS, NMR, OP	4,6-Dimethylcurvulinic acid	[64]
Penicillium expansum 091006 from the mangrove plant *Excoecaria agallocha* EtOAc, AC HPLC, UV, IR, HRESIMS, TLC, NMR, OP, ECD	Phenolic bisabolane sesquiterpenoid and diphenyl ether units, expansols A and B, (S)-(+)-11-dehydrosydonic acid,(7S,11S)-(+)-12-acetoxysydonic acid, (S)-(+)-sydonic acid, diorcinol,(S)-(+)-2-[3-hydroxy-4-(2-methoxy-6-methylheptan-2-yl)benzyl]-5-(3-hydroxy-5-methylphenoxy)-3-methylphenol, S-(+)-2-[3-hydroxy-4-(2-hydroxy-6-methylheptan-2-yl)benzyl]-5-(3-hydroxy-5-methylphenoxy)-3-methylphenol, (S)-(+)-3-hydroxy-4-(2-hydroxy-6-methylhept-6-en-2-yl)benzoic acid, and 4-[(2S,6S)-7-acetoxy-2-hydroxy-6-methylheptan-2-yl]-3-hydroxybenzoic acid	[71]
Scopulariopsis sp. EtOAc, MeOH, W, H HPLC-PDA, RP-HPLC, LC-MS, HRESIMS, TLC, NMR, OP	12-Dimethoxypinselin; 12-O-acetyl-AGI-B4, 11,12-dihydroxysydonic acid; 1-hydroxyboivinianic acid	[72]
ZSDS1-F11 from the sponge *Phakellia fusca* EtOAc, AC TLC, CC, HRESIMS, NMR	Phenolic bisabolane sesquiterpenoid and diphenyl ether units; expansols A–F; diorcinol	[73]
Cyanobacteria		
Anabaena C5 E, US HPLC-MS/MS	Quinic acid; catechin $EC_{50\ DPPH}$ = 0.1 mg/mL; FRAP = 11.4 mg ASE/g	[74]
Arthrospira S1, S2 EtOH, US HPLC-MS/MS	Catechin $EC_{50,\ DPPH}$ = 0.1 mg/mL; FRAP = 15.1–21.0 mg ASE/g	[74]
Calothrix sp. SI-SV MeOH, W HPLC-UV/VIS	Rutin; tannic acid; orcinol; phloroglucinol; protocatechuic acid $EC_{50,\ ABTS}$ = 65.79 µg/mL, $EC_{50,\ DPPH}$ = 69.38 µg/mL	[75]
Leptolyngbya sp. MeOH, W HPLC-UV/VIS	Rutin; tannic acid; orcinol; phloroglucinol; protocatechuic acid SI-SM ($EC_{50,\ ABTS}$ = 63.45, $EC_{50,\ DPPH}$ = 67.49 µg/mL)	[75]
Nostoc commune MeOH, W RP-HPLC-DAD	Gallic and chlorogenic acids	[76]
Nostoc sp EtOH, US HPLC-MS/MS	Gallic acid; chlorogenic acid; quinic acid; catechin; epicatechin; kaempferol; rutin; apiin $IC_{50,\ DPPH}$ = 0.04–9.47 mg/mL; FRAP = 8.4–13.7 mg ASE/g	[74]

Table 1. *Cont.*

Marine Organism Extraction Chemical Analysis	Phenolic Compounds Antioxidant Activity (When Provided)	Ref.
Microalgae		
Ankistrodesmus sp. MeOH, W RP-HPLC-DAD	Protocatechuic acid DPPH scavenging$_{10\ mg/mL}$ = 29%	[76]
Euglena cantabrica MeOH, W RP-HPLC-DAD	Gallic acid; protocatechuic acid; chlorogenic acid; (+) catechin; (−)epicatechin DPPH scavenging$_{10\ mg/mL}$ = 71%	[76]
Nannochloropsis sp. EtOH, MeOH, H RP-HPLC-UV HPLC-ESI-MS/MS	Phenolic acids: chlorogenic; caffeic; gallic; protocatechuic; hydroxybenzoic; syringic; vanillic; ferulic	[27]
Spirulina sp. RP-HPLC-UV HPLC-ESI-MS/MS	Protocatechuic; gallic; chlorogenic; vanillic; hydroxybenzoic; syringic; vanillic acids	[27]
Spirogyra sp. MeOH, W RP-HPLC-DAD	Gallic acid DPPH scavenging$_{10\ mg/mL}$ = 62%	[76]
Seagrasses		
Cymodocea nodosa MeOH, CH$_2$Cl$_2$ HPLC-DAD, LC/MS-ESI, NMR	Diosmetin 7-sulfate; caftaric acid; coutaric acid	[77]
Halodule wrightii, Thalassia testudinum Agar, W HPLC	p-hydroxybenzoic acid; ferulic acid; p-coumaric acid; syringic acid; gallic acid	[78]
Halophila stipulacea MeOH, EtOAc, Hexane HR-LC-MS/MS GNPS	Luteolin; apigenin; matairesinol; cirsimarin; spiraeoside; 2,4-dihydroxyheptadec-16-ynyl acetate; 3-hydroxy-4-methoxycinnamic acid, alpha-cyano-4-hydroxycinnamic	[79]
Posidonia oceanica (L.) EtOH, W, Formic acid HPLC-ESI-MS/MS	Procyanidin B2; procyanidin C2; isorhamnetin-3-O-glucoside; quercetin-3-O-glucoside; quercetin-3-O-malonylglucoside; isorhamnetin-3-O-malonylglucoside EC$_{50\ DPPH}$ = 32 μg/mL	[80]
Ruppia cirrhosa (Petagna) Grande, *Ruppia maritima* L. MeOH, W, EtOAc HPLC-DAD, HR-LCMS-ESI+/TOF, NMR	Chicoric acid; quercetin 3-O-β-ᴅ-(6''-O-malonyl)-glucopyranoside; quercetin 3-O-β-ᴅ-galactopyranoside; quercetin 3-O-β-ᴅ-glucopyranoside; quercetin 3-O-β-ᴅ-(6'-O-malonyl)galactopyranoside; isorhamnetin 3-O-β-ᴅ-galactopyranoside; isorhamnetin 3-O-β-ᴅ-glucopyranoside; isorhamnetin 3-O-β-ᴅ-(6''-O-malonyl) galactopyranoside; isorhamnetin 3-O-β-ᴅ-(6''-O-malonyl)-glucopyranoside EC$_{50\ DPPH}$ = 23–176 μg/mL	[81]

Table 1. *Cont.*

Marine Organism / Extraction / Chemical Analysis	Phenolic Compounds / Antioxidant Activity (When Provided)	Ref.
Syringodium isoetifolium / MeOH / HPLC-EI-MS	Caftaric acid; 2 3-(4-Hydroxyphenyl)]lactic acid; caffeic acid; caffeoyl-4'-O-phenyllactate; 3-phenyllactic acid; 4-coumaric acid; chicoric acid; DPPH = 5.4 mg TE/g; ABTS = 9.6 mg TE/g; CUPRAC = 18.7 mg TE/g; FRAP = 9.5 mgTE/g; Chelating ability = 9.17 mg EDTAE/g	[82]
Thalassia testudinum / AC, W, AA	3,4-Dihydroxybenzoic acid; p-hydroxybenzoic acid; p-coumaric acid; vanillin	[83]
T. testudinum / EtOH, W / RP-HPLC / LC-MS, NMR	3,4-Dihydroxybenzoic acid, p-hydroxybenzoic acid, p-coumaric acid and vanillin	[84]
Zostera asiatica and *Z. marina* / HPLC-MS	Rosmarinic acid; luteolin; 7,3'-disulfate luteolin; ROS scavenger; protecting or enhancing endogenous antioxidants; metal chelation	[85]
Z. marina / Hexane, AC / HPLC-MS, NMR	Deoxycymodienol; isotedarene A	[86]
Z. marina / SPA, IPA / HPLC-MS/MS	3-Hydroxyhexanoic acid; 4-hydroxynonenoic acid; p-coumaric acid; caffeic acid; ferulic acid; zosteric acid; apigenin; luteolin; diosmetin; apigenin-7-sulfate; rosmarinic acid; luteolin-7-sulfate; diosmetin-7-sulfate; kaempferol-7,4'-dimethylether-3-O-sulfate	[5]
Z. noltii / MeOH / HPLC, NMR	Rosmarinic acid; caffeic acid; zosteric acid	[87]
Z. noltei / MeOH / HPLC-PDA-MS-ESI-QTOF, NMR	Rosmarinic acid; apigenin-7-O-glucoside; luteolin; apigenin; diosmetin; acacetin; luteolin-7-sulfate; apigenin-7-sulfate; diosmetin-7-sulfate; acacetin-7-sulfate	[88]
Zostera noltei leaves / MeOH, W / HPLC-DAD, LC-MS, NMR	Apigenin 7-sulfate; diosmetin 7-sulfate	[89]
Zostera noltii, Z. marina	Apigenin 7-sulphate; luteolin 7-sulphate; diosmetin 7-sulphate; rosmarinic acid; luteolin 7-glucoside; apigenin 7-glucoside; apigenin; luteolin 7-(6''-malonyl) glucoside; apigenin 7-(6''-malonyl) glucoside	[81]
Zostera muelleri / MeOH, AA / RP-HPLC	Proanthocyanidins; gallic acid; rosmarinic acid	[90]
	Sponges	
Didiscus aceratus / MeOH, CH$_2$Cl$_2$, H / HRESIMS, NMR	(S)-(+)-Curcuphenol; 10β-hydroxycurcuphenol; 10α-hydroxycurcudiol; dicurcuphenols A–E; dicurcuphenol ether F	[91]

Table 1. Cont.

Marine Organism Extraction Chemical Analysis	Phenolic Compounds Antioxidant Activity (When Provided)	Ref.
Hyrtios erectus MeOH, EtOAc HRAPCIMS, HRESIMS, NMR	Phenolic alkenes; erectuseneols A–F	[92]
Myrmekioderma sp. MeOH, CH$_2$Cl$_2$, EtOAc, BuOH, hexane HRESIMS, NMR	1-(2,4-Dihydroxy-5-methylphenyl)ethan-1-one; (1'Z)-2-(1',5'-dimethylhexa-1',4'-dieny1)-5- methylbenzene-1,4-diol; 1,8-epoxy-1(6),2,4,7,10-bisaborapentaen-4-ol; 6-(3-hydroxy-6-methyl-1,5- heptadien-2-yl)-3-methylbenzene-1,4-diol; 4-hydroxy-3,7-dimethyl-7-(3-methylbut-2- en-1-yl)benzofuran-15-one; 6-(2-methoxy-6-methylhept-5-en-2-yl)-3-methylbenzene-1,4-diol; 9-(3,3-dimethyloxiran-2-yl)-1,7-dimethyl-7-chromen-4-ol	[93]
Myrmekioderma sp. MeOH, EtOAc HRAPCIMS, HRESIMS, NMR	(R)-Biscurcudiol; (S)-biscurcudiol; myrmekiodermaral; myrmekiodermanol; myrmekioperoxide A; myrmekioperoxide B (4); myrmekiodermaral; (+)-curcudiol; (+)-dehydrocurcudiol; abolene; abolene epimer at C-5'; (+)-oxoabolene; (+)-curcuphenol; 5'α-hydroxycurcudiol; 5'β- hydroxycurcudiol; curcuepoxide A; curcuepoxide B	[94]

AA: ascorbic acid; AC: acetone; ASE: ascorbic acid equivalents; BuOH: butanol; CA: chelating ability; CAA: antioxidant assay for cellular antioxidant activity [95]; CLPAA: cellular lipid peroxidation antioxidant activity assay [95]; EA: ethyl acetate; ECD: electronic circular dichroism; EtOAc: ethyl acetate; EtOH: ethanol; GNPS: global natural product social molecular networking; H: hexane; HRAPCIMS: high resolution atmospheric pressure chemical ionization mass spectrometry; HRESIMS: high resolution electrospray ionization mass spectrometry; IPA: isopropanol; LPIA: lipid peroxidation inhibition assay [96]; MeCN: acetonitrile; MeOH: methanol; OP: polarimetry; PDA: photodiode array; PGU: phloroglucinol units; RP: reducing power; RP-HPLC: reversed phase HPLC; SPA: solid phase adsorption; SRSA: superoxide radical scavenging assay [96]; TAA: total antioxidant capacity [97]; TLC: thin-layer chromatography; US: ultrasound; W: water.

2.2.3. Macroalgae

Bromophenols

Among the halogenated secondary metabolites synthesized by seaweeds, brominated ones are more usual due to the availability of chloride and bromide ions in seawater; iodine and fluorine are less frequent. Whereas iodination can be found in brown algae, bromine or chlorine metabolites are more abundant in red and in green seaweeds [98]. The most abundant bromophenolic compounds found in macroalgae are bromophenols and their transformation products bromoanisoles, hydroxylated and methoxylated bromodiphenyl ethers and polybrominated dibenzo-p-dioxins [2,37]. Other brominated compounds have also been identified in macroalgae, such as brominated sesquiterpenes [99].

Some bromophenols identified in seaweeds are shown in Tables 2 and 3. Specifically, 2,4,6-tribromophenol is widely distributed, coming from environmental contaminants, pesticides and from marine organisms, which produce it as a defense against predators and biofouling. Machado et al. [100] found bromoform (1.7 mg/g), dibromochloromethane (15.8 µg/g), bromochloroacetic acid (9.8 µg/g) and dibromoacetic acid (0.9 µg/g) in *Asparagopsis taxiformis*, and Bidleman et al. [37] found bromoanisols at more than 1000 pg/g in *Ascophyllum nodosum*, *Ceramium tenuicorne*, *Ceramium virgatum*, *Fucus radicans*, *Fucus serratus*, *Fucus vesiculosus*, *Saccharina latissima*, *Laminaria digitata*, and *Acrosiphonia/Spongomorpha sp*. The presence of these compounds can be associated with off-flavors. Among the bromophenols identified in prawn species, probably obtained from marine algae and bryozoan from the diet, are 2-and 4-bromophenol, 2,4-dibromophenol, 2,4,6-tribromophenol and 2,6-dibromophenol. The latter confers an iodoform off-flavor at 60 ng/kg and it was found in prawn, reaching more than 200 µg/kg in the eastern king prawn. This off-flavor can be reduced by handling and processing [101]. Kim et al. [102] reported that 3-bromo-4,5-dihydroxybenzaldehyde exerted antioxidant effects in skin cells subjected to oxidative stress, by increasing the protein and mRNA levels of glutathione synthesizing enzymes, enhancing the production of reduced glutathione in HaCaT cells and protecting cells against oxidative stress via the activation of the NF-E2-related factor.

Their extraction can be achieved with organic solvents, i.e., methanol or methanol-dichloromethane [37], but yields can vary with other factors. Seasonal variations and different profiles among species, locations and environmental conditions have been observed [40], their production being induced by environmentally stressing conditions, such as the presence of herbivores and the elevated levels of light and salinity [2].

Simple Phenolics

The presence of benzoic and cinnamic acids has been reported, particularly in brown seaweeds, which also present flavonoids [20,103]. Brown seaweeds present higher contents of benzoic and cinnamic acids (1 mg/g) than red (0.2 3 mg/g) and green (0.01–0.9 mg/g) seaweeds [26,104,105]. Higher values (1–9 mg/g) have been reported for gallic acid in green and red seaweeds [106]. These authors reported catechin content up to 14 mg/g in red seaweeds and up to 11.5 mg/g in green ones, whereas in brown seaweeds reached up to 11 mg/g. Phloroglucinol derivatives are the major phenolics in brown seaweeds, and flavonoids account for 35% of the total, the most abundant being gallic, chlorogenic acid, caffeic acid, ferulic acid [20].

The phenolic levels correlated positively with elevated irradiance exposure and temperature and their content differs among different parts of the seaweed. Extracts of the thallus were more active than extracts of the receptacles, and the solvent was also important, the best being acetone, ethanol, and water. The drying stage should also be optimized, since degradation may occur, i.e., dried material provided lower yield and less active extracts than frozen ones [24].

Table 2. Phenolic compounds found in brown seaweeds.

Seaweed / Extraction / Chemical Analysis	Compounds / Antioxidant Activity (When Provided)	Ref.
A. nodosum, F. spiralis; MeOH, AC, Hexane; UPLC, MS, NMR	Phlorotannins (4–6, 9–12 PGU)	[107]
A. nodosum, Fucus spiralis, F. vesiculosus, Pelvetia canaliculata, Saccharina longicruris; MeOH, W; UPLC, HRMS	Phlorotannins (3–50 PGU)	[108]
Carpophyllum flexuosum, Carpophyllum plumosum, Ecklonia radiata; W, MAE; HPLC-DAD-ESI-MS, NMR	Bifuhalol, bifuhalol dimer, bifuhalol trimer, hydroxytrifuhalol, trifuhalol, tetrafuhalol; DPPH = 2.7–37.4 mg GAE/g; FRAP = 4.4–62.1 mg GAE/g	[109]
Cystoseira barbata; TFA, W; LC-QTOF-MS	Phloroglucinol, rutin, phlorofucofuroeckol, 3-O-rutinosyl-kaempferol, catechin–catechin-O-gallate, gallocatechin, gallocatechin-O-glucuronide, 1-hydroxy-2-(β-ᴅ-glucopyranosyloxy)-9,10 anthraquinone, 2-O-(6,9,12-octadecatrienoyl)glyceryl β-galactopyranoside, chlorogenic acid butyl ester, phloroglucinol, quercetin; $EC_{50, DPPH}$ = 11.7 µg/mL; $EC_{50, OH}$ = 11.4 µg/mL; $EC_{50, RP}$ = 51 g/mL; $EC_{50, CA}$ = 40 g/mL	[110]
C. barbata; AC, MeOH, W; UHPLC-DAD-QTOF-MS	Fucophlorethol and eckol derivatives (3–7 PGU); $EC_{50, DPPH}$ = 14 µg/mL; $EC_{50, ABTS}$ = 0.5 µM Trolox; $EC_{50, RP}$ = 16–35 µg/mL	[111]
Cystoseira nodicaulis, Cystoseira tamariscifolia, Cystoseira usneoides, F. spiralis; AC, Hexane, W; HPLC-DAD-ESI-MS^n	Fucophloroethol, fucodiphloroethol, fucotriphloroethol, 7-phloroeckol, phlorofucofuroeckol, bieckol, dieckol; $EC_{50, SRSA}$ = 0.93–4.02 mg/mL; $EC_{50, LPIA}$ = 2.32–>9.1 mg/mL	[96]
Cystoseira nodicaulis, F. serratus, F. vesiculosus, Himanthalia elongata; EtOH, W; UPLC-ESI-MS	Phlorotannin (3–16 PGU); $EC_{50, DPPH}$ = 4–28 µg/mL; FRAP = 101–307 µg TE/mg	[21]
Durvillaea antarctica, Lessonia spicata; EtOH, EE, EtOAc, W; HPLC-MS-MRM	Phlorotannin (3–8 PGU), flavonoids; $EC_{50, DPPH}$ = 0.97–1.24 mg/mL; FRAP = 2.95–6.20 mM TE/kg; ORAC = 4.75–25.9 µM TE/g	[112]
Eisenia bicyclis; EtOH; HPLC-PDA	Eckol, phlorofucofuroeckol-A, dieckol, 6,6'-bieckol, 8,8'-bieckol	[113]
Ecklonia cava; EtOH; UPLC-PDA	Phloroglucinol, eckol, eckstolonol, triphlorethol-A, dieckol	[114]

Table 2. *Cont.*

Seaweed Extraction Chemical Analysis	Compounds Antioxidant Activity (When Provided)	Ref.
E. cava EtOH, US HPLC-DAD-ESI/MS, NMR	Dieckol, phlorofucofuroeckol-A, 2,7-phloroglucinol-6,6-bieckol, pyrogallol-phloroglucinol-6,6-bieckol	[115]
E. cava EtOH, W RP-HPLC	Dieckol ABTS = 1.3 g VCE/g; DPPH = 0.4 g VCE/g	[116]
Ecklonia stolonifera EtOH, W HPLC-PDA, NMR	2-phloroeckol, dioxinodehydroeckol, eckol, phlorofucofuroeckol B, 6,6′-bieckol, dieckol, 974-B, phlorofucofuroeckol A	[117]
F. vesiculosus MeOH, W Q-ToF-MS, UPLC-TQD-MS/MS-MRM	Phlorotannins (3-18 PGU) $EC_{50, DPPH}$ = 18.2 µg/mL	[118]
F. vesiculosus AC, EtOAc, EtOH, MeOH, W HPLC-DAD-ESI/MS[a]	Fucodiphlorethol A, trifucodiphlorethol isomers, phlorotannins (3-10 PGU) $EC_{50, DPPH}$ = 2.79–4.23 µg/mL; Fe^{2+}-CA = 25.1–47.6%; RP = 17.8–910.7 mg ASEs/g	[119]
F. vesiculosus AC, W UPLC-DAD-ESI/MS[a]	Fucols, fucophlorethols, fuhalols, phlorotannin derivatives (3-22 PGU), fucofurodiphlorethol, fucofurotriphlorethol, fucofuropentaphlorethol	[120]
Halidrys siliquosa AC, W MALDI-TOF-MS, NMR	Diphlorethol, triphlorethol, trifuhalol, tetrafuhalol $EC_{50, DPPH}$ = 0.02–1.00 mg/mL; $EC_{50, RP}$ = 0.06–0.62 mg/mL; $EC_{50, NBT}$ = 0.66–2.44 mg/mL; ORAC = 5.39 µmol TE/mg; BCB = 0.21–1.50 g/mL	[121]
H. elongata MeOH, W HPLC-DAD, HPLC-ESI-MS/MS	Phloroglucinol, gallic acid, chlorogenic acid, caffeic acid, ferulic acid, hydroxybenzaldehyde, kaempferol, myricetin, quercetin $EC_{50, DPPH}$ = 14.5 µg/mL	[20]
Hydroclathrus clathratus, Padina minor, Padina sp., Sargassum oligosystum, Sargassum aff. batuanens, Sargassum sp. MeOH, W GC-MS-EI-SIM	2,4,6-tribromophenol; 2,4,6-tribromoanisol; 2′-hydroxy-2,3′,4,5′-tetrabromodiphenyl ether; 2′-methoxy-2,3′,4,5′-tetrabromodiphenyl ether; 6-hydroxy-2,2′,4,4′-tetrabromodiphenyl ether; 6-methoxy-2,2′,4,4′-tetrabromodiphenyl ether; 2′,6-dihydroxy-2,3′,4,5′-tetrabromodiphenyl ether; 2′,6-dimethoxy-2,3′,4,5′-tetrabromodiphenyl ether; 2,2′-dihydroxy-3,3′,5,5′-tetrabromodiphenyl; 2,2′-dimethoxy-3,3′,5,5′-tetrabromodiphenyl	[122]
L. digitata MeOH, W RP-UPLC-UV-MS[a], MALDI-TOF-MS, NMR	Di-fuhalols (6-7 PGU), fucols (3-7 PGU), fucophlorethols (3-16 PGU), fuhalols (4-5 PGU), phlorethols (3-18 PGU)	[123]

Table 2. Cont.

Seaweed Extraction Chemical Analysis	Compounds Antioxidant Activity (When Provided)	Ref.
Leathesia nana CH$_2$Cl$_2$, EtOH, W HRESIMS, NMR	2,2',3,3'-tetrabromo-4,4',5,5'-tetrahydroxydiphenylmethane; 3-bromo-4-(2,3-dibromo-4,5-dihydroxybenzyl)-5-methoxymethylpyrocatechol; 2,3,3'-tribromo-4,4',5,5'-tetrahydroxyl-1'-ethyloxymethyldiphenyl methane; 2,3-dibromo-4,5-dihydroxybenzaldehyde; 2,3-dibromo-4,5- dihydroxybenzyl alcohol; 2,3-dibromo-4,5-dihydroxybenzyl methyl ether; 2,3-dibromo-4,5-dihydroxybenzyl ethyl ether; 3,5-dibromo-4-hydroxybenzaldehyde; 3,5-dibromo-4-hydroxybenzoic acid; 3-bromo-4,5-dihydroxybenzoic acid methyl ester, 3-bromo-5-hydroxy-4-methoxybenzoic acid, 3-bromo-4-hydroxybenzoic acid EC$_{50}$, DPPH = 14.5 µg/mL	[124]
Lessonia trabeculate MeOH, W, MWE HPLC-DAD-ESI-MS/MS	Phlorotannins derivatives (2–3 PGU), gallocatechin derivative, p-coumaric acid derivative	[103]
Padina tetrastromatica PE, MeOH, CHCl$_3$, MEK, Soxhlet HPLC-UV, UPLC-MS/MS	Fucophlorethol (2–18 PGU) EC$_{50}$, DPPH = 17 µg/mL	[125]
Sargassum fusiforme EtOH, W UPLC-DAD-ESI-MS/MS	Eckol, dieckol, dioxinodehydroeckol, fuhalols (2–12 PGU), phlorethols/fucols/fucophlorethols (2–11 PGU), eckols (2–8 PGU) EC$_{50}$, DPPH = 15–150 µg/mL; FRAP = 1.29 mg TE/mg	[126]
Sargassum muticum EtOH, W, PHLE LC × LC-DAD-ESI-MS/MS	Decafuhalol, dihydroxytetrafuhalol, dihydroxypentafuhalol, dihydroxyhexafuhalol, dihydroxyheptafuhalol, dihydroxyoctafuhalol, dihydroxynonafuhalol, heptaphlorethol, hexafuhalol, hexaphlorethol, hydroxytetrafuhalol, hydroxypentafuhalol, hydroxyhexafuhalol, nonafuhalol, octafuhalol, pentafuhalol, pentaphlorethol, tetrafuhalol, trifuhalol, trihydroxyhexafuhalol, trihydroxyheptafuhalol, trihydroxyoctafuhalol ABTS = 0.65–2.29 mmol TE/g	[127]
Silvetia compressa EtOH, US HPLC-DAD, HPLC-TOF-MS	Dihydroxytetrafuhalol, dieckol, eckol derivative, eckstolonol, 7-phloroeckol (3 PGU), dihydroxypentafuhalol, phlorofucofuroeckol A, pentafuhalol, trifuhalol	[128]

Phlorotannins

These are structural components of the cell wall in brown algae, and also play a function in macroalgal chemical defense comparable to those of secondary metabolites, such as protection from UV radiation and defense against grazing [4]. The species and cultivation conditions affect the composition. Lopes et al. [7] found five and six ringed phloroglucinol oligomers in wild grown and aquaculture-grown *F. vesiculosus*, trimers and tetramers in extracts from *Fucus guiryi*, *F. serratus* and *F. spiralis*. Fucophlorethols are dominant in *Fucus sp*, exhibiting molecular weights ranging from 370 to 746 Da, and relatively low degree of polymerization (3–6 phloroglucinol units, PGU). Moreover, isomers of fucophlorethol, dioxinodehydroeckol, difucophlorethol, fucodiphlorethol, bisfucophlorethol, fucofuroeckol, trifucophlorethol, fucotriphlorethol, tetrafucophlorethol, and fucotetraphlorethol were identified.

Heffernan et al. [21] reported that most low molecular weight (LMW) phlorotannins presented 4–16 monomers of phloroglucinol. The level of isomerization differed among macroalgal species and *F. vesiculosus* showed up to 61 compounds with 12 PGU. Species-specific phenolic profiles, with varying degrees of composition, polymerization and isomerization have been described and the antiradical activity observed was not only due to higher phlorotannin concentrations but to their geometric arrangement and the position of the free hydroxyl groups [4]. *F. vesiculosus* low molecular weight fractions were predominantly composed of phlorotannins between 498 and 994 Da (4–8 PGUs), in *P. canaliculata* most structures presented 9–14 PGUs, whereas in *H. elongata* most phlorotannins were composed of 8–13 PGU.

The influence of the solvent has been reported in a number of studies either as extractant or as fractionation medium, i.e., Murugan and Iyer [129] found higher ferrous ion chelation and growth inhibition of MG-63 cells by methanolic and aqueous extracts from *Caulerpa peltata*, *Gelidiella acerosa*, *Padina gymnospora*, and *S. wightii*. However, the higher extraction of phenols and flavonoids was found with chloroform and ethyl acetate, as well as the DPPH radical scavenging and growth inhibitory activities in cancer cells. Aravindan et al. [130] selected dichloromethane and ethyl acetate fractions from *Dictyota dichotoma*, *Hormophysa triquerta*, *Spatoglossum asperum*, *Stoechospermum marginatum* and *P. tetrastromatica* for their high levels of phenolics, antioxidants and inhibitors of pancreatic tumorigenic cells (MiaPaCa-2, Panc-1, BXPC-3 and Panc-3.27) growth. The use of intensification techniques can enhance the extraction yields. Kadam et al. [16] reported that under ultrasound-assisted extraction of *A. nodosum* in acidic media, the extraction of high molecular weight phenolic compounds was facilitated. However, other bioactives were also solubilized in a short time and crude solvent extracts contained several non-phenolic components, such as carbohydrates, amino acids and pigments. Further purification strategies have been tried, i.e., a multistep scheme with successive precipitation of lipophilic compounds and further chromatographic fractionation [24], solvent partition and membrane fractionation [21], solvent partition and column chromatography [19,22], adsorption, washing and further elution [131], or chromatography and then membrane processing by ultrafiltration and dialysis [132].

Table 3. Phenolic compounds found in red and green seaweeds.

Seaweed / Extraction / Chemical Analysis	Compounds / Antioxidant Activity (When Provided)	Ref.
Green seaweeds		
Caulerpa lentillifera, C. taxifolia, Chaetomorpha crassa, Chara sp., Chlorodesmis sp., Cladophora sp. / MeOH, W / GC-MS-EI-SIM	2,4,6-Tribromophenol; 2,4,6-tribromoanisol; 2'-hydroxy-2,3',4,5'-tetrabromodiphenyl ether; 2'-methoxy-2,3',4,5'-tetrabromodiphenyl ether; 6-hydroxy-2,2',4,4'-tetrabromodiphenyl ether; 6-methoxy-2,2',4,4'-tetrabromodiphenyl ether; 2',6-dihydroxy-2,3',4,5'-tetrabromodiphenyl ether; 2,2'-dihydroxy-3,3',5,5'-tetrabromodiphenyl; 2,2'-dimethoxy-3,3',5,5'-tetrabromodiphenyl	[122]
Dasycladus vermicularis / MeOH / UPLC-MS/MS	4-(Sulfooxy)phenylacetic acid; 4-(sulfooxy)benzoic acid	[133]
Red seaweeds		
Acanthophora specifera, Ceratodictyon spongiosum, Gracilaria edulis, Halymenia sp., Hydropuntia edulis, Jania adhaeren, Jania sp., Kappaphycus alvarezii / MeOH, W / GC-MS-EI-SIM	2,4,6-Tribromophenol; 2,4,6-tribromoanisol; 2'-hydroxy-2,3',4,5'-tetrabromodiphenyl ether; 2'-methoxy-2,3',4,5'-tetrabromodiphenyl ether; 6-hydroxy-2,2',4,4'-tetrabromodiphenyl ether; 6-methoxy-2,2',4,4'-tetrabromodiphenyl ether; 2',6-dihydroxy-2,3',4,5'-tetrabromodiphenyl ether; 2,2'-dihydroxy-3,3',5,5'-tetrabromodiphenyl; 2,2'-dimethoxy-3,3',5,5'-tetrabromodiphenyl	[122]
Asparagopsis taxiformis / W, MeOH, CH$_2$Cl$_2$, H / GC-MS	Bromoform, dibromochloromethane, bromochloroacetic acid, dibromoacetic acid	[100]
Bostrychia radicans / MeOH, H, EtOAc / GC-MS, NMR	N,4-dihydroxy-N-(2'-hydroxyethyl)-benzamide; N,4-dihydroxy-N-(2'-hydroxyethyl)-benzeneacetamide; methyl 4-hydroxymandelate; methyl 2-hydroxy-3-(4-hydroxyphenyl)-propanoate	[134]
C. tenuicorne / H + DEt + 2-P / GC-MS, ECNI	Phenols, hydroxylated, and methoxylated penta- and hexabrominated diphenyl ethers	[135]
C. tenuicorne / H + DEt + 2-P / GC-MS, ECNI	Hydroxylated polybrominated diphenyl ethers; 2'-hydroxy-2,3',4,5'-tetrabromodiphenyl ether; 6-hydroxy-2,2',4,4'-tetrabromodiphenyl ether	[136]
Laurencia nipponica, Odonthalia corymbifera, Polysiphonia morrowii / A, W, MeOH / LC-MS, NMR	3,5-Dibromo-4-hydroxybenzaldehyde; 3-bromo-4,5-dihydroxybenzyl ether; 3-bromo-4,5-dihydroxybenzyl alcohol; 5-(2,3-dibromo-4,5-dihydroxybenzyloxy)methyl)-3,4-dibromobenzene-1,2-diol; 5-(2-bromo-3,4-dihydroxy-6-(hydroxymethyl) benzyl)-3,4-dibromobenzene-1,2-diol	[137]

Table 3. Cont.

Seaweed Extraction Chemical Analysis	Compounds Antioxidant Activity (When Provided)	Ref.
O. corymbifera, Neorhodomela aculeata, Symphyocladia latiuscula A, W, MeOH LC-MS, NMR	n-Butyl 2,3-dibromo-4,5-dihydroxybenzyl ether; 3-bromo-4-(2,3-dibromo-4,5-dihydroxybenzyl)-5-methoxymethylpyrocatechol; 2,3-dibromo-4,5- dihydroxybenzyl alcohol; 2,3-dibromo-4,5-dihydroxybenzyl methyl ether; bis-(2,3,6-tribromo-4,5-dihydroxybenzyl) ether; 2,3,6-tribromo-4,5-dihydroxybenzyl methyl ether; 2,2',3,3'-tetrabromo-4,4',5,5'-tetrahydroxydiphenylmethane; 5-(2-bromo-3,4-dihydroxy-6-(hydroxymethyl)benzyl)-3,4-dibromobenzene-1,2-diol; 5-((2,3-dibromo-4,5-dihydroxybenzyloxy)methyl)-3,4-dibromobenzene-1,2-diol	[138]
Odonthalia corymbifera MeOH, EtOAc NMR	Odonthalol, odonthadione	[139]
Polysiphonia decipiens 3:1 MeOH:CH2Cl2 NMR	α-O-Methyllanosol; lanosol; 5-(2-bromo-3,4-dihydroxy-6-(hydroxymethyl)benzyl)-3,4-dibromobenzene-1, 2-diol; 5-(2-bromo-3, 4-dihydroxy-6-(methoxymethyl)benzyl)-3, 4-dibromobenzene-1, 2-diol; rhodomelol; polysiphonol	[140]
Polysiphonia morrowii W, MeOH, CH2Cl2 NMR	3-bromo-4,5-dihydroxybenzyl methyl ether; 3-bromo-4,5-dihydroxybenzaldehyde	[141]
P. morrowii W, MeOH ESI-MS, NMR	bis (3-Bromo-4,5-dihydroxybenzyl) ether	[142]
Rhodomela confervoides EtOH NMR	3-(2,3-Dibromo-4,5-dihydroxybenzyl) pyrrolidine-2,5-dione; methyl 4-(2,3-dibromo-4,5-dihydroxybenzylamino)-4-oxobutanoate; 4-(2,3-dibromo-4,5-dihydroxybenzylamino)-4-oxobutanoic acid; 3-bromo-5-hydroxy-4-methoxybenzamide; 2-(3-bromo-5-hydroxy-4-methoxyphenyl)acetamide; 3-bromo-4,5-bis(2,3-dibromo-4,5-dihydroxybenzyl) pyrocatechol; methyl 1-(2-(2,3-dibromo-4,5-dihydroxybenzyl)-3-bromo-4,5- dihydroxybenzyl)-5-oxopyrrolidine-2-carboxylate; 5-(2,3-dibromo-4,5-dihydroxybenzyloxy)methyl)-3,4-dibromobenzene-1,2-diol; 5-(2-bromo-3,4-dihydroxy-6-(hydroxymethyl)benzyl)-3,4-dibromobenzene-1,2-diol; 5-(2-bromo-6-(methoxymethyl) benzyl)-3,4-dibromobenzene-1,2-diol; 5-(2-bromo-6-(ethoxymethyl)-3,4-dihydroxybenzyl)-3,4-dibromobenzene-1,2-diol; 5-(2,3-dibromo-4,5-dihydroxybenzyl)- 3,4-dibromobenzene-1,2-diol; 1-(2,3-dibromo-4,5-dihydroxybenzyl)-5-oxopyrrolidine-2-carboxylic acid; methyl 1-(2,3-dibromo-4,5-dihydroxybenzyl)-5-oxopyrrolidine-2-carboxylate EC50, DPPH = 5.2–23.6 μmol/L; ABTS = 2.1–3.6 mmol TE/L	[143]
Symphyocladia latiuscula EtOH NMR	2,3-Dibromo-4,5-dihydroxybenzyl methyl ether; 3,5-dibromo-4-hydroxybenzoic acid; 2,3,6-tribromo-4,5-dihydroxymethylbenzene; 2,3,6-tribromo-4,5-dihydroxybenzaldehyde; 2,3,6-tribromo-4,5-dihydroxybenzyl methyl ether; bis(2,3,6-tribromo-4,5-dihydroxyphenyl)methane; 1,2-bis(2,3,6-tribromo-4,5-dihydroxyphenyl)-ethane; 1-(2,3,6-tribromo-4,5-dihydroxybenzyl)-pyrrolidin-2-one	[144]

Table 3. *Cont.*

Seaweed Extraction Chemical Analysis	Compounds Antioxidant Activity (When Provided)	Ref.
S. latiuscula EtOH, EtOAc, W HRMS, NMR, MS	1-[2,5-dibromo-3,4-dihydroxy-6-(2,3,6-tribromo-4,5-dihydroxybenzyl)benzyl]pyrrolidin-2-one; methyl 4-[(2,3,6-tribromo-4,5-dihydroxybenzyl)[(2,3,6-tribromo-4,5-dihydroxybenzyl)carbamoyl] amino]butanoate; methyl 4-[(2,5-dibromo-3,4-dihydroxybenzyl)[(2,3,6-tribromo-4,5-dihydroxybenzyl)carbamoyl]amino] butanoate; 2,5-dibromo-3,4-dihydroxy-6-(2,3,6-tribromo-4,5-dihydroxybenzyl)benzyl methyl ether $EC_{50, DPPH}$ = 14.5, 20.5 µg/L	[145,146]
S. latiuscula EtOH, EtOAc, W ESIMS, NMR	Bromocatechol conjugates (symphyocladins)	[147,148]
S. latiuscula MeOH, CH$_2$Cl$_2$, EtOAc, BuOH, W ESIMS, NMR	2,3,6-tribromo-4,5-dihydroxybenzyl alcohol; 2,3,6-tribromo-4,5-dihydroxybenzyl methyl ether; bis-(2,3,6-tribromo-4,5-dihydroxybenzyl) ether	[149]
Vertebrata lanosa MeOH, EtOAc ESIMS, NMR	2,3-Dibromo-4,5-dihydroxybenzylaldehyde; 2,2'-tribromo-3',4,4',5-tetrahydroxy-6'-hydroxymethyldiphenylmethane; bis(2,3-dibromo-4,5-dihydroxylbenzyl) ether; 5,5''-oxybis(methylene)bis (3-bromo-4-(2',3'-dibromo-4',5'-dihydroxylbenzyl)benzene-1,2-diol) ORAC = 0.08–0.33 µg TE/mL; CAA$_{inhib. 10 µg/mL}$ = 68%; CLPAA$_{inhib. 10 µg/mL}$ = 100%	[95]
V. lanosa	Methylrhodomelol; lanosol; lanosol methyl ether; 2-amino-5-(3-(2,3-dibromo-4,5-dihydroxybenzyl)ureido)pentanoic acid; 3-bromo-4-(2,3-dibromo-4,5-dihydroxybenzyl)-5-methoxymethylpyrocatechol; 5-((2,3-dibromo-4,5-dihydroxybenzyloxy)methyl)-3,4-dibromobenzene-1,2-diol; 2,2',3,3'-tetrabromo-4,4',5,5'-tetrahydroxydiphenylmethane	[150]

ABTS: 2,2'-azino-bis (3-ethylbenzothiazoline-6-sulphonic acid); ASE: ascorbic acid equivalents CA: chelating ability; BCBM: β-carotene bleaching; CAA: antioxidant assay for cellular antioxidant activity [95]; CLPAA: cellular lipid peroxidation antioxidant activity assay [95]; DCM: dichloromethane; DEtE: diethylether; DPPH: 2,2-diphenyl-1-picrylhydrazyl; ECNI: electron capture negative ionization; EtOH: ethanol; EtOAc: ethyl acetate; FRAP: ferric reducing antioxidant power; H: hexane; LPIA: lipid peroxidation inhibition assay; MeOH: methanol; NBT: superoxide anion scavenging test; ORAC: oxygen radical absorbance capacity; 2-P: 2-propanol; PGU: phloroglucinol units; RP: reducing power; SRSA: superoxide radical scavenging assay [96]; TE: trolox equivalents; TEAC: trolox equivalent antioxidant capacity; W: water.

18

Kirke et al. [118] reported that low molecular weight phlorotannin fractions (<3 kDa) from *F. vesiculosus* in powder form remained stable under storage for 10 weeks, when exposed to temperature and oxygen. Although, when suspended in an aqueous matrix, this fraction underwent oxidation when exposed to atmospheric oxygen and 50 °C, and both the DPPH radical scavenging activity and the content of phlorotannins with 6–16 PGUs decreased.

Since other compounds found in the crude seaweed extracts could be responsible for the biological activities, the correlation between phenolic content and antioxidant properties was not always found in seaweed extracts. These relationships should also consider the type of activity assayed since some of them share the same mechanisms [151]. Furthermore, not only the phenolic content is determinant, but also their structure. In brown seaweeds, the classical correlation of phenolic content and radical scavenging has been established with the antiradical properties and molecular weight. Phlorotannin-enriched fractions from water and aqueous ethanolic extracts of *A. nodosum* and *Pelvetia canaliculata* contain predominantly larger phlorotannins (DP 6–13) compared to *F. spiralis* (DP 4–6) [25]. The 3.5–100 kDa and/or >100 kDa fractions from the cold water and aqueous ethanolic extracts showed higher phenolic content, radical scavenging abilities and ferric reducing antioxidant power (FRAP) than the <3.5 kDa, which could enhance their activity after a reverse-phase flash chromatography fractionation [152]. In a study on *F. vesiculosus*, Bogolitsyn et al. [22] concluded that the highest radical scavenging activity was observed for average molecular weights from 8 to 18 kDa and the activity decreased with increasing molecular weight from 18 to 49 kDa. This effect has been ascribed to the formation of intramolecular and intermolecular hydrogen bonds between hydroxyl groups, causing conformational changes in phlorotannin molecules and, therefore, mutual shielding and a decrease in the availability of active centers. *Ascophylllum nodosum* purified oligophenolic fraction was more active than the crude fraction as ABTS scavengers, and the fraction containing phenolic compounds with a MW ≥50 kDa was the most active and showed higher correlation with the content of phenolic compounds [132]. Whereas the radical scavenging activity and reducing power showed correlation, particularly in brown seaweeds, the chelating properties did not, and were higher in green seaweed extracts, because the major activity could come from the saccharidic fraction [13]. The FRAP activity displayed a stronger correlation with the phlorotannin content than the radical scavenging capacity, as well as the phenolic content, molecular weight and structural arrangement [4]. However, other authors did not find any significant correlation between the total phenolic content of the extracts and the inhibition of red blood cell hemolysis and lipid peroxidation [153], or the antioxidant activity (DPPH and β-carotene bleaching assays) [154].

2.2.4. Seagrasses

Compared to algae, seagrasses are scarcely exploited [11,33]. The worldwide distributed *Zostera* genus produces large amounts of leaf material. This is not utilized, representing an abundant waste which could be proposed to recover valuable compounds and therefore compensate the costs of cleaning beaches and shorelines used for recreational purposes [10].

Seagrasses are a rich source of (poly)phenolics, including simple and sulfated phenolic acids, such as zosteric acid, and condensed tannins [33,83]. Rosmarinic acid and caffeic acid (0.4–19.2 mg/g) were the major phenolic components in leaves and roots-rhizomes of eelgrass (*Zostera marina* L.), and higher concentrations have been found during spring in the younger leaves and roots-rhizomes [9,78]. Rosmarinic acid was also reported as an active phenolic in the methanolic extracts from detritus of *Z. noltii* and *Z. marina* (2.2–18.0 and 1.3–11.2 mg/g, respectively) [10]. Extraction yields, seasonally dependent for the two species, vary in the range of 9.3–19.7% (g/g dw) for *Z. noltii*, and 9.6–31% for *Z. marina*; near 85% of the rosmarinic acid was recovered from the crude methanolic extract using ethyl acetate [10]. Chicoric acid was the major compound in *Ruppia* sp, with 30 mg/g; twice the total flavonoid content [81]. High concentrations of phenolics in methanolic *Zostera* extracts correspond to higher growth inhibition of the toxic red tide dinoflagellate *A. catenella* [23]. In sc-CO_2 extracts with ethanol or methanol cosolvents, the phenolic content and radical scavenging capacity correlated well with the cytotoxicity on tumoral cell lines; this high activity might be due to the high content of phenylpropanoids [155]. The supercritical CO_2 extraction of phenolic compounds from *Zostera marina* residues using 20% ethanol as co-solvent enhanced the solubilization of polar compounds (chicoric, p-coumaric, rosmarinic, benzoic, ferulic and caffeic) [29,155], reaching phenolic contents comparable to those found in the ethanolic and methanolic Soxhlet extracts; the DPPH radical-scavenging activities were also similar.

2.2.5. Sponges

Despite being a rich source of highly bioactive compounds [93], there are few studies in the literature regarding the extraction and identification of polyphenols in sponges. Methanol and dichloromethane were normally used for extraction, while new phenolic compounds have been identified using HRAPCIMS, HRESIMS and NMR. Bisabolenes are particularly interesting polyphenolic compounds found in sponges. These phenolics are characterized by a C-7 absolute stereochemistry. All sponge bisabolenes possess a unique 7S configuration, while other marine and terrestrial bisabolenes possess a 7R configuration [91]. (S)-(+)-curcuphenol, a member of this family commonly found in sponges, presents several biological activities [91].

3. Bioactive Properties of Marine Phenolics

Epidemiological, clinical and nutritional studies strongly support the evidence that dietary polyphenols play important roles in human health. Their regular consumption has been associated with a reduced risk of different chronic diseases, including cardiovascular diseases (CVDs), cancer and neurodegenerative disorders [156]. Marine polyphenols have also attracted much attention because, similar to other polyphenols, they are bioactive compounds with potential health benefits in numerous human diseases due to their enzyme inhibitory effect and antimicrobial, antiviral, anticancer, antidiabetic, antioxidant, or anti-inflammatory activities; however, most of the findings are based on in vitro assays and animal testing on rodents.

Studies demonstrating the multi-targeted protective effect of marine phenolics, focused on the most prevalent diseases such as type 2 diabetes mellitus, obesity, metabolic syndrome, Alzheimer's disease and cancer, are included in this section. In addition, the influence of marine phenolics on gut human microbiota and other infectious have been also addressed (Tables 4–10).

3.1. Type 2 Diabetes Mellitus

Type 2 diabetes mellitus (T2DM) is one of the most common non-communicable diseases in the world, which can be attributed to hyperglycemia characterized by a high glucose concentration circulating in the blood, and has a marked impact on the quality of life [157]. This disease leads to higher risk of premature death and is associated with several health problems such as vision loss, kidney failure, leg amputation, nerve damage, heart attack and stroke [158]. Due to its chronic nature, T2DM is also associated with several comorbidities such as metabolic syndrome (MetS), overweight and obesity, hypertension, non-alcoholic hepatic steatosis, coronary disease, and neuropathy, among others [159].

Phlorotannins of edible seaweeds are involved in various antidiabetic mechanisms: inhibition of starch-digesting enzymes α-amylase and α-glucosidase, protein tyrosine phosphatase 1B (PTP1B) enzyme inhibition, modulation of glucose-induced oxidative stress and reduction in glucose levels and lipid peroxidation, among others [160,161]. There are a few recent reviews that summarize the huge number of in vitro studies along with minor number of in vivo studies focused in evaluating the antidiabetic activity of polyphenols [35,36,160,162] or bioactive components of seaweeds [161,163]. Key in vitro studies along with the recent in vivo studies about antidiabetic activity of marine polyphenols are detailed below (Table 4).

Alpha-amylase, located in the pancreas, and α-glucosidase, at the brush border of intestinal cells, are two key enzymes involved in carbohydrate metabolism [164]. These enzymes break down carbohydrates into monosaccharides that are absorbed into the bloodstream, resulting in a rise in blood glucose following a meal. Oral glucosidase inhibitor drugs are the common clinical treatment for T2DM; however, long-term use can result in side effects such as renal tumors and hepatic injury [164]. Hence, looking for alternative natural products with no side effects is an active research area. Most brown seaweeds belonging to the genus *Ecklonia* and family *Lessoniaceae* have been reported to exhibit antidiabetic activities [160]. Five isolated phlorotannins from *E. cava*, fucodiphloroethol G, dieckol, 6,6'-bieckol, 7-phloroeckol, phlorofucofuroeckol-A, have shown a marked α-glucosidase inhibition with 19.5 μM, 10.8 μM, 22.2 μM, 49.5 μM and 19.7 μM, respectively, as well as some α-amylase inhibitory effect with IC_{50} values of >500 μM, 125 μM, >500 μM, 250 μM and >500 μM, respectively [165]. Phlorotannins extracted from *A. nodosum* [166,167], *Alaria marginata* and *Fucus distichus* [168] are also able to inhibit α-amylase and α-glucosidase, while those of *F. vesiculosus* [160] and *L. trabeculate* [103] inhibit α-glucosidase activity, and *L. trabeculate* [103] inhibits lipase activity (Table 4). Generally, seaweed extracts and isolated compounds exhibited more inhibitory potency towards α-glucosidase compared to α-amylase (see IC_{50} values in Table 4), which is desirable since high inhibition of α-amylase activity has been suggested to cause abnormal fermentation of undigested carbohydrates by the colonic microbiota [169]. This promising inhibitory activity towards the enzymes involved in the digestion of carbohydrates has led to the development of polyphenol-rich extracts from seaweeds as alternative drugs to treat T2DM. Catarino et al. [120] obtained crude extracts and semi-purified phlorotannins from *F. vesiculosus* containing fucols, fucophlorethols, fuhalols and several other phlorotannin derivatives, tentatively identified as fucofurodiphlorethol, fucofurotriphlorethol and fucofuropentaphlorethol. These extracts showed the potential to control the activities of α-amylase, pancreatic lipase, and particularly α-glucosidase, for which a greater inhibitory effect was observed compared to the pharmaceutical drug acarbose (IC_{50}~4.5 − 0.82 μg/mL against 206 μg/mL, respectively). Park et al. [170] isolated minor phlorotannin derivatives from *E. cava* that effectively inhibited the activity of α-glucosidase, with IC_{50} values ranging from 2.3 to 59.8 μM; they obtained the kinetic parameters of the receptor–ligand binding by a fluorescence-quenching study. In the same line, Lopes et al. [8] isolated phlorotannins from four edible *Fucus* species (*F. guiryi*, *F. serratus*, *F. spiralis* and *F. vesiculosus*). These were chemically characterized using mass spectrometry-based techniques (HPLC–DAD–ESI/MS and UPLC–ESI–QTOF/MS). The isolated phlorotannins showed inhibitory activity against α-amylase and α-glucosidase, being particularly important in the activity of the latter, with IC_{50} values significantly lower (between 2.48 and 4.77 μg/mL) than those obtained for the pharmacological inhibitors acarbose and miglitol (between 56.43 and 1835.37 μg/mL). *F. guiryi* and *F. serratus* were the most active of the tested *Fucus* species. In addition, xanthine oxidase activity, an enzymatic system usually overexpressed in diabetes and responsible for producing deleterious free radicals, was also inhibited, related with the antioxidant activity associated to phlorotannins [8].

Protein tyrosine phosphatase 1B (PTP1B) is a major negative regulator of insulin signaling and is localized on the cytoplasmic surface of the endoplasmic reticulum in hepatic, muscular and adipose tissues. Due to its ubiquity in the insulin-targeted tissues and its role in insulin resistance development [142], inhibition of PTP1B activity would be a target for the treatment of T2DM and obesity. Ezzat et al. [171] reviewed the in vitro studies focused on evaluating the inhibitory activity of PTP1B marine polyphenols. Xu et al. [172] studied the inhibitory activity of a marine-derived bromophenol compound (3,4-dibromo-5-(2-bromo-3,4-dihydroxy-6-(ethoxymethyl)benzyl)benzene- 1,2-diol) isolated from the red alga *Rhodomela confervoides* in insulin-resistant C2C12 myotubes. This bromophenol has the ability to inhibit PTP1B activity (IC_{50} 0.84 μM), permeate into cells and bind to the catalytic domain of PTP1B in vitro, activate insulin signaling and potentiate insulin sensitivity in C2C12 myotubes as well as enhance glucose uptake. Similarly, 3-bromo-4,5-bis(2,3-dibromo-4,5-dihydroxybenzyl)-1,2-benzenediol isolated from the red alga *Rhodomela confervoides* was able to activate insulin signaling and prevent palmitate-induced insulin resistance by intrinsic PTP1B inhibition (IC_{50} 2.0 μM). Moreover, this compound also activated the fatty acid oxidation signaling in palmitate-exposed C2C12 myotubes [173].

Glycated insulin is commonly found in T2DM patients and is less effective in controlling glucose homeostasis and stimulating glucose uptake than non-glycated insulin [174]. Non-enzymatic protein glycation is an irreversible modification between reducing sugars and primary amino groups and leads to the production of advanced glycation end-products (AGEs) [175], whose accumulation causes various diabetic complications such as nephropathy, retinopathy and atherosclerosis as well as stimulates the development of neurodegenerative diseases such as Alzheimer's disease (AD) [176]. The inhibition of AGEs formation is another approach being explored in managing hyperglycemia using seaweeds. Crude phlorotannins contained in the Japanese *Lessoniaceae* exhibited an inhibitory effect on the formation of fluorescence bound AGEs (IC_{50} 0.43–0.53 mg/mL), and among the purified phlorotannins (phlorofucofuroeckol-A, eckol, phloroglucinol, fucofuroeckol A, dieckol and 8,8′-bieckol), phlorofucofuroeckol A showed the highest inhibitory activity (IC_{50} 4.1–4.8 × 10^2 µM) against fluorescent AGEs formation, being about 15 times more active than the reference drug aminoguanidine hydrochloride [177]. Further studies carried out with methanolic extracts from brown algae *Padina pavonica*, *Sargassum polycystum*, and *Turbinaria ornata*, rich in phlorotannins, inhibited the glucose-induced protein glycation and formation of protein-bound fluorescent AGEs (IC_{50} 15.16 µg/mL, 35.25 µg/mL and 22.7 µg/mL, respectively). Furthermore, brown algal extracts containing phlorotannins exhibited protective effects against AGEs formation in *Caenorhabditis elegans* (a species of nematode) with induced hyperglycemia [178]. From five phlorotannins isolated from *E. stolonifera*, only phlorofucofuroeckol-A inhibited, in a dose-dependent form, the induced non-enzymatic insulin glycation of D-ribose and D-glucose (IC_{50} 29.50 µM and 43.55 µM, respectively) [179]. These authors used computational analysis to find that phlorofucofuroeckol-A interacts with the Phe1 in insulin chain-B, blocking D-glucose access to the glycation site of insulin.

The need to secrete increasing amounts of insulin to compensate for progressive insulin resistance and the hyperglycemia-induced oxidative stress lead to an eventual deterioration of pancreatic β-cells [180]. Lee et al. [181] confirmed the protective effect of octaphlorethol A, a novel phenolic compound isolated from *Ishige foliacea*, against streptozotocin (STZ)-induced pancreatic β-cell damage investigated in a rat insulinoma cell line (RINm5F pancreatic β-cells). Thus, octaphlorethol A reduced the intracellular reactive oxygen species (ROS) and generation of thiobarbituric acid reactive substances (TBARs), extensively produced by STZ-treated pancreatic β-cells. The oxidative stress involved in diabetes-associated pathological damages reduces antioxidant enzyme activities (catalase (CAT), superoxide dismutase (SOD), and glutathione peroxidase (GPx)), and octaphlorethol A treatment increased the enzyme activity due to its antioxidant potency. The phlorotannins isolated from *E. cava*, 6,6-bieckol, phloroeckol, dieckol and phlorofucofuroeckol inhibited high glucose-induced ROS and cell death in zebrafish. Particularly, the antioxidant activity of dieckol significantly reduced heart rates, ROS, nitric oxide (NO) and lipid peroxidation generation in high glucose-induced oxidative stress. Dieckol also reduced overexpression of inducible nitric oxide synthase (iNOS) and cyclooxygenase-2 (COX-2) [182]. A recent study addressed the efficacy of an extract of the red seaweed *Polysiphinia japonica* on preserving cell viability and glucose-induced insulin secretion in a pancreatic β-cell line, Ins-1, treated with palmitate [183]. However, the tested extract contained, in addition to polyphenols, other components such as carbohydrates, lipid and proteins; hence, the described bioactivities may not be due only to polyphenols.

Glucose uptake and disposal mainly occurs in the skeletal muscle, playing an important role in the energy balance regulation [184]; marine polyphenols are also involved in this mechanism. Lee et al. [185] confirmed that octaphlorethol A from *Ishige foliacea* increased glucose uptake in skeletal muscle cells (differentiated L6 rat myoblast). Furthermore, this compound increased glucose transporter 4 (Glut4) translocation to the plasma membrane, in a process depending on the protein kinase B (Akt) and AMP-activated protein kinase (AMPK) activation, a therapeutic target for treatment of hyperglycemia, which is associated with insulin resistance [186].

Table 4. Effect of marine phenolics in the prevention of type 2 diabetes mellitus (T2DM).

Compounds/Marine Source	Test Model	Outcome	Ref.
Five isolated phlorotannins from *E. cava* (fucodiphlorethol G, dieckol, 6,6'-bieckol, 7-phloroeckol, phlorofucofuroeckol-A)	In vitro assay: α-glucosidase and α-amylase inhibitory activity	Inhibition of α-glucosidase (IC$_{50}$ values ranged from 10.8 μM for dieckol to 49.5 μM for 7-phloroeckol) and α-amylase (IC$_{50}$ values ranged from 125 μM for dieckol to <500 μM for the rest of compounds, except 7-phloroeckol with a value of 250 μM) activities	[165]
Methanolic extract isolated from *A. nodosum* rich in phlorotannins	In vitro assay: α-glucosidase and α-amylase inhibitory activity	Inhibition of α- glucosidase (IC$_{50}$~20 μg/mL GAE) and α-amylase (IC$_{50}$~0.1 μg/mL GAE) activities	[166]
Cold aqueous and ethanolic extracts of *A. nodosum* and *F. vesiculosus* rich in phlorotannins	In vitro assay: α-glucosidase and α-amylase inhibitory activity	Inhibition of α- glucosidase (IC$_{50}$~0.32–0.50 μg/mL GAE for *F. vesiculosus*) and α-amylase (IC$_{50}$~44.7–53.6 μg/mL GAE for *A. nodosum*) activities	[167]
Methanolic extract from *Alaria marginata* and *Fucus distichus* rich in phlorotannins	In vitro assay: α-glucosidase and α-amylase inhibitory activity	Inhibition of α- glucosidase (IC$_{50}$~0.89 μg/mL) and α-amylase (IC$_{50}$~13.9 μg/mL) activities	[168]
Polyphenol-rich extracts from *L. trabeculate*	In vitro assay: α-glucosidase and lipase activity	Inhibition of α-glucosidase and lipase activities (IC$_{50}$ < 0.25 mg/mL)	[103]
Crude extract and semi-purified phlorotannins from *F. vesiculosus* composed by fucols, fucophlorethols, fuhalols and several other phlorotannin derivatives	In vitro assay: α-glucosidase, α-amylase and pancreatic lipase inhibitory activity	Inhibition of α-amylase (IC$_{50}$~28.8–2.8 μg/mL), α-glucosidase (IC$_{50}$~4.5–0.82 μg/mL) and pancreatic lipase (IC$_{50}$~45.9–19.0 μg/mL) activities	[120]
Phlorotannin derivatives from *E. cava*	In vitro assay: α-glucosidase inhibitory activity	Inhibition of α-glucosidase activity (IC$_{50}$~2.3–59.8 μM) Kinetic parameters of receptor–ligand binding	[163]
Phlorotannin-targeted extracts from four edible *Fucus* species (*F. guiryi*, *F. serratus*, *F. spiralis* and *F. vesiculosus*)	In vitro assay: α-glucosidase and α-amylase inhibitory activity	Inhibition of α-glucosidase (IC$_{50}$~2.48–4.77 μg/mL), α-amylase (IC$_{50}$~23.31–253.31 μg/mL) and xanthine oxidase (IC$_{50}$~157.66–800.08 μg/mL) activities	[8]
Marine-derived bromophenol compound (3,4-dibromo-5-(2-bromo-3,4-dihydroxy-6-(ethoxymethyl)benzyl)benzene-1,2-diol) isolated from *Rhodomela confervoides*	In vitro: insulin resistant C2C12 cells treated with bromophenol (0.1–0.5 μM for phenol)	Inhibition of PTP1B activity (IC$_{50}$~0.84 μM) Activation of insulin signaling and potentiate insulin sensitivity	[172]
3-Bromo-4,5-bis(2,3-dibromo-4,5-dihydroxybenzyl)-1,2-benzenediol isolated from the red alga *Rhodomela confervoides*	In vitro: palmitate-induced insulin resistance in C2C12 cells treated with bromophenol (0.5–2.0 μM for phenol)	Inhibition of PTP1B activity (IC$_{50}$~2 μM) Activation of insulin signaling and prevent palmitate-induced insulin resistance	[173]
Phlorofucofuroeckol-A, eckol, phloroglucinol, fucofuroeckol A, dieckol and 8,8'-bieckol isolated and crude phlorotannins from *Lessoniaceae*	In vitro assay: human and bovine serum albumin models	Inhibition of AGEs formation, crude phlorotannins showed IC$_{50}$~0.43–0.53 mg/mL, and among the purified phlorotannins, phlorofucofuroeckol A was the most active (IC$_{50}$~4.1–4.8 μM)	[177]
Methanolic extract from *P. pavonica* and *Turbinaria ornate* rich in phlorotannins	In vitro assay: BSA-glucose assay In vivo: *Caenorhabditis elegans* with induced hyperglycemia	Inhibition of AGEs formation (IC$_{50}$~15.16 μg/mL, 35.25 μg/mL and 22.70 μg/mL, respectively) Inhibition of AGEs formation	[178]
Phlorofucofuroeckol-A isolated from *E. stolonifera*	In vitro assay for non-enzymatic insulin glycation	Inhibition of AGEs formation (IC$_{50}$ 29.50-43.55 μM for D-ribose and D-glucose-induced insulin glycation, respectively)	[179]
Octaphlorethol A isolated from *Ishige foliacea*	In vitro: STZ-induced pancreatic β-cell damage (RINm5F pancreatic β-cells) (12.5–50.0 μg/mL for phenol)	Decreased the death of STZ-treated pancreatic β-cells Decreased the TBARS and ROS Increased the activity of antioxidant enzymes	[181]
6,6-Bieckol, phloroeckol, dieckol and phlorofucofuroeckol isolated from *E. cava*	In vivo: high glucose-stimulated oxidative stress in Zebrafish, a vertebrate model (10–20 μM of phenols)	Inhibition of high glucose-induced ROS and cell death Dieckol reduced the heart rates, ROS, NO and lipid peroxidation Dieckol reduced the overexpression of iNOS and COX-2	[182]
Extract isolated from the red seaweed *Polysiphonia japonica*	In vitro: palmitate-induced damage in β-cells (Ins-1 cells) (1–10 μg/mL of extract)	Inhibited the palmitate-induced damage in β-cells Preserved the glucose-induced insulin secretion in β-cells	[183]

Table 4. *Cont.*

Compounds/Marine Source	Test Model	Outcome	Ref.
Octaphlorethol A from *Ishige foliacea*	In vitro: rat myoblast L6 cells (6.25–50 µM of phenol)	Increased the glucose uptake Increased the Glut4 translocation to the plasma membrane, via Akt and AMPK activation	[185]
Dieckol *isolated* from *E. cava*	In vivo: STZ-induced diabetic mice (acute; 100 mg/kg bw of dieckol administered orally)	Delayed the absorption of dietary carbohydrates	[187]
2,7″-Phloroglucinol-6,6′-bieckol from *E. cava*	In vivo: STZ-induced diabetic mice (acute, 10 mg/kg bw of phenol administered orally)	Delayed the absorption of dietary carbohydrates Inhibition of α-glucosidase and α-amylase activities (IC$_{50}$ 23.35 µM and 6.94 µM, respectively)	[188]
Polyphenol-rich seaweed extract from *F. vesiculosus*	In vivo: 38 healthy adults (acute, 500 mg and 2000 mg of phenol)	No change in postprandial blood glucose and insulin levels	[189]
Dieckol isolated from brown seaweed *E. cava*	In vivo: a T2DM mouse model (C57BL/KsJ-db/db) (10 and 20 mg/kg bw of phenol for 14 days administered intraperitoneal injection)	Diminished the fasting blood glucose and insulin levels Diminished the body weight Decreased the TBARS Increased the activity of antioxidant enzymes in liver tissues Increased the levels of AMPK and Akt phosphorylation in muscle tissues	[190]
Polyphenol-rich extracts from brown macroalgae *L. trabeculata*	In vitro assay: α-glucosidase and lipase inhibitory activities	Inhibition of α-glucosidase and lipase activities (IC$_{50}$ < 0.25 mg/mL)	[103]
	In vivo: high-fat diet and STZ-induced diabetic rats (200 mg/kg/day bw of phenol for 4 weeks by gavage)	Diminished the fasting blood glucose and insulin levels Improved the serum lipid profile Improved the antioxidant stress parameters	
Water-ethanolic extract of green macroalgae *Enteromorpha prolifera* rich in flavonoids	In vivo: STZ-induced diabetic rats (150 mg/kg/day bw of phenol for 4 weeks by gavage)	Diminished the fasting blood glucose and improved oral glucose tolerance Hypoglycemic effect by increasing IRS1/PI3K/Akt and suppressing JNK1/2 in liver	[191]
Dieckol-rich extract of brown algae *E. cava*	In vivo: 8 pre-diabetic adults (1500 mg per day for 12 weeks)	Decreased the postprandial glucose, insulin, and C-peptide levels	[192]

GAE: gallic acid equivalents; PTP1B: protein tyrosine phosphatase 1B; AGEs: advanced glycation end-products; ROS: reactive oxygen species; TBARs: thiobarbituric acid reactive substances; NO: nitric oxide; iNOS: inducible nitric oxide synthase; COX-2: cyclooxygenase-2; Glut4: glucose transporter 4; Akt: protein kinase B; AMPK: AMP-activated protein kinase; PI3K: phosphoinositide 3-kinase; IRS1: Insulin receptor substrate 1; JNKs: c-Jun N-terminal kinases.

Since postprandial blood glucose is a stronger predictor of cardiovascular events than fasting blood glucose in T2DM [193], polyphenol-rich extracts from seaweeds have been evaluated for their postprandial effect. After oral administration of soluble starch with dieckol (100 mg/kg bw), isolated from *E. cava*, a significant reduction in the postprandial blood glucose level in both normal mice and STZ-induced diabetic mice [187] were observed. Likewise, a phlorotannin constituent of *E. cava* (2,7''-phloroglucinol-6,6'-bieckol) inhibited α-glucosidase and α-amylase activities (IC_{50} values of 23.35 and 6.94 μM, respectively), which was more effective than that observed with the positive control acarbose (IC_{50} values of 130.04 and 165.12 μM, respectively). In addition, this phlorotannin alleviated postprandial hyperglycemia in diabetic mice treated with 10 mg/kg bw [188]. A randomized cross-over trial carried out by Murray et al. [189] evaluated the impact of a single dose of a polyphenol-rich seaweed extract from *F. vesiculosus* on postprandial glycemic control in 38 healthy adults. Neither low (500 mg) nor high (2000 mg) doses of the polyphenol-rich brown seaweed affected the postprandial blood glucose and insulin levels in healthy volunteers.

The in vivo chronic treatment with polyphenol-rich extracts from seaweeds showed an important activity in the attenuation of T2DM. The antidiabetic activity of dieckol isolated from brown seaweed *E. cava* was evaluated in a T2DM mouse model (C57BL/KsJ-db/db). Dieckol was administrated daily at doses of 10 and 20 mg/kg bw for 14 days. Results showed a significant reduction in blood glucose and serum levels as well as body weight, when compared to the untreated group [190]. Furthermore, reduced TBARs and increased activity of antioxidant enzymes (SOD, CAT and GPx) in liver tissues, as consequence of the antioxidant potency of phlorotannins, and increased levels of AMPK and Akt phosphorylation in muscle tissues, which play a vital role in the glucose homeostasis, were observed in the dieckol treated group. Another recent study showed the capacity of a polyphenol-rich extracts from the brown macroalgae *Lessonia trabeculata* to attenuate hyperglycemia in high-fat diet and STZ-induced diabetic C57BL/6J rats treated for 4 weeks (200 mg/kg bw/day). Lower fasting blood glucose and insulin levels, as well as a better serum lipid profile and antioxidant stress parameters compared with the diabetic control group, were observed [103]. Similarly, a water-ethanolic extract of green macroalgae *E. prolifera* rich in flavonoids showed antidiabetic activity, by improving oral glucose tolerance and insulin sensitivity, decreasing fasting blood glucose levels and protecting kidney and liver from high sucrose–fat diet on STZ-induced diabetic mice treated with 150 mg/kg bw/day of the assayed extract for 4 weeks. This flavonoid-rich fraction revealed a hypoglycemic effect as confirmed by activation of the IRS1/PI3K/Akt and inhibition of the c-Jun N-terminal kinases (JNK)1/2 insulin pathway in liver [191].

In pre-diabetic human subjects, the efficacy and safety of a dieckol-rich extract from *E. cava* was evaluated by the development of a double-blind, randomized, placebo-controlled clinical trial. The daily consumption of 1500 mg of the dieckol-rich extract decreased postprandial glucose, insulin, and C-peptide levels after 12 weeks, but there was no significant difference between the supplemented and placebo groups [192].

Strong evidence regarding the antidiabetic activity of several marine polyphenols have been obtained (Table 4). They are involved in different and complementary mechanisms (Figure 2), although most of the studies are in vitro or in vivo with animals, not with humans as desirable.

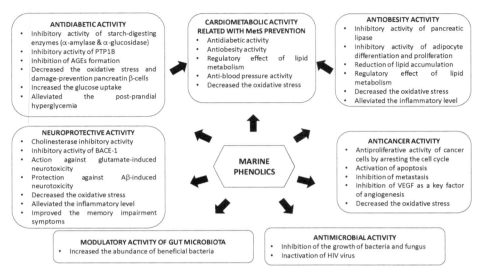

Figure 2. Mechanisms of action of marine phenolics.

3.2. Obesity

Obesity—defined as the excessive or abnormal accumulation of body fat in the adipose tissue, energy imbalance, and lipogenesis—results from modern lifestyles characterized by high intakes of fat, sugar, and calories, in addition to poor exercise and physical activity [194]. The molecular mechanism of obesity mediated by cytokines, adiponectin, and leptin has been correlated with increasing inflammation and oxidative stress, and leads to the development of metabolic diseases including certain types of cancer, hyperglycemia, T2DM, high blood pressure, as well as liver, heart, and gallbladder diseases [195–197]. Consequently, researchers have been exploring functional materials of plant origin that contain antioxidants and other properties to combat obesity and its comorbidities, as an alternative to conventional approaches such as surgery and antiobesity drugs.

Pancreatic lipase is a key enzyme for triglyceride absorption in the small intestine, which hydrolyses triglycerides into glycerol and fatty acids. Pancreatic lipase inhibitors hinder fat digestion and absorption and are a potential therapeutic target for the treatment of diet-induced obesity in humans. In an ongoing search for new pancreatic lipase inhibitors from natural sources, polyphenols isolated from seaweeds have repeatedly shown inhibitory activity against this enzyme, such as a methanolic extract of the marine brown algae *E. bicyclis*. Bioassay-guided isolation of this methanolic extract using a pancreatic lipase inhibitory test led to the identification of six known phlorotannins: eckol, fucofuroeckol A, 7-phloroeckol, dioxindehydroeckol, phlorofucofuroeckol A, and dieckol. Among them, fucofuroeckol and 7-phloroeckol showed the most potent inhibitory effect on pancreatic lipase activity (IC$_{50}$ values of 37.2 and 12.7 μM, respectively) [198]. More recently, Austin et al. [199] evaluated the inhibitory lipase activity of a polyphenol-rich extract from the edible seaweed *A. nodosum*. This crude extract showed higher inhibitory activity than the known commercial product, Orlistat. Additionally, a phlorotannin-enriched fraction obtained from the crude extract was even more potent than the un-purified extract. Although the purified extract also contained polysaccharides such as alginate that might contribute to the total inhibitory lipase activity, apparently these were less effective than phlorotannins (Table 5).

Obesity is related to adipogenesis, which is the process of pre-adipocyte differentiation into adipocytes. This process plays a central role in keeping lipid homeostasis and energy balance, by storing triglycerides (TG) and releasing free fatty acids in response to changing energy demands. Adipogenesis is regulated by multiple processes, including pre-adipocytes proliferation, differentiation, as well as fatty acid oxidation and synthesis, which are controlled by several factors [196,200]. Thus, the inhibitory effect of adipocyte differentiation and proliferation has been suggested to be an important strategy for preventing or treating obesity. Dieckol from *E. cava* [201], phloroglucinol, eckol, dieckol, dioxinodehydroeckol, and phlorofucofuroeckol A from *E. stolonifera* [202] and 6,6'-bieckol, 6,8'-bieckol, 8,8'-bieckol, dieckol and phlorofucofuroeckol A from *E. bicyclis* [203] exhibited antiobesity activity by suppressing the differentiation of 3T3-L1 pre-adipocytes cells in a dose-dependent manner. These phlorotannins were able to down-regulate the expression of the proliferator activated receptor gamma (PPARγ) and the CCAAT/enhancer-binding protein alpha (C/EBPα) [201–203]. The activation of C/EBPα promote differentiation of preadipocytes through cooperation with PPARγ resulting in transactivation of adipocyte-specific genes such as fatty acid binding protein (FABP) and fatty acid synthase (FAS). Sterol regulatory element-binding protein 1 (SREBP1) is the earliest transcription factor, which also appears to be involved in adipocyte differentiation, and increases the expression of several lipogenic genes, including acyl-coA carboxylase (ACC) and FAS. Therefore, over expression of these transcription factors can accelerate adipogenesis. In this sense, dieckol, moreover, down-regulated the expression of the SREBP1 and that of the FABP4 by AMPK activation [201], the latter also related to the obesity control. In addition, 6'6'-bieckol down-regulated the sterol regulatory element binding protein-1c (SREBP-1c), the FAS and the ACC [203]. Similarly, Kong et al. [204] characterized the antiadipogenic activity of triphlorethol-A, eckol and dieckol isolated from *E. cava* in differentiating 3T3-L1 pre-adipocytes by measuring glycerol release level and adipogenic-related gene expression. These phlorotannins increased the glycerol secretion and reduced the glucose consumption level of adipocytes. In addition, phlorotannins down-regulated the expression of PPARγ, C/EBPα, SREBP-1c, as well as FABP4, FAS, acyl-CoA synthetase-1 (ACS1), fatty acid transport protein-1 (FATP1) and leptin. FATP1 has been reported to take part in fatty acid utilization along with FABP4 [205]. Leptin is a hormone related to food intake and body weight reduction. Obese subjects present leptin resistance, i.e., despite this enzyme being found in high levels in these subjects, it is unable to exercise any anorexigenic effect [206]. Phlorotannins also increased mRNA expression of hormone-sensitive lipase while they suppressed perilipin and tumor necrosis factor alpha (TNFα) expressions. Kim et al. [207] demonstrated that an extract containing eckol, dieckol and phlorofucofuroeckol-A from *E. cava* inhibited adipogenesis in 3T3-L1 adipocytes, shown by the significant reduction in glucose utilization and TG accumulation without showing cytotoxicity. This suppressive effect may be mediated by decreasing the expression levels of C/EBPα, SREBP-1c, adipocyte fatty acid binding protein (A-FABP), FAS and adiponectin (Table 4). Karadeniz et al. [208] also confirmed the antiadipogenic effect of triphlorethol-A, eckol and dieckol isolated from *E. cava* on 3T3-L1 pre-adipocytes, by reducing lipid accumulation and suppressing the expression of adipogenic differentiation markers. Considering that adipocytes and osteoblasts are derived from a common mesenchymal stem cell precursor, molecules that lead to osteoblastogenesis inhibit adipogenesis and vice versa. Thus, these authors also observed that the isolated phlorotannins successfully enhanced the osteoblast differentiation evaluated in MC3T3-E1 pre-osteoblasts, by increasing the alkaline phosphatase activity along with raising the osteoblastogenesis indicators and intracellular calcification. These results showed the potential of the selected phlorotannins for mitigating obesity and osteoporosis, which are closely related [208].

A complete study developed by Choi et al. [209] demonstrated that dieckol, a major phlorotannin in *E. cava*, suppressed lipid accumulation in 3T3-L1 cells, high-fat diet-fed zebrafish and mice (15 mg/kg bw/day and 60 mg/kg bw/day for 11 weeks). Furthermore, the findings suggested that dieckol was able to inhibit early adipogenic events by suppressing cell cycle progression, and played an important role in regulating AMPKα, ERK, and Akt signaling pathways to inhibit lipid accumulation. Recently, Ko et al. [210] demonstrated that the treatment of 3T3-L1 pre-adipocytes with 5-bromo-3,4-dihydroxybenzaldehyde isolated from the red alga *Polysiphonia morrowii* could inhibit intracellular lipid accumulation and TG levels by downregulating protein expression of adipogenic-specific factors such as PPARγ, C/EBPα, SREBP-1, FABP4, FAS, leptin, and adiponectin through phosphorylation of AMPK and ACC (Table 5).

Visceral obesity is characterized by chronic local and systemic inflammation [195]. It is well established that an increase in pro-inflammatory cytokines may be related to enlarged adipose tissue, and dysregulation of lipid metabolism, ultimately leading to insulin resistance. Thus, a phlorotannin fraction from the brown alga *Fucus distichus* decreased mRNA expression of acute and chronic inflammatory biomarkers via TRL attenuation in RAW 264.7 macrophages. Additionally, *F. distichus* fractions decreased lipid accumulation in 3T3-L1 adipocytes up to 55% and increased free glycerol concentrations, by increasing in adiponectin and uncoupling protein 1 (UCP-1) and decreasing in leptin mRNA expression [211]. Compared with lean adipocytes, hypertrophic adipocytes had higher expression of inflammatory cytokines (e.g., TNFα, interleukin (IL)-1β) and of receptors for advanced glycation end-products (RAGEs) and RAGE ligands (e.g., AGE, HMGB1, S100b, free fatty acids (FFAs)) [212]. Choi et al. [213] demonstrated the antiobesity effect of pyrogallol-phloroglucinol-6,6-bieckol (PPB) contained in *E. cava* by reducing the expression of RAGE and the secretion of ligands in a mouse model of diet-induced obesity that consumed PPB (2.5 mg/kg bw/day) for 4 weeks. In addition, this phlorotannin reduced the number of activated macrophages and inflammatory cytokine levels (TNFα and IL-1β).

The regulatory effect of marine phenolics on lipid metabolism has also been evaluated. Typically, dyslipidemia of obesity consists of increased fasting plasma TG and FFAs, decreased HDL-cholesterol (HDL-C) and normal or slightly increased LDL-cholesterol (LDL-C). Yeo et al. [214] demonstrated that oral administration of polyphenol extracts of the marine brown algae *E. cava* and dieckol effectively suppressed body weight gain and reduced total cholesterol (TC), TG and LDL-C levels in high-fat diet-induced obese mice treated with 1.25, 2.5 and 5.0 mg extract/mouse or 0.5, 1.0 and 2.0 mg dieckol/mouse for 4 weeks. The antihyperlipidemic effect was related to the inhibition of 3-hydroxyl-methyl glutaryl coenzyme A (HMGCoA) reductase activity, which is involved in the metabolic pathway that produces cholesterol and other isoprenoids. Likewise, Park et al. [215] confirmed the antiobesity activity of a polyphenol-rich fraction of the brown alga *E. cava* in high fat diet-induced obese mice. Oral administration of a polyphenol extract (200 mg/kg bw/day) for 8 weeks was effective in reducing body weight gain, body fat, and hyperglycemia, as well as in improving glucose tolerance. The mRNA expression of inflammatory cytokines (TNF-α and IL-1β) and macrophage marker gene (F4/80) was decreased in treated obese mice. These authors compared the efficacy of *E. cava* from different areas in Korea: that from Gijang was consistently more effective than that from Jeju due to its higher amounts of polyphenols and richness in 8,8′-bieckol, the major component in Gijang extract. In agreement with Park et al. [215], a later study developed by Eo et al. [216] reported that the treatment with a polyphenol-rich extract of *E. cava* containing dieckol, 2,7″-phloroglucino-6,6′-bieckol, pyrogallo-phloroglucinol-6,6′-bieckol and phlorofucofuro-eckol A (100 mg/kg bw/day or 500 mg/kg bw/day, 5 times a week for 12 weeks) was able to reduce body weight gain, adipose tissue mass, plasma lipid levels (TC and TG), hepatic fat depositions, insulin resistance and plasma leptin/adiponectin ratio of diet-induced obese mice. Moreover, polyphenol supplementation selectively ameliorated hepatic protein levels associated with hepatic lipogenesis (SREBP-1c, PPARα, FAS, and LPL), fatty acid β-oxidation (p-ACC and CPT1A), inflammation (TNF-α, IL-1β and NFkB) as well as enhancing the antioxidant defense system by activating the AMPK and SIRT1 signaling pathway (Table 4). Ding et al. [217] demonstrated the strong effect of diphlorethohydroxycarmalol, the most abundant bioactive compound in *Ishige okamurae*, against high-fat levels in diet-induced obese mice through in vivo regulation of multiple pathways. Oral administration of this polyphenol (25 and 50 mg/kg bw/day for six weeks) significantly reduced adiposity and body weight gain and improved lipid profile (lowered TG and LDL-C and increased HDL-C levels). This compound reduced hepatic lipid accumulation, by the reduction in expression levels of the critical enzymes for lipogenesis (SREBP-1c, FABP4, and FAS). In addition, diphlorethohydroxycarmalol reduced the adipocyte size and the expression levels of key adipogenic-specific proteins and lipogenic enzymes such as PPARγ, C/EBPα, SREBP-1c, FABP4, and FAS, which regulate the lipid metabolism in the epididymal adipose tissue. Finally, diphlorethohydroxycarmalol stimulated the phosphorylation of AMPK and ACC in both liver and epididymal adipose tissue.

Clinical trials also have demonstrated the potential of marine phenolics to prevent obesity. The efficacy of a polyphenol-rich extract from *E. cava* (low dose-72 mg/day or high dose-144 mg/day) was tested in 97 overweight adults enrolled in a randomized, double-blind, placebo-controlled clinical trial with parallel-group design. Results demonstrated that the polyphenol-rich extract consumed for 12 weeks lowered body fat and serum lipid levels (TC and LDL-C) [218].

All these data together highlight the potential of marine phenols in the prevention and treatment of obesity (Table 5, Figure 2), although more studies, especially clinical trials, would reinforce their use in the management of obesity.

Table 5. Effect of marine phenolics in the prevention of obesity.

Compounds/Marine Source	Test Model	Outcome	Ref.
Methanolic extract of *E. bicyclis* (eckol, fucofuroeckol A, 7-phloroeckol, dioxindehydroeckol, phlorofucofuroeckol A, and dieckol)	In vitro: assay of pancreatic lipase activity	Inhibition of pancreatic lipase activity; fucofuroeckol A and 7-phloroeckol were the most potent (IC$_{50}$ values of 37.2 and 12.7 μM, respectively)	[198]
Polyphenol-rich extract (crude) from the edible seaweed *A. nodosum* and phlorotannin-enriched fraction from crude extract	In vitro: assay of pancreatic lipase activity	Inhibition of pancreatic lipase activity Evaluated the interaction between phlorotannins and polysaccharides on inhibitory lipase activity and phlorotannins were more effective	[199]
Dieckol isolated from *E. cava*	In vitro: 3T3-L1 pre-adipocytes cells (25–100 μM of phenol)	Suppression of pre-adipocytes differentiation Down-regulated the expression of PPARγ, C/EBPα, SREBP1 and FABP4 by AMPK activation	[201]
Phloroglucinol, eckol, dieckol, dioxinodehydroeckol, and phlorofucofuroeckol A isolated from *E. stolonifera*	In vitro: 3T3-L1 pre-adipocytes cells (12.5–100 μM of phenol)	Suppression of pre-adipocytes differentiation Down-regulated the expression of PPARγ and C/EBPα	[202]
6,6'-Bieckol, 6,8'-bieckol, 8,8'-bieckol, dieckol and phlorofucofuroeckol A isolated from *E. bicyclis*	In vitro: 3T3-L1 pre-adipocytes cells (10–50 μg/mL of phenol)	Suppression of pre-adipocytes differentiation Down-regulated the expression of PPARγ, C/EBPα, SREBP-1c, FAS and ACC	[195]
Triphlorethol-A, eckol and dieckol from *E. cava*	In vitro: 3T3-L1 pre-adipocytes cells (5 μM of phenol)	Increased the glycerol secretion and reduced glucose consumption level Down-regulated the expression of PPARγ, C/EBPα, SREBP-1c as well as FABP4, FATP1, FAS, leptin and ACSL1	[204]
Extract from *E. cava* containing eckol, dieckol and phlorofucofuroeckol-A	In vitro: 3T3-L1 adipocytes cells (50 μg/mL of extract)	Inhibited the glucose utilization and TG accumulation Down-regulated the expression of C/EBPα, SREBP-1c, A-FABP, FAS and adiponectin	[207]
Triphlorethol-A, eckol and dieckol isolated from *E. cava*	In vitro: 3T3-L1 pre-adipocytes cells (1–20 μM of phenol) — In vitro: MC3T3-E1 cells (1–20 μM of phenol)	Suppressed the lipid accumulation and expression of adipogenic differentiation markers — Enhanced the osteoblast differentiation by increasing alkaline phosphatase activity and raising intracellular calcification	[208]
Dieckol from *E. cava*	In vitro: 3T3-L1 pre-adipocytes cells (25–100 μM of phenol) — In vivo: high-fat diet-fed zebrafish (1–4 μM of phenol) — In vivo: high-fat diet-fed mice (15 mg/kg bw/day or 60 mg/kg bw/day, 11 weeks administered orally)	Suppressed the lipid accumulation in the three models Inhibited the early adipogenic events by suppressing cell cycle progression Regulated the AMPKα, ERK, and Akt signaling to inhibit lipid accumulation	[209]
5-Bromo-3,4-dihydroxybenzaldehyde isolated from *Polysiphonia morrow*	In vitro: 3T3-L1 pre-adipocytes cells (25–100 μM)	Inhibited the intracellular lipid accumulation and triglyceride levels Down-regulated the expression of PPARγ, C/EBPα, SREBP-1, FABP4, FAS, leptin, and adiponectin by AMPK and ACC activation.	[210]
Phlorotannin fraction from *Fucus distichus*	In vitro: murine macrophage RAW 264.7 cells (12.5–50 μg/mL of extract) — In vitro: 3T3-L1 adipocytes cells (12.5–50 μg/mL of extract)	Anti-inflammatory activity via TLR attenuation in macrophages — Decreased the lipid accumulation in 3T3-L1 adipocytes cells	[211]
Pyrogallol-phloroglucinol-6,6-bieckol from *E. cava*	In vivo: mouse model of diet-induced obesity (2.5mg/kg bw/day for 4 weeks administered orally)	Reduced the expression of RAGE and the secretion of ligands Reduced the inflammatory cytokine level (TNFα and IL-1β)	[213]

Mar. Drugs **2020**, 18, 501

Table 5. *Cont.*

Compounds/Marine Source	Test Model	Outcome	Ref.
Polyphenol extracts from E. cava and dieckol	In vivo: high-fat diet-induced obese mice (5.0 mg, 2.5 mg and 1.25 mg extract/mouse; 2.0 mg, 1.0 mg and 0.5 mg dieckol/mouse for 4 weeks administered orally)	Suppressed the body weight gain; Reduced the TC, TG and LDL-C levels	[14]
	In vitro: 3T3-L1 pre-adipocytes cells	Inhibited the lipid accumulation; Inhibition of HMGCoA reductase activity	
Polyphenol-rich fraction of E. cava from Gijang (Korea)	In vivo: high-fat diet-induced obese mice (200 mg/kg bw for 8 weeks by oral intubation)	Reduced the body weight gain, body fat and hyperglycemia; Reduced the mRNA expression of inflammatory cytokines (TNF-α and IL-1β) and macrophage marker gene (F4/80)	[15]
Polyphenol-rich fraction of E. cava containing dieckol, 2,7''-phloroglucino-6,6'-bieckol, pyrogallo-phloroglucinol-6,6'-bieckol and phlorofucofuro-eckol A	In vivo: high-fat diet-induced obese mice (100mg/kg bw/day or 500 mg/kg bw/day, 5 times a week for 12 weeks by gavage)	Reduced the body weight gain, body fat, plasma lipid levels (TC and TG), insulin resistance and plasma leptin/adiponectin ratio; Ameliorated the hepatic protein levels: hepatic lipogenesis (PPARγ, SREBP-1c, FAS and LPL), fatty acid β-oxidation (p-ACC and CPT1A), inflammation (TNF-α, NFkB and IL-1β) and antioxidant defense system	[16]
Diphloroethohydroxycarmalol isolated from Ishige okamurae	In vivo: high-fat diet-induced obese mice (25 mg/kg bw/day or 50 mg/kg bw/day for 6 weeks administered orally)	Reduced the body weight gain, body fat and hepatic lipid accumulation, and improved lipid profile; Reduced the hepatic lipid accumulation by reduction in expression level of SREBP-1c, FABP4, and FAS; Reduced the adipocyte size by down-regulation of enzyme expression (PPARγ, C/EBPα, SREBP-1c, FABP4, and FAS)	[17]
Polyphenol-rich extract from E. cava	In vivo: 97 overweight adults (low dose-72 mg/day or high dose-144 mg/day for 12 weeks)	Decreased the body fat and serum lipid levels (TC and LDL-C)	[18]

PPARγ: proliferator activated receptor gamma; C/EBPα: CCAAT/enhancer-binding protein alpha; SREBP1: sterol regulatory element binding protein 1; FABP4: fatty acid binding protein 4; AMPK: AMP-activated protein kinase; SREBP-1c: sterol regulatory element binding protein-1c; FAS: fatty acid synthase; ACC: acyl-CoA carboxylase; FATP1: fatty acid transport protein-1; ACSL1: adipose acyl-CoA synthetase 1; ERK: extracellular signal-regulated kinase; Akt: protein kinase B; TLR: toll-like receptor; RAGE: receptor for advanced glycation end-products; TNFα: tumor necrosis factor alpha; IL-1β: interleukin 1b; TC: total cholesterol; TG: triglycerides; LDL-C: LDL-cholesterol; HMGCoA: 3-hydroxyl-methyl gluttaryl coenzyme A; ACC: acetyl-CoA carboxylase; CPT1A: carnitine palmitoyltransferase I; NFkB: nuclear factor kappa B.

3.3. Metabolic Syndrome.

Metabolic syndrome (MetS) is not a disease but a metabolic disorder that includes hypertension, obesity, glucose dysregulation and dyslipidemia [219]. A person has MetS when three or more of the following five cardiovascular risk factors have been diagnosed: (i) central obesity (waist circumference: men ≥102 cm; women ≥88 cm); (ii) elevated TG (≥150 mg/dL); (iii) diminished HDL-C (men <40 mg/dL; women <50 mg/dL) (or treated for dyslipidemia); (iv) systemic hypertension (≥130/≥85 mm Hg) (or treated for hypertension); (v) elevated fasting glucose (≥100 mg/dL) (or treated for hyperglycemia) [220]. MetS appears to be two times more frequent in women than in men, and menopause contributes to its rapid acceleration [221]. A recent study examined prospectively the association between habitual dietary iodine and seaweed consumption and the incidence of MetS among 2588 postmenopausal women 40 years or older in the Korean Multi-Rural Communities Cohort (MRCohort) [222] for an average time of 2.85 years (between 2 and 4 years). Results showed an inverse association between seaweed consumption with MetS incidence. The unmeasured bioactives of seaweed, such as polysaccharides, peptides, carotenoids and polyphenols, make it difficult to understand the real involvement of marine phenolics in the observed effects (Table 6).

Table 6. Effect of marine phenolics in the prevention of metabolic syndrome (MetS).

Compounds/Marine Source	Test Model	Outcome	Ref.
Dietary iodine and seaweed consumption	In vivo: 2588 postmenopausal women for 2.85 years (between 2 and 4 years)	Inverse association between seaweed consumption with MetS incidence	[222]
Bioactive fraction of *Sargassum wightii*	In vitro: assays of ACE enzyme activity and antioxidant activity (DPPH, ABTS and FRAP)	Inhibition of ACE activity (IC$_{50}$ 56.96 μg/mL) and improved the antioxidant potency determined	[223]
An extract I from *E. cava* and pyrogallol-phloroglucinol-6,6-bieckol	In vivo: two mice models, high-fat diet-induced obese mice and high-cholesterol and saline diet-induced hypertension mice (70 mg extract or 500 mg extract or 2.5 mg pure phenol/kg bw/day for 4 weeks administered orally) In vitro: VSMC cells, an endothelial cell line	Reduced the blood pressure and serum lipoprotein levels in vivo Reduced the adhesion molecule expression, endothelial cell death and excessive migration and proliferation of VSMCs in vitro, as well as in the obese and hypertension mouse models	[224]
Ethanolic extract from *Ulva lactuca* enriched in phlorotannins	In vivo: hypercholesterolemic mice (250 mg/kg body weight for 4 weeks by gavage)	Improved the heart oxidative stress, plasma biochemical parameters and index of atherogenesis Down-regulated the expression of pro-inflammatory cytokines (TNFα, IL-1β and IL-6) in the heart	[225]
Food supplement from *K. alvarezii*	In vivo: rats fed for 8 weeks on high-carbohydrate, high-fat diet, alone or supplemented with 5% (w/w) algae	Reduced the body weight, adiposity, systolic blood pressure and plasma lipid levels Improved the heart and liver structure	[226]

ACE: angiotensin-1 converting enzyme; DPPH: 2,2-diphenyl-1-picrylhydrazyl; ABTS: 2,2'-azino-bis (3-ethylbenzothiazoline-6-sulphonic acid); FRAP: ferric reducing antioxidant power: VSMC: human vascular smooth muscle cell line; TNFα: tumor necrosis factor alpha; IL-1β: interleukin 1β; IL-6: interleukin 6.

The aforementioned in vitro and in vivo studies have evidenced the involvement of marine phenolics in regulating lipid metabolism, hyperglycemia and obesity; these studies were reviewed by Gomez-Guzman et al. [227]. Additionally, hypertension, which is a strong independent risk factor for stroke and coronary heart disease, is also a cardiovascular risk factor in patients with MetS [228]. Angiotensin-I converting enzyme (ACE) is a zinc-containing metalloproteinase that catalyzes the conversion of angiotensin I to angiotensin II, a potent vasoconstrictor involved in the pathogenesis of hypertension. ACE also facilitates the degradation of the vasodilator bradykinin. This enzyme has a crucial role in the control of blood pressure and its inhibition has become a major target for hypertension control. Seca et al. [229] recently reviewed several marine polyphenols that have been reported to inhibit ACE activity. A bioactive fraction of the brown algae *Sargassum wightii* with optimal antioxidant and ACE inhibition activities (IC_{50} 56.96 µg/mL) was characterized by Vijayan et al. [223]. An in vivo study evaluated the efficacy of a polyphenol-rich extract from *E. cava* as well as its major component (pyrogallol-phloroglucinol-6,6-bieckol) for improving blood circulation in diet-induced obese and diet-induced hypertension mouse models [224]. After four weeks of administering 70 mg and 500 mg of extract/kg bw or 2.5 mg of phenol/kg bw, the study found a reduction in blood pressure and in serum lipoprotein levels in the obese and hypertension mouse models. A reduced expression of adhesion molecules and endothelial cell death as well as a reduction in excessive migration and proliferation of vascular smooth cells was also observed (Table 6).

In vivo studies with animals supplemented with diet-induced MetS have also evidenced the potential of seaweed polyphenols to prevent metabolic disorders. An ethanolic extract from *U. lactuca* enriched in phlorotannins was tested against hypercholesterolemia and other risk factors involved in CVD. Treatment of hypercholesterolemic mice with *U. lactuca* extract (250 mg/kg body weight) for 4 weeks revealed a cardioprotective effect by improving heart oxidative stress, plasma biochemical parameters, and index of atherogenesis. Additionally, a reduction in gene expression of proinflammatory cytokines (TNFα, IL-1β and IL-6) in the heart of *U. lactuca*-supplemented animals was also observed [225]. *Kappaphycus alvarezii*, a red seaweed, was tested as a food supplement to prevent diet-induced MetS in rats. Rats were randomly divided and fed for 8 weeks with control diet or high-fat/high-carbohydrate diet supplemented with 5% (*w/w*) algae. *Kappaphycus*-treated rats showed normalized body weight and adiposity, lower systolic blood pressure, improved heart and liver structure, and lower plasma lipids [226]. The hypotensive activity of marine polyphenols, in addition to their antidiabetic, antilipidemic and antiobesity activities, turns this group of compounds into allies to combat MetS and related cardiovascular complications (Table 6 and Figure 2).

3.4. Neurodegenerative Diseases

Seaweed-derived phenols have been described to possess neuroprotective properties [230]. Although this pathology has been less explored than those described above, knowing the role of phenolic constituents of seaweed as neuro-active compounds has gained tremendous interest in the last decade. Alzheimer's disease (AD) is the most common form of irreversible dementia, and its neuropathological hallmarks are characterized by amyloid plaques and neurofibrillary tangles composed of aggregated amyloid-β peptides (Aβ) and microtubule-associated protein tau, respectively [231]. Although the exact mechanisms of Aβ-induced neurotoxicity are still unclear, it has been reported that pathological deposition of Aβ leads to cholinergic dysfunction, glutamate excitotoxicity, beta-amyloid aggregation, oxidative stress, apoptosis and neuro-inflammation, inducing the progressive degeneration of *cognitive functions* in AD patients (Figure 2).

AD development has been linked with an impaired cholinergic pathway which is caused by upregulation of acetylcholinesterase (AChE) and butyrylcholinesterases (BChE) as well as rapid depletion of acetylcholine (AChE) [232]. In addition, BACE-1 (β-site amyloid precursor protein cleaving enzyme 1) is the major β-secretase for generation of Aβ by neurons [233] and its inhibition could block one of the earliest pathologic events in AD. The activity of some phlorotannins, in particular eckols from *E. cava* [234] and *E. bicyclis* [235], showed an inhibitory effect against AChE and BChE activities, higher than the currently used anti-AD drugs. Recently, aqueous extracts of some seaweeds (*G. beckeri*, *G. pristoides*, *Ulva rigida* and *Ecklonia maxima*), composed mainly by phloroglucinol, catechin and epicatechin 3-glucoside, showed high antioxidant potency, inhibitory activity of AChE and BChE enzymes and Aβ aggregation [236] (Table 7). The study by Olasehinde et al. [237] revealed that aqueous-ethanolic extracts of *G. pristoides*, *E. maxima*, *U. lactuca* and *G. gracilis* containing phlorotannins, flavonoids and phenolic acids exhibited a strong inhibitory activity of BACE-1, AChE and BChE enzymes, as well as hampered Aβ aggregation. Choi et al. [234] showed that phlorofucofuroeckol isolated from *E. cava* also reduced BACE-1 activity (IC$_{50}$ values in Table 7).

Table 7. Effect of marine phenolics in the prevention of Alzheimer's disease (AD).

Compounds/Marine Source	Test Model	Outcome	Ref.
Phlorotannin-rich extract from *E. cava* (dieckol, 6,6'-bieckol, 8,8'-bieckol, eckol and phlorofucofuroeckol-A)	In vitro: assays of AChE, BChE and BACE-1 activities — In vitro: Jurkat clone E1–6 cells (GSK3β activity at 50 μM)	Inhibition of AChE and BChE activities (IC$_{50}$ 16.0–96.3 μM and 0.9–29.0 μM, respectively) Inhibition of BACE-1 activity (18.6–58.3% at 1 μM) Inhibition of GSK3β activity (14.4–39.7% at 50 μM)	[234]
Phlorotannin-rich extract from *E. bicyclis* (eckols)	In vitro: assays of AChE and BChE activities	Inhibition of AChE and BChE activities (IC$_{50}$ 2.78 and 3.48 μg/mL, respectively)	[235]
Aqueous extracts of *Gracilaria beckeri*, *Gelidium pristoides*, *U. rigida* and *E. maxima* composed by phloroglucinol, catechin and epicatechin 3-glucoside	In vitro: assays of AChE and BChE activities	High antioxidant potency Inhibition of AChE and BChE activities (IC$_{50}$ 49.41 and 52.11 μg/mL, respectively, for *E. maxima*) Inhibition of Aβ aggregation	[236]
Aqueous-ethanolic extracts from *E. maxima*, *G. pristoides*, *Gracilaria gracilis*, and *Ulva lactuca* containing phlorotannins, flavonoids and phenolic acids	In vitro: assays of AChE, BChE and BACE-1 activities	Inhibition of AChE and BChE activities (IC$_{50}$ 1.74–2.42 and 1.55–2.04 mg/mL, respectively) Inhibition of BACE-1 activity (IC$_{50}$ 0.052–0.062 mg/mL) Inhibition of Aβ aggregation	[237]
Phlorofucofuroeckol isolated from *E. cava*	In vitro: Glutamate-stimulated PC12 cells (10 μM of phenol)	Increased the cell viability and attenuated glutamate excitotoxicity Inhibited the apoptosis in a caspase-dependent manner Regulated the production of ROS and attenuated mitochondrial dysfunction	[238]
Phloroglucinol isolated from *E. cava*	In vitro: Aβ-induced neurotoxicity in HT-22 cells (10 μg/mL) — In vivo: 5XFAD mice, model of AD (acute, 1.2 μmol of phenol bilaterally delivery)	Reduced the Aβ-induced ROS accumulation in HT-22 cells Ameliorated the reduction in dendritic spine density — Attenuated the impairments in cognitive dysfunction	[239]
Eckmaxol from *E. maxima*	In vitro: Aβ oligomer-induced neurotoxicity in SH-SY5Y cells (5–20 μM of phenol)	Prevented the Aβ oligomer-induced neurotoxicity Inhibition of GSK3β and ERK signaling pathway	[240]
E. cava rich in phlorotannins (eckol, 8,80-bieckol and dieckol)	In vitro: Aβ 25–35-induced damage in PC12 Cells (1–50 μM of phenol)	Inhibition of pro-inflammatory enzymes preventing Aβ production and neurotoxicity on the brain	[241]
Phlorotannin-rich fraction from *Ishige foliacea*	In vivo: scopolamine-induced amnesic mice (50 and 100 mg/kg bw/day of extract orally administered for 6 weeks)	Inhibition of AChE activity in the brain Improved the status antioxidant Prevented the memory impairment via regulation of ERK–CREB–BDNF pathway	[242]

AD: alzheimer's disease; AChE: acetylcholinesterase; BChE: butyrylcholinesterase; BACE-1: beta-site amyloid precursor protein cleaving enzyme 1; Aβ: amyloid-β peptides; GSK3β: glycogen synthase kinase 3β; ROS: reactive oxygen species; ERK: extracellular signal-regulated kinase; BDNF: brain-derived neurotrophic factor.

Glutamate is an important neurotransmitter responsible for memory, learning and cognitive function. However, excessive glutamate release from the presynaptic terminals has also been suggested as a mechanism for increased Aβ production via NMDA receptor-mediated Ca^{2+} influx [243]. Hence, administering biological active compounds capable of protecting the brain cells against glutamate excitotoxicity may be an appealing therapeutic intervention. Phlorofucofuroeckol isolated from *E. cava* increased cell viability in glutamate-stimulated PC12 cells, attenuating glutamate excitotoxicity. Kim et al. [238] showed that phlorofucofuroeckol inhibits glutamate-induced apoptotic cell death in a caspase-dependent manner, regulates the production of ROS and attenuates mitochondrial dysfunction.

No drug has been developed yet to combat Aβ aggregation, although some marine phenolics have shown the ability to attenuate Aβ-induced neurotoxicity in AD models. Phloroglucinol isolated from *E. cava* reduced ROS generation caused by Aβ-induced neurotoxicity in HT-22 cells. Yang et al. [239] have shown that phloroglucinol ameliorated the reduction in dendritic spine density induced by Aβ treatment in rat primary hippocampal neuron cultures. Administration of phloroglucinol to the hippocampal region attenuated the impairments in cognitive dysfunction 5XFAD mice, an animal model of AD [239]. Eckols from *E. cava* demonstrated it is able to inhibit glycogen synthase kinase 3β (GSK3β), which inhibits the biosynthesis of amyloid precursor proteins and is related to the formation of hyperphosphorylated tau and the generation of Aβ [234]. Likewise, eckmaxol isolated from *Ecklonia maxima* also prevented Aβ oligomer induced neurotoxicity in SH-SY5Y cells, via the inhibition of glycogen synthase kinase 3β (GSK3β) and the ERK signaling pathway [240]. Neurodegenerative disorders are often characterized by a wide range of diverse and intertwined neuro-inflammatory processes, leading to primary or secondary central nervous system damage. A recent study showed that eckols, 8,8′-bieckol and dieckol, were able to inhibit TNFα, IL-1β and prostaglandin E2 (PGE2) production at protein level, related to the down-regulation of proinflammatory enzymes, iNOS and COX-2, through the negative regulation of the NF-kB pathway in Aβ$_{25-35}$-stimulated PC12 cells, preventing the neurotoxicity on the brain. Especially, dieckol showed strongest anti-inflammatory effects via suppression of p38, ERK and JNK [241]. Um et al. [242] assessed the neuroprotective activity of a phlorotannin-rich fraction from *Ishige foliacea* on mice with scopolamine-induced memory impairment. A supplementation of 50 and 100 mg/kg of the phlorotannin-rich fraction for 6 weeks improved the memory impairment symptoms of the rodents, reduced AChE activity in their brain and improved their antioxidant status by decreasing lipid peroxidation levels and increasing glutathione levels and SOD activity. Additionally, the phlorotannin-rich supplementation up-regulated the expression levels of: brain-derived neurotrophic factor (BDNF), tropomyosin receptor kinase B, phosphorylated ERK and cyclic AMP-response element-binding protein (CREB). Therefore, the phlorotannin-rich fraction prevented the memory impairment via regulation of the ERK–CREB–BDNF pathway.

In summary, marine phenolics show potential to prevent or delay the consequences of AD (Table 7 and Figure 2), although it is still a little explored pathology and more in vitro studies need to be undertaken. In addition, the blood–brain barrier represents a challenge for the bioavailability of these compounds, and although there are a few studies confirming that dietary polyphenols may cross the blood–brain barrier [244], it is necessary to confirm the results derived from in vitro models in in vivo studies.

3.5. Cancer

Cancer represents a group of diseases related to the abnormal proliferation of any of the different kinds of cells in the body with the potential to spread to other parts of the body [245]. The side effects of antineoplastic drugs and chemotherapy motivate the search for natural products that could be used as new therapeutic agents with more efficacy, specificity and without adverse effects. Among the bioactive compounds present in marine sources, polyphenols have been demonstrated to have potent anticancerogenic activity, which has been recently reviewed [36,246,247]. An association of dietary seaweed intake (gim, miyeok and dashima) with single-nucleotide polymorphisms (SNPs; rs6983267, rs7014346, and rs719725) and colorectal cancer risk in a Korean population has been established.

Colorectal cancer risk and c-MYC rs6983267 association was derived from an analysis of 923 patients and 1846 controls [248]. Furthermore, an inverse association between dietary seaweed intake (gim, miyeok, and dashima) and colorectal cancer risk was observed, suggesting that dietary seaweed may have a positive benefit as a chemotherapeutic or chemopreventive agent for colorectal cancer risk associated with the rs6983267 genotype. Although phenolic compounds are not the only bioactives present in the consumed dietary seaweeds, it is well known that they can contribute to preventing or slowing down carcinogenic processes through the different mechanisms that are discussed next (Table 8 and Figure 2).

Polyphenol-rich extracts, as well as isolated phlorotannins and bromophenols, have been extensively described as inhibitors of cancer cell proliferation (Table 8). Aqueous extracts derived from brown *Cystoseira crinita* showed a significant antiproliferative activity against colon (HCT15) and breast (MCF7) human tumor cell lines, and these were associated with the total phenolic content and the antioxidant activity of the extracts [249]. Likewise, a phlorotannin-rich extract from *A. nodosum* inhibited the viability of colon carcinoma HT29 cells [250]. Montero et al. [127] evaluated five purified hydroalcoholic extracts of *S. muticum* from the North Atlantic coast. Their results revealed that the *S. muticum* sample with the highest level of total phlorotannins presented the highest antiproliferative activity against HT29 adenocarcinoma colon cancer cells (IC_{50}~53–58 μg/mL after 24 h of treatment). Phlorotannins isolated from the brown alga *E. maxima* (phloroglucinol, eckol, 7-phloroeckol and 2-phloroeckol) showed antiproliferative activity in HeLa, H157 and MCF7 cancer cell lines, with eckol being the most bioactive tested phlorotannin (IC_{50} < 50 μg/mL against HeLa and MCF7 cells after 24 h of treatment) [251]. Namvar et al. [252] evaluated the antiproliferative activity against five human cancer cell lines (MCF-7, MDA-MB-231, HeLa, HepG2, and HT-29) of seaweed alcoholic extracts of red (*Gracillaria corticata*), green (*Ulva fasciata*) and brown (*Sargassum ilicifolium*). All the extracts showed a dose-dependent antiproliferative activity against all the cancer cell lines, although *G. corticata* had the greatest inhibition activity against MCF-7 cell line (IC_{50} value of 30 μg/mL after 24 of treatment). Lopes-Costa et al. [253] reported that phloroglucinol not only reduced the growth of two colorectal cancer cell lines (HCT116 and HT29), but also intensified the activity of 5-fluorouracil, one of the most commonly used chemotherapeutic drugs to treat colorectal cancer. Two polybrominated diphenyl ethers, 3,4,5-tribromo-2-(2′,4′-dibromophenoxy)-phenol and 3,5-dibromo-2-(2′,4′-dibromophe-noxy)-phenol, which were isolated from *Dysidea* sp., an Indonesian marine sponge, showed antiproliferative activity against PANC-1 cells under glucose-starved conditions. The first bromophenol might act by inhibiting complex II in the mitochondrial electron transport chain [254].

There are many studies focusing on the isolation of seaweed extracts rich in bioactive compounds, along with their chemical characterization and antiproliferative activity. Zenthoefer et al. [24] determined the cytotoxic potential of different extracts from *F. vesiculosus* L. against human pancreatic cancer cells (Panc89 and PancTU1) and the most active extract (IC_{50} value of 72 μg/mL against Panc89 and 77 μg/mL against PancTU1 cells after 72 h of treatment) was characterized by H-1—NMR spectroscopy, identifying two chemical structures belonging to the phlorotannin group. Bernardini et al. [255] explored the chemical composition of French Polynesian *P. pavonica* extract by spectrophotometric assays (total phenolic compounds, tannin content and antioxidant activity) and GC–MS analysis to obtain extracts with improved antiproliferative and pro-apoptotic activities against two osteosarcoma cell lines, SaOS-2 and MNNG. Likewise, extracts of three brown marine macroalgae *Dictyota dichotoma*, *P. pavonica* and *Sargassum vulgare* were tested for improving their antioxidant, antimicrobial and cytotoxic activities on human colon carcinoma LS174 cells, human lung carcinoma A549 cells, malignant melanoma FemX cells and chronic myelogenous leukemia K562 cells [256]. Sevimli-Gur and Yesil-Celiktas [155] extracted detached leaves of *Posidonia oceanica* and *Zostera marina* with CO_2, with ethanol as co-solvent, to obtain phenolic acids with cytotoxic properties on breast, cervix, colon, prostate and neuroblastoma tumor cells. *Z. marina* extract showed the best IC_{50} values of 25, 20 and 8 μg/mL after 48 h in neuroblastoma, colon and cervix cancer cell lines, respectively. In the same line, optimized extraction, preliminary chemical characterization, and evaluation of the in vitro antiproliferative activity of phlorotannin-rich fraction from brown seaweed, *Cystoseira sedoides*, was developed by Abdelhamid et al. with promising results (IC_{50} value of 78 μg/mL after 72 h of treatment) [257]. Another recent example was carried out by Abu-Khudir et al. [258], who evaluated the antioxidant, antimicrobial and cytotoxic effect of crude extracts of the Egyptian brown seaweeds, *Sargassum linearifolium* and *Cystoseira crinita*, against a group of cancer cells—the latter with a strong cytotoxic activity against MCF-7 cells (IC_{50} value of 18 μg/mL after 48 h). They observed an increased mRNA and protein expression of the pro-apoptotic

Bax and the marker of autophagy Beclin-1, a reduced expression of the anti-apoptotic Bcl-2, as well as revealed the ability of these extracts to induce apoptosis and autophagy in MCF-7 cells. Finally, Premarathana et al. [259] carried out a preliminary screening of the cytoxicity activity on a mouse fibroblast (L929) cell line of twenty-three different seaweed species in Sri Lanka (Table 8). Crude extracts of brown and red seaweed species showed high mortality rate compared to green seaweeds and *Jania adherens* showed a remarkable cytotoxic effect on L929 cell line (51% cell viability compared with control after 24 h).

Activation of apoptosis, programmed cell death, is an important target in cancer therapy. Namvar et al. [252] demonstrated the ability of an alcoholic extract from the red seaweed *Gracillaria corticata* to induce apoptosis in human breast cancer cells (MCF-7), as well as *Sargassum linearifolium* and *Cystoseira crinite*, as already mentioned [258]. Dieckol suppressed ovarian cancer cell (SKOV3) growth by inducing caspase-dependent apoptosis via ROS production and the regulation of Akt and p38 signaling pathways [260]. A phlorotannin-rich extract from *E. cava*, mainly composed of dieckol, was assessed in terms of cisplatin responsiveness, and in its effects on A2780 and SKOV3 ovarian cancer cell lines, as well as on a SKOV3-bearing mouse model [261]. They found that dieckol may improve the efficacy of platinum drugs for ovarian cancer, by enhancing cancer cell apoptosis via the ROS/Akt/NFκB pathway and reducing nephrotoxicity. Phlorofucofuroeckol A, a phlorotannin present in the brown alga *E. bicyclis*, exhibited antiproliferative and proapoptotic properties in human cancer cells (LoVo, HT-29, SW480 and HCT116) by activating the transcription factor 3 (ATF3)-mediated pathway in human colorectal cancer cells [262]. Park et al. [263] showed that an ethanolic extract of *Hizikia fusiforme* decreased the viability of B16F10 mouse melanoma cells and induced apoptosis through activation of extrinsic and intrinsic apoptotic pathways and ROS-dependent inhibition of the PI3K/Akt signaling pathway. No chemical characterization of the tested extract was carried out, and other bioactive compounds present in the extract, apart from polyphenols, could have contributed to the observed effect.

Metastasis is an important cellular marker of cancer progression and has been associated with an increase in the activity of matrix metalloproteinases (MMPs), which are needed to degrade connective tissues. A polyphenol-rich extract of *E. cava* showed a potent inhibitory effect on the metastatic activity of A549 human lung carcinoma cells, including the suppressions of migration and invasion. This polyphenol-rich extract down-regulated MMP-2 activity through the inhibition of the PI3K/Akt signaling pathway [264]. Phloroglucinol, isolated from the brown alga *E. cava*, diminished the population of breast cancer cell lines (MCF7, SKBR3 and BT549) in tumors, by inhibiting KRAS and its downstream PI3K/Akt and RAF-1/ERK signaling pathways. Furthermore, phloroglucinol increased sensitization of breast cancer cells to conventional therapy (chemotherapy and ionizing radiation) [265]. The same research group also confirmed the effectiveness of phloroglucinol against metastasis of breast cancer through downregulation of SLUG by the inhibition of PI3K/Akt and RAS/RAF-1/ERK signaling pathways [266]. Phloroglucinol was also effective against metastasis of breast cancer cells, drastically suppressing their metastatic ability in lungs, and extending the survival time of mice. In agreement with in vitro data, phloroglucinol also exhibited breast anticancer activity at 25 mg/kg bw, either by decreasing tumor growth or by suppressing the metastatic ability of breast cancer cells that spread to the lungs, contributing in both cases to an increase in survival time in mice [266].

Angiogenesis has a crucial role in tumor growth and metastasis and is also related to an aggressive tumor phenotype where vascular endothelial growth factor (VEGF) is the most important component. Qi et al. [267] demonstrated that bis(2,3-dibromo-4,5-dihydroxybenzyl) ether treatment repressed angiogenesis in human endothelial cells (HUVECs) and in zebrafish embryos via inhibiting the VEGF signal systems. Dieckol modulated the expression of key molecules that regulate apoptosis, inflammation, invasion, and angiogenesis. Daily administration of dieckol isolated from *E. cava* (40 mg/kg for 15 weeks) to rats with N-nitrosodiethylamine(NDEA)-induced hepatogenesis regulated xenobiotic-metabolizing enzymes and by modulating Bcl-2 family proteins induced apoptosis via the regulation of mitochondrial release of cytochrome c and the activation of caspases [268].

Anti-inflammatory activity of dieckol was associated with inhibition of the nuclear factor-kappa B (NF-κB) and COX2. In addition, dieckol treatment inhibited invasion by decreasing proliferating cell nuclear antigen (PCNA) expression and angiogenesis by changing MMP-2 and MMP-9 activities and VEGF expression. Li et al. [269] found that dieckol exhibited antiangiogenic activity by inhibiting the proliferation and migration of EA.hy926 cells through mitogen-activated protein kinase (MAPK), extra-cellular signal regulated kinase (ERK) and p38 signaling pathways (Table 8).

The antioxidant activity of phlorotannins and bromophenols offers a complementary mechanism to mitigate cancerous processes as observed in a few studies already discussed. Zhen et al. [270] associated the protective effects of eckol against PM2.5-induced cell damage on human HaCaT keratinocytes with a reduced ROS generation, ensuring the stability of molecules, and maintaining a steady mitochondrial state. In addition, eckol protected cells from apoptosis by inhibiting the MAPK signaling pathway. An interesting study carried out by Zhang et al. [271] investigated the in vivo antitumor effect and the mechanisms involved in a sarcoma 180 (S180) xenograft-bearing animal model supplemented with low-dose (0.25 mg/kg), middle-dose (0.5 mg/kg) and high-dose (1.0 mg/kg) of eckol. The pro-apoptosis and antiproliferation activities of eckol were manifested by the increased TUNEL-positive apoptotic cells, the upregulated Caspase-3 and Caspase-9 expression, and the downregulated expression of Bcl-2, Bax, EGFR and p-EGFR in eckol-treated transplanted S180 tumors. Eckol stimulated the mononuclear phagocytic system, recruited and activated DCs, promoted the tumor-specific Th1 responses, increased the CD4+/CD8+ T lymphocyte ratio, and enhanced cytotoxic T lymphocyte responses in the eckol-treated animals; this suggests its potent stimulatory property on innate and adaptive immune responses.

Despite the promising anticancer activity described for marine phenolics, no human studies have been conducted to directly confirm their efficacy against cancer. DNA damage results in an increased rate of genetic mutations that often lead to the development of cancer [272]. The anticancerogenic activity of seaweeds was indirectly verified in a clinical trial. A modest improvement in DNA damage was observed in an obese group after consuming 100 mg/day for 8 weeks of a (poly)phenol-rich extract of the brown algae *A. nodosum*.

In summary, there are many in vitro studies—in addition to in vivo studies—using animal models that demonstrate the potential of marine polyphenols to block carcinogenic mechanisms (Table 8 and Figure 2). Given the prevalence of this pathology, the next step would be to test their efficacy in human trials.

Table 8. Effect of marine phenolics on the prevention of cancer.

Compounds/Marine Source	Test Model	Outcome	Ref.
Dietary seaweed intake (gim, miyeok, and dashima)	In vivo: 923 colorectal cancer patients and 1846 controls	Association between c-MYC rs6983267 and colorectal cancer risk Inverse association between dietary seaweed intake and colorectal cancer risk	[248]
Aqueous extract derived from brown *Cystoseira crinita*	In vitro: HCT15 and MCF7 cells (25–250 µg/mL for extracts)	Antiproliferative activity (IC$_{50}$ of 10.5–26.4 µg/mL on HCT15 and 17.9–29.5 µg/mL for 24 h) associated with phenolic content and antioxidant activity	[249]
Phlorotannin-rich extract from *A. nodosum*	In vitro: HT29 cells (100–500 µg/mL for extracts)	Antiproliferative activity	[250]
Ethanolic extract from *S. muticum* rich in phlorotannins	In vitro: HT29 cells (12.5–100 µg/mL for extracts)	Antiproliferative activity (IC$_{50}$ of −53.5–57.9, 55.0–57.8 and 59.4–74.0 µg/L for 24, 48 and 72 h of treatment of *S. muticum* extracts)	[127]
Phlorotannins isolated from *Ecklonia maxima* (phloroglucinol, eckol, 7-phloroeckol and 2-phloroeckol)	In vitro: HeLa, H157 and MCF7 cells (6.25–500 µg/mL for phenol)	Antiproliferative activity: eckol was the most active of all the tested phlorotannins against HeLa and MCF7 cells after 24 of treatment (IC$_{50}$ < 50 µg/mL).	[251]
Alcoholic extract from red (*Gracillaria corticata*), green (*Ulva fasciata*) and brown (*Sargassum ilicifolium*) seaweeds	In vitro: MCF-7, MDA-MB-231, HeLa, HepG2 and HT-29 cells (15–300 µg/mL for extracts)	Antiproliferative activity: *G. corticata* extract had the greatest activity against MCF-7 cells (IC$_{50}$ of 30, 37, 53, 102 and 250 µg/mL on MCF-7, HeLa, MDA-MB-231, HepG2 and HT-29 cells, respectively, after 24 h of treatment). *G. corticata* extract induced the apoptosis in human breast cancer cells	[252]
Phloroglucinol	In vitro: HCT116 and HT29 cells (10–300 µM of phenol)	Antiproliferative activity Intensified the 5-fluorouracil activity	[253]
3,4,5-Tribromo-2-(2′,4′-dibromophenoxy)-phenol (1) and 3,5-dibromo-2-(2′,4′-dibromophenoxy)-phenol (2) isolated from marine sponge *Dysidea* sp.	In vitro: PANC-1 cells under glucose-starved conditions (1–100 µM of phenol)	Antiproliferative activity (IC$_{50}$ values of 2.1 and 3.8 µM for 1 and 2, respectively, after 12 h) Inhibition of the complex II in the mitochondrial electron transport chain	[254]
Different extracts from *F. vesiculosus* L. rich in phlorotannins	In vitro: Panc89 and PancTU1 cells (0.8–500 µg/mL for crude extracts and 0.16–200 µg/mL for fractions)	Antiproliferative activity (IC$_{50}$ of 72 µg/mL against Panc89 and of 77 µg/mL against PancTU1 cells after 72 h of treatment for the most active crude extract)	[24]
Extract from *P. pavonica*	In vitro: SaOS-2 and MNNG cells (0.5–2.5 µg/mL for extract)	Antiproliferative (IC$_{50}$ value of 152.2 and 87.75 µg/mL for SaOS-2 and MNNG cells, respectively, after 24 h) and pro-apoptotic activities	[255]
Extracts of three brown marine macroalgae *Dictyota dichotoma*, *Padaina pavonia* and *Sargassum vulgare*	In vitro: LS174, A549, FemX, K562 cells (12.5–200 µg/mL for extract)	Characterization of the cytotoxic activity *D. dichotoma* showed the strongest cytotoxic activity of all the tested extracts (IC$_{50}$ values ranging from 9.76 to 50.96 µg/mL after 72 h)	[256]
Extracts from detached leaves of *Posidonia oceanica* and *Zostera marina*	In vitro: MCF-7, MDA-MB-231, SK-BR-3, HT-29, HeLa, PC-3 and Neuro 2A cells, as well as African green monkey kidney (VERO) cells (6.25–100 µg/mL for extract)	Characterization of the cytotoxic activity *Z. marina* extract showed the best IC$_{50}$ values of 25, 20 and 8 µg/mL after 48 h in neuroblastoma, colon and cervix cancer cell lines, respectively	[155]
Phlorotannin-rich fraction from *Cystoseira sedoides*	In vitro: MCF-7 cells (10–200 µg/mL for extract)	Characterization of the antiproliferative activity (IC$_{50}$ value of 78 µg/mL after 72 h)	[257]
Crude extracts from two Egyptian brown seaweeds, *Sargassum linearifolium* and *Cystoseira crinita*	In vitro: a panel of cancer cells such as MCF-7 cells, among others (0.01–2000 µg/mL for extract)	Characterization of the cytotoxic activity. *C. crinita* cold methanolic extract showed a strong cytotoxic activity against MCF-7 cells (IC$_{50}$ value of 18 µg/mL after 48 h) Induced the apoptosis and autophagy in MCF-7 cells	[258]

Table 8. *Cont.*

Compounds/Marine Source	Test Model	Outcome	Ref.
Aqueous seaweed extracts of 23 different species in Sri Lanka	In vitro: L929 cells (10–100 μg/mL for extract)	Antiproliferative activity Crude extracts of brown and red seaweeds species have shown high mortality rate compared to green seaweeds *Jania adherens* showed a remarkable cytotoxic effect on L929 cell line (51% cell viability compared with control after 24 h)	[259]
Ethanolic extract from *E. cava* whose main component was dieckol	In vitro: A2780 and SKOV3 cells	Cytotoxic effects on A2780 and SKOV3 ovarian cancer cells (IC_{50} ranging from 84 to 100 μg/mL for extract and from 77 to 169 μM for phenols, with dieckol being the most active of all, after 24 h) Induced the apoptosis on SKOV3 cells via Akt and p38 signaling pathways	[260]
Phlorotannin-rich extract from *E. cava* rich in dieckol	In vitro: A2780 and SKOV3 cells (50–100 μg/mL) In vivo: SKOV3-bearing mouse model (75 and 150 mg/kg bw for extract and 50 and 100 mg/kg bw for dieckol was given orally three times/week for 4 weeks)	Phlorotannin-rich extract may improve the efficacy of cisplatin for ovarian cancer by enhancing cancer cell apoptosis via the ROS/Akt/NFκB pathway	[261]
Phlorofucofuroeckol A present in *E. bicyclis*	In vitro: LoVo, HT-29, SW480 and HCT116 cells (25–100 μM of phenol)	Antiproliferative and pro-apoptotic properties Induced the apoptosis on colorectal cancer cells by ATF3 signaling pathway	[262]
Ethanolic extract of *H. fusiforme*	In vitro: B16F10 cells (25–400 μg/mL of extract)	Cytotoxic activity Induced the apoptosis through activation of extrinsic and intrinsic apoptotic pathways and ROS-dependent inhibition of the PI3K/Akt signaling pathway	[263]
Phlorotannin-rich extract from *E. cava* rich in phenolic compounds	In vitro: A549 cells (12.5–50 μg/mL of extract)	Inhibition of metastatic activity including suppression of migration and invasion Down-regulated the MMP-2 activity via PI3K/Akt	[264]
Phloroglucinol isolated from *E. cava*	In vitro: MCF7, SKBR3 and BT549 cells (10–100 μM of phenol) In vivo: MDA-MB231 breast cancer cells implanted into mammary fat pads of NOD-scid gamma (NSG) mice, treated with phloroglucinol 4 times on alternate days (25 mg/kg bw by intratumoral injections)	Antiproliferative effect by KRAS inhibition and its downstream PI3K/Akt and RAF-1/ERK signaling pathways Increased the sensitization of breast cancer cells to conventional therapy	[265]
Phloroglucinol isolated from *E. cava*	In vitro: BT549 and MDA-MB-231 cells (10–100 μM of phenol) In vivo: GFP-labeled metastatic MDA-MB231 cells transplanted into mammary fat pads of NSG mice, treated with phloroglucinol 4 times on alternate days (25 mg/kg bw by intraperitoneal injection)	Inhibited the metastatic ability of breast cancer cells Decreased the expression of SLUG, EMT master regulator through inhibition of PI3K/Akt and Ras/Raf-1/ERK Inhibited the in vivo metastatic ability of breast cancer cells	[266]
Bis(2,3-dibromo-4,5-dihydroxybenzyl) ether	In vitro: HUVEC cells (12.5–50 μM of phenol) In vivo: Zebrafish embryos model (6.25–25 μM of phenol)	Repressed the angiogenesis in both in vitro and in vivo models by inhibiting the VEGF signal systems	[267]
Dieckol from *E. cava*	In vivo: N-nitrosodiethylamime-induced hepatocarcinogenesis rats (40 mg/kg bw/day for 15 weeks administered orally)	Regulated the xenobiotic-metabolizing enzymes Induced the apoptosis by mitochondrial pathway Inhibited the invasion by decreasing PCNA expression Inhibited the angiogenesis by changing MMP-2 and MMP-9 activity and VEGF expression Anti-inflammatory activity by inhibiting NF-kB and COX2	[268]
Dieckol	In vitro: EA.hy926 cells (10–100 μM of phenol)	Antiangiogenic activity by inhibiting the proliferation and migration of cells through MAPK, ERK and p38 signaling pathways	[269]

Table 8. *Cont.*

Compounds/Marine Source	Test Model	Outcome	Ref.
Eckol	In vitro: on human HaCaT keratinocytes against PM2.5-induced cell damage (30 μM of phenol for 17 days)	Decreased the ROS generation Protected the cells from apoptosis by inhibiting MAPK signaling pathway	[270]
Eckol	In vivo: sarcoma 180 (S180) xenograft-bearing animal model supplemented with low dose (0.25 mg/kg bw), middle dose (0.5 mg/kg bw) and high dose (1.0 mg/kg bw) of phenol administered orally	Proapoptotic and antiproliferative activities by improving the immune response	[271]
Polyphenol-rich extract from *A. nodosum*	In vivo: 80 overweight or obese population (100 mg/day of extract for 8 weeks)	Improvements in DNA damage in the obese subset	[30]

ROS:reactive oxygen species; Akt: protein kinase B; NFkB: nuclear factor kappa B; ATF3: transcription factor 3; MMP: metalloproteinase; PI3k: phosphoinositide 3-kinase; VEGF: vascular endothelial growth factor; PCNA: proliferating cell nuclear antigen; COX-2: cyclooxygenase-2; MAPK: mitogen-activated protein kinase; ERK: extracellular signal-regulated kinase; DNA: deoxyribonucleic acid.

3.6. Human Gut Microbiota

The human intestine contains an intricate ecological community of dwelling bacteria, referred to as gut microbiota, which plays a pivotal role in host homeostasis. Multiple factors could interfere with this delicate balance, including genetics, age, antibiotics, as well as environmental factors, particularly diet, thus causing a disruption of microbiota equilibrium (dysbiosis). Growing evidence supports the involvement of gut microbiota dysbiosis in gastrointestinal and extra-intestinal cardiometabolic diseases, namely obesity and diabetes [273]. Even though, seaweeds and microalgae are excellent sources of prebiotics such as fucoidans, alginates, carrageenans and exopolysaccharides that can be partially fermented. We will focus next on marine polyphenol studies that explore their influence on gut microbiota (Table 9).

Table 9. Effect of marine phenolics in human gut microbiota.

Compounds/Marine Source	Test Model	Outcome	Ref.
Food supplement from *Kappaphycus alvarezii*	In vivo: rats fed for 8 weeks on high-carbohydrate, high-fat diet, alone or supplemented with 5% (*w/w*) algae	Improved the cardiovascular, liver and metabolic biomarkers in obese rats. Modulated the balance between *Firmicutes* and *Bacteroidetes* in the gut	[226]
Polyphenol-rich extracts from brown macroalgae *L. trabeculate*	In vivo: high-fat diet and STZ-induced diabetic rats (200 mg/kg/day bw of phenol for 4 weeks by gavage)	Attenuated the hyperglycemia in diabetic rats. Increased the short-chain fatty acid contents in fecal samples. Enhanced the abundance of *Bacteroidetes*, *Odoribacter* and *Muribaculum*. Decreased the abundance of *Proteobacteria* as well as *Firmicutes/Bacteroidetes* ratio	[103]
Water-ethanolic extract of green macroalgae *Enteromorpha prolifera* rich in flavonoids	In vivo: STZ-induced diabetic rats (150 mg/kg/day bw of phenol for 4 weeks by gavage)	Showed the antidiabetic activity on diabetic mice. Modulated the balance between *Firmicutes* and *Bacteroidetes* in the gut and increased the abundance of the *Lachnospiraceae* and *Alisties* bacteria involved in the prevention of T2DM	[191]
Water-soluble compounds from *Nitzschia laevis* extract	In vivo: high-fat diet obese mice (50 mg/kg/day bw of extract for 8 weeks by gavage)	Prevented obesity in mice. Protected the gut epithelium and positively reshaped the gut microbiota	[274]

Several studies showed that polyphenol-rich extracts had a positive effect on regulating the dysbiosis of the microbial ecology in rats. The red seaweed *K. alvarezii* tested as a food supplement demonstrated its capacity to improve cardiovascular, liver, and metabolic biomarkers in obese rats. *Kappaphycus* also modulated the balance between *Firmicutes* and *Bacteroidetes* in the gut, which could serve as a potential mechanism to reverse MetS through selective inhibition of obesogenic gut bacteria and promote healthy gut bacteria [226]. Polyphenol-rich extracts from *L. trabeculate* attenuated hyperglycemia in high-fat diet and STZ-induced diabetic rats [103], as aforementioned. Higher *Bacteroidetes*, *Odoribacter* and *Muribaculum* abundances, lower *Proteobacteria* abundances, as well as a reduced *Firmicutes/Bacteroidetes* ratio, were observed in the polyphenol supplemented group in comparison with untreated diabetic rats. In addition, rats supplemented with polyphenols showed higher amounts of short-chain fatty acids in fecal samples compared with the un-supplemented diabetic group. In their study discussed above, Yan et al. [191] showed the antidiabetic activity of a water-ethanolic extract of the green macroalgae *E. prolifera*. This extract, which was rich in flavonoids, significantly modulated the balance between *Firmicutes* and *Bacteroidetes* and increased the abundance of the *Lachnospiraceae* and *Alisties* bacteria involved in the prevention of T2DM. Guo et al. [274] demonstrated the efficacy of administering 50 mg/kg/day for 8 weeks of a *Nitzschia laevis* extract in preventing obesity in mice fed with a high-fat diet. This extract protected the gut epithelium and positively reshaped the gut microbiota composition against the damaging effect of a high-fat diet. The *Nitzschia laevis* extract was a mixture of bioactive compounds, including carotenoids and polyphenols; therefore, the specific functional ingredient(s) of this product and their potential synergistic effect (if any) are yet to be defined.

3.7. Infectious Diseases

Apart from the dietary and lifestyle-related diseases, marine phenolics are involved in the prevention of other pathological processes due to their multiple bioactivities (enzyme inhibitory effect and antimicrobial, antiviral, anticancer, antidiabetic, antioxidant, and anti-inflammatory activities, among others). Special attention should be focused on infectious diseases caused by bacteria, viruses, and fungi that continue to grow despite the development of antibiotics in the 1940s. In the western world, the issue is not the availability of antimicrobial treatments, but the developed immunity of microorganisms to pharmaceutical drugs and disinfectants. Natural products are an important source of new drugs. Approximately 80 antibacterial drugs, which were approved from 1981 to 2014, either were natural products or directly derived from them [275]. Therefore, bacterial and fungal infections and the emerging multidrug resistance are driving interest into fighting these microorganisms with natural products, which have generally been considered complementary to pharmacological therapies, and marine phenolics can be an appealing alternative (Table 10).

Lopes et al. [131] found that in vitro phlorotannin purified extracts from ten brown algal species, collected along the Portuguese west coast, were shown to be less effective against fungi and Gram-negative bacteria than Gram-positive bacteria. *F. spiralis* and *C. nodicaulis* were the most effective species (MIC = 3.9 mg/mL), followed by *C. usneoides*, *S. vulgare* (MIC = 7.8 mg/mL), and *C. tamariscifolia* (MIC = 31.3 mg/mL) against *Trichophyton rubrum*. Likewise, *C. nodicaulis* extracts were the most effective against *C. albicans* (MIC = 7.8 mg/mL). *Cystoseira* sp, and *F. spiralis* were the most active against *Staphylococcus* and against *M. luteus* (with minimum inhibitory concentration (MIC) values of 2.0–3.9 mg/mL). These effects could be related to their content in phlorotannins of the purified extracts, although their microbial activity is not truly relevant considering the MIC values. Rajauria et al. [276] reported that aqueous methanolic extracts isolated from the Irish brown seaweed *H. elongata* showed the highest antimicrobial activity against the Gram-positive bacteria *L. monocytogenes* and *E. faecalis*, and against the Gram-negative *P. aeruginosa* and *S. abony*. These authors related the antimicrobial activity with their polyphenol content and antioxidant activity.

Steele et al. [83] reported a "pseudo-induction" of plant phenolic acids (p-hydroxybenzoic acid, p-coumaric acid and vanillin) caused by changing the pattern of rearrangements of resources in plant tissues as a response of turtlegrass *Thalassia testudinum* to infection with *Labyrinthula* sp. The eelgrass *Zostera marina* possesses defensive mechanisms possibly associated with surface metabolites for surface protection and fouling control against marine epiphytic yeasts. The major constituents of eelgrass leaf surfaces and whole tissues were rosmarinic acid, p-coumaric acid, caffeic acid, ferulic acid, zosteric acid, apigenin-7-sulfate, luteolin-7-sulfate, diosmetin-7-sulfate (the most abundant) and their desulfated forms, as well as kaempferol-7,4′-dimethylether-3-O-sulfate. Papazian et al. [5] confirmed the existence of a selective chemical defense system in eelgrass which involved surface-associated phenolics and fatty acids to control growth and settlement of the microfouling yeasts *Cryptococcus fonsecae* and *Debaryomyces hansenii*. In addition, the antioxidant and cytotoxic capacities of desulfated flavonoids were enhanced compared to their sulfated compounds [5].

Free phenolic acid extracts from *Nannochloropsis* sp. (chlorogenic, gallic, protocatechuic, hydroxybenzoic, syringic, vanillic and ferulic acids) and *Spirulina* sp. (chlorogenic, hydroxybenzoic, protocatechuic and gallic acids) were efficient in reducing the mycelial growth rates of *Fusarium*. Moreover, synthetic mixtures of phenolic acids from both microalgae were less efficient than the natural extracts (EC_{50} values of 49.6 μg/mL and 33.9 μg/mL for *Nannochloropsis* and *Spirulina* phenolic acid extracts, respectively) to inhibit fungal growth, indicating that no purification is required [27]. Maadame et al. [50] evaluated the antimicrobial activities of nine marine microalgae from Moroccan coastlines (*Nannochloropsis gaditana*, *Dunaliella salina*, *Dunaliella* sp., *Phaeodactylum tricornutum*, *Isochrysis* sp., *Navicula* sp., *Chaeotoceros* sp., *Chlorella* sp. and *Tetraselmis* sp.). Ethanolic extracts of the selected microalgae were evaluated against bacteria (*Escherichia coli*, *Pseudomonas aeruginosa* and *Staphylococcus aureus*), yeast (*Candida albicans*) and fungus (*Aspergillus niger*). *Tetraselmis* sp. and *Nannochloropsis gaditana* extracts exhibited an inhibitory effect against the three types of bacteria while extracts from *Dunaliella salina*, *Phaeodactylum tricornutum* and *Isochrysis* sp. showed inhibitory activity only against the first two strains. *Tetraselmis* sp. was the most active of all the marine microalgae tested with MIC of 2.6 to 3.0 μg/mL of extract, indicative of high antimicrobial activity. All the tested extracts modestly inhibited the growth of *Candida albicans*, although *N. gaditana* showed the highest activity with MIC of 4.0 mg/mL of extract. None of them were able to inhibit *Aspergillus niger*. The observed antimicrobial activities were linked to fatty acid, carotenoid, and phenolic content of the extracts.

Sushanth and Rajashekhar [59] found that the extracts of four marine microalgae (*Chaetoceros calcitrans*, *Skeletonema costatum*, *Chroococcus turgidus* and *Nannochloropsis oceanica*) possessed effective inhibitory activity against *Staphylococcus aureus*, *Streptococcus pyogenes* and *Bacillus subtilis*. A hexane extract of *Chroococcus turgidus* showed significant inhibition activity against *Escherichia coli*, followed by an ethanol extract of *Skeletonema costatum* against *Streptococcus pyogenes*. Antifungal activity was found only in *Skeletonema costatum* and *Chroococcus turgidus* (Table 10).

Recently, Besednova et al. [277] have reviewed the activity of marine algal metabolites as promising therapeutics for the prevention and treatment of human immunodeficiency virus infection and acquired immunedeficiency syndrome (HIV/AIDS), discussing some studies focused on phlorotannins. Diphlorethohydroxycarmalol isolated from *Ishige okamurae* exhibited inhibitory effects on HIV-1 reverse transcriptase (RT) and integrase (IC$_{50}$ values of 9.1 μM and 25.2 μM, respectively), although it did not show an inhibitory activity against HIV-1 protease [278]. Specifically, 6,6'-bieckol isolated from *E. cava* showed a strong inhibition against HIV-1 induced syncytia formation, lytic effects and viral p24 antigen production [279]. In addition, 6,6'-bieckol selectively inhibited the activity of HIV-1 RT enzyme and HIV-1 entry. Another compound of this group, 8,4''-dieckol isolated from *E. cava* [280], also showed similar results as those reported by Artan et al. [279]. Therefore, there is enough evidence to support the antimicrobial activity of marine phenolics, which encourages the research community to continue exploring their application through the development of animal and human studies.

Table 10. Effect of marine phenolics on the prevention of infectious diseases.

Compounds/Marine Source	Test Model	Outcome	Ref.
Phlorotannins purified extracts isolated from ten brown algal species (Cystoseira tamariscifolia, C. nodicaulis, C. usneoides, Sargassum vulgare, F. spiralis, Halopteris filicina, Stypocaulon scoparium, Cladostephus spongiosus, P. pavonica and Saccorhiza polyschides) from Portugal	In vitro broth microdilution assay	Less effective against fungi than bacteria. Phlorotannin extracts were more effective against Gram-positive than Gram-negative bacteria. Cystoseira species and F. spiralis were the most active against Staphylococcus and M. luteus (minimum MIC of 2.0 mg/mL). F. spiralis and C. nodicaulis extracts were the most effective against the studied fungi (MIC = 3.9 mg/mL)	[131]
Aqueous methanolic extracts isolated from Irish brown seaweed H. elongata	In vitro broth microdilution assay	High antimicrobial activity against the Gram-positive bacteria, L. monocytogenes and E. faecalis. High antimicrobial activity against the Gram-negative bacteria, P. aeruginosa and S. abony	[276]
Turtlegrass Thalassia testudinum	Inoculations of healthy turtlegrass blades with Labyrinthula sp.	The emergence of Labyrinthula sp. lesions on turtlegrass blades causes a "pseudo-induction" of plant phenolic acids as carbon resources over-accumulate in tissues located above wound sites	[83]
Extracts isolated from Eelgrass Zostera marina, whose leaf surface contained hydroxycinnamic acids, flavones and flavanols	In vitro bioassays against microbial foulers	Involvement of surface-associated phenolic compounds to control yeasts	[5]
Free phenolic acid extracts from Nannochloropsis sp. and Spirulina sp., as well as pure compounds	In vitro antifungal activity of phenols	Antifungal activity of phenolic acid extracts of the microalgae. Higher activity of the natural free phenolic acid extracts (EC_{50} values of 49.6 µg/mL and 33.9 µg/mL for Nannochloropsis sp. and Spirulina sp., respectively) than the synthetic mixtures	[27]
Ethanolic extracts isolated from nine marine microalgae (Nannochloropsis gaditana, Dunaliella sp., Phaeodactylum tricornutum, Isochrysis sp., Navicula sp., Chaetoceros sp., Chlorella sp. and Tetraselmis sp.)	In vitro broth microdilution assay	Variable inhibitory activity against Escherichia coli, Pseudomonas aeruginosa and Staphylococcus aureus (Tetraselmis sp. was the most active of all those tested with MIC of 2.6 to 3.0 mg/mL of extract). Inhibition of the growth of Candida albicans (N. gaditana showed the highest activity with a MIC of 4.0 mg/mL of extract). Aspergillus niger (fungus) was resistant to the effects of the extracts. Activity of the extracts was due to the presence of fatty acids, carotenoids and phenols	[50]
Methanol, ethanol and hexane extracts from four marine microalgae (Chaetoceros calcitrans, Skeletonema costatum, Chroococcus turgidus and Nannochloropsis oceanica)	In vitro disc diffusion method	Inhibitory activity against Staphylococcus aureus, Streptococcus pyogenes and Bacillus subtilis. Antifungal activity only in Skeletonema costatum and Chroococcus turgidus	[59]
Diphlorethohydroxycarmalol isolated from Ishige okamurae	In vitro antiviral enzyme assay	Inhibited the activity of HIV-1 reverse transcriptase and integrase with IC_{50} values of 9.1 µM and 25.2 µM, respectively	[278]
8,4'''-Dieckol isolated from E. cava	In vitro: H9, H9/HIV-1IIIB, CEM-SS, C8166 cells (1–50 µM of phenol)	Inhibited the activity of HIV-1 reverse transcriptase (RT) enzyme (91% inhibition ratio at 50 µM) and HIV-1 entry. Exhibited the inhibitory effects against HIV-1 induced syncytia formation, lytic effects and viral p24 antigen production	[280]
6,6'-Bieckol isolated from E. cava	In vitro: H9, H9/HIV-1IIIB, CEM-SS, C8166 cells (0.1–30 µM of phenol)	Inhibited the activity of HIV-1 RT enzyme (EC_{50} 1.07 µM) as well as HIV-1 entry. Exhibited the inhibitory effects against HIV-1 induced syncytia formation (EC_{50} 1.72 µM), lytic effects (EC_{50} 1.23 µM) and viral p24 antigen production (EC_{50} 1.26 µM)	[279]

MIC: minimum inhibitory concentration; HIV-1: human immunodeficiency virus-1; RT: reverse transcriptase; AIDS: acquired immunodeficiency syndrome.

4. Conclusions

Marine organisms represent a widely available and renewable source of bioactives, many of them found exclusively in this environment. Phenolics are among the most active families, but contrarily to those found in terrestrial sources, marine phenolics are much less studied. Advances in the analysis of their complex and diverse structure are desirable. These tools allow their characterization, needed both for commercialization and for the study of the structure activity relationships. Classical reverse phase (RP) chromatography is the most used approach but slightly ineffective since the hydrophobic stationary phase of RP columns weakly retain these compounds that, in addition to the close polar nature among the extensively isomerized phlorotannins, make their right resolution difficult. Thus, MS^n coupled to chromatographic techniques is widely used based on their mass-to-charge ratio (m/z) and fragmentation patterns (m/z of precursor and product ions, respectively). Quadrupole time-of-flight (qTOF) and triple quadrupole (QqQ) analyzers have been widely used to this aim. Given the high complexity of marine phenolics, MS^n spectrums help only partially to identify the polymerization degree and structure of phlorotannins. Coupling NMR and tandem mass spectrometry (MS^n) with liquid chromatography is another strategy used to identify and characterize the chemical structure of this group of compounds. No less important are the advances in clean and efficient extraction methods, as well as the fractionation and purification strategies, which can promote the rational utilization of these compounds as bioactive components in functional foods, nutraceuticals and medicines. This is especially relevant since other compounds (carbohydrates, pigments, or toxic heavy metals) can be co-extracted with marine phenolics. Among the isolation techniques assayed, classical solid–liquid extraction using organic solvents is the most studied method. Alternatively, pressurized hot liquid extraction (PHLE) is a more recent option to obtain pure phlorotannins and bromophenols extracts, with lower environmental impact than solid-liquid extraction, but difficult to scale up to industrial production.

Although deficiencies in polyphenol intake do not result in specific diseases, adequate consumption of polyphenols could confer health benefits, especially related to the prevention of non-communicable diseases. The reviewed studies have revealed the multi-targeted protective effect of marine phenolics against the most prevalent diseases, such as T2DM, obesity, metabolic syndrome, Alzheimer's, or cancer, along with infectious diseases, among others. The modulatory activity of human gut microbiota has been also described, although few studies are currently available, and it would be desirable to expand them to address this aspect in depth.

Many studies have demonstrated the involvement of polyphenols in various multifactorial mechanisms underlying several diseases, due to their enzyme inhibitory effect along with their antidiabetic, antiobesity, antihypertensive, anti-inflammatory, anticancer, antimicrobial, or antiviral activities. This is an important difference compared to the available drugs used to treat most of the diseases, i.e., the ability of marine phenolics intervening in multiple pathways involved in the pathological processes. This reinforces their consideration in the pharmaceutical and cosmeceutical industries as drug substitutes. This step must be supported by the development of human studies since current understanding on the bioactivity of marine phenolics is almost exclusively based on the data available from the in vitro assays or cellular and animal models; hence, they cannot be extrapolated without reliable human clinical data.

The majority of the reported clinical trials aimed to ascertain the effect of marine phenolics on obesity and diabetes and there is not one on cancer or Alzheimer's. Regarding the polyphenol types, phlorotannins bioactivity was much more explored than bromophenols; particularly eckols and their derivatives have shown to be promising. Therefore, it is essential to design clinical trials to confirm the current knowledge about the bioactivity of marine phenols, rule out adverse effects, and study their metabolism and bioavailability for their study is almost un-existent so far.

In conclusion, marine organisms represent an important polyphenol source with promising beneficial properties to ameliorate the prevalent non-communicable diseases such as diabetes, obesity, cancer, and neurodegenerative pathologies.

Author Contributions: Conceptualization, R.M., J.R.P.-C. and H.D.; writing—original draft preparation, R.M., J.R.P.-C. and H.D.; writing—review and editing, R.M., J.R.P.-C. and H.D. All authors have read and agreed to the published version of the manuscript.

Funding: Financial support from the Xunta de Galicia (Centro singular de investigación de Galicia accreditation 2019–2022), the European Union (European Regional Development Fund–ERDF), (Ref. ED431G2019/06) and from FONDECYT Regular project number 1180571, is gratefully acknowledged.

Acknowledgments: We appreciate the technical support of Javiera Pérez-Manríquez on managing the references. Lisa Gingles edited the English text.

Conflicts of Interest: The authors declare no conflict of interest. The funders had no role in the writing of the manuscript, or in the decision to publish the results.

Abbreviatures

A2780	Human ovarian carcinoma cell line
A549	Adenocarcinomic human alveolar basal epithelial cell line
Aβ	Amyloid-beta peptides
ABTS	2, 2′-Azino-bis (3-ethylbenzothiazoline-6-sulphonic acid)
ACC	Acetyl-CoA carboxylase
ACE	Angiotensin-I converting enzyme
AChE	Acetylcholinesterase
ACh	Acetylcholine
ACSL1	Adipose acyl-CoA synthetase 1
AD	Alzheimer's disease
A-FABP	Adipocyte fatty acid binding protein
AGEs	Advanced glycation end-products
AIDS	Acquired immunedeficiency syndrome
Akt	Protein kinase B
AMPK	AMP-activated protein kinase
ASE	Ascorbic acid equivalents
ATF3	Transcription factor 3
B16F10	Murine melanoma cell line
BACE-1	Beta-site amyloid precursor protein cleaving enzyme 1
BCBM	β-Carotene bleaching
BChE	Butyrylcholinesterases
BDNF	Brain-derived neurotrophic factor
BT549	Human triple negative breast cancer cell line
C2C12	Mouse myoblast cell line
C8166	Human cancer cell line
CA	Chelating ability
CAA	Antioxidant assay for cellular antioxidant activity
CAT	Catalase
CEM-SS	Human lymphoblastic leukemia cell line
CLPAA	Cellular lipid peroxidation antioxidant activity assay
COSY	Homonuclear correlation spectroscopy
GC–MS	Gas chromatography–mass spectrometry
Glut4	Glucose transporter 4
C/EBPα	CCAAT/enhancer-binding protein alpha
COX-2	Cyclooxygenase-2
CPT1A	Carnitine palmitoyltransferase I
CVD	Cardiovascular disease
DNA	Deoxyribonucleic acid
DPPH	2,2-Diphenyl-1-picrylhydrazyl
E1-6	Human Jurkat clone cell line
EA.hy926	Human umbilical vein endothelial cell line
ERK	Extracellular signal-regulated kinase

FABP4	Fatty acid binding protein 4
FAS	Fatty acid synthase
FATP1	Fatty acid transport protein-1
FemX	Human malignant melanoma cell line
FFAs	Free fatty acids
FRAP	Ferric reducing antioxidant power
GAE	Gallic acid equivalents
Glut4	Glucose transporter 4
GPx	Glutathione peroxidase
GR	Glutathione reductase
GSK3β	Glycogen synthase kinase 3β
H157	Human oral squamous cell carcinoma cell line
H9	Human embryonic stem cell line
HCT-15	Human colon adenocarcinoma cell line
HCT-116	Human colon cancer cell line
HDL-C	HDL-cholesterol
HeLa	Human cervical cervix cancer cell line
HepG2	Human hepatocellular carcinoma cell line
HIV-1	Human immunodeficiency virus-1
HMGCoA	3-Hydroxyl-methyl glutaryl coenzyme A
HPLC–DAD–ESI/MS	High performance liquid chromatography–Diode array–Electrospray ionization–Mass spectrometry
HSQC	Heteronuclear single-quantum coherence spectroscopy
HT-22	Mouse hippocampal neuronal cell line
HT29	Human colon adenocarcinoma cell line
HUVEC	Human umbilical vein endothelia cells
IL	Interleukin
Ins-1	Rat insulinoma cell line
IR	Infrared spectroscopy
IRS1	Insulin receptor substrate 1
JNKs	c-Jun N-terminal kinases
K562	Human myelogenous leukemia cell line
L6	Rat skeletal myoblast cell
L929	Mouse fibroblasts cell line
LC–ESI–MS/MS	Liquid chromatography–Electrospray ionization–tandem Mass spectrometry
LDL-C	LDL-cholesterol
LMW	Low molecular weight
LoVo	Human colon cancer cell line derived from supraclavicular lymph node metastasis
LPIA	Lipid peroxidation inhibition assay
LS174	Human colon adenocarcinoma cell line
MAPK	Mitogen-activated protein kinase
MC3T3-E1	Mouse osteoblastic cell line
MetS	Metabolic syndrome
MMPs	Matrix metalloproteinase
MNNG	Human osteosarcoma cell line
NBT	Superoxide anion scavenging test
MCF-7	Human breast adenocarcinoma cell line
MDA-MB-231	Human breast adenocarcinoma cell line
MIC	Minimum inhibitory concentration
NFkB	Nuclear factor kappa B
NMR	Nuclear magnetic resonance
Neuro 2A	Mouse neuroblastoma cell line
NO	Nitric oxide
NOESY	Nuclear overhauser spectroscopy

iNOS	Inducible nitric oxide synthase
ORAC	Oxygen radical absorbance capacity
PANC-1	Human pancreatic carcinoma cell line
Panc89	Human pancreatic carcinoma cell line
PancTU1	Human pancreatic carcinoma cell line
PC12	Rat neuronal cell line
PC3	Human prostate cancer cell line
PCNA	Proliferating cell nuclear antigen
PGE2	Prostaglandin E2
PGU	Phloroglucinol units
PI3K	Phosphoinositide 3-kinase
PPARγ	Proliferator activated receptor gamma
PTP1B	Protein tyrosine phosphatase 1B
RAGE	Receptor for advanced glycation end-products
RAW 264.7	Murine macrophage cell line
RINm5F	Rat insulinoma cell line (pancreatic β-cells)
ROS	Reactive oxygen species
RP	Reducing power
RRLC-ESI-MS	Rapid resolution liquid chromatography coupled to mass spectrometry detection with negative ion electrospray ionization
RT	Reverse transcriptase
SaOS-2	Human osteosarcoma cell line
SKBR3	Human breast cancer cell line
SK-N-SH	Human neuroblastoma cell line
SKOV3	Human ovarian carcinoma cell line
SOD	Superoxide dismutase
SREBP1	Sterol regulatory element binding protein 1
SREBP-1c	Sterol regulatory element binding protein -1c
SRSA	Superoxide radical scavenging assay
STZ	Streptozotocin
SW480	Human colon cancer cell line
3T3-L1	Mouse adipocyte cell line
T2DM	Type 2 diabetes mellitus
TAA	Total antioxidant capacity
TBARs	Thiobarbituric acid reactive substances
TC	Total cholesterol
TEAC	Trolox equivalent antioxidant capacity
TG	Triglycerides
TLR	Toll-like receptor
TNFα	Tumor necrosis factor alpha
TOCSY	Total correlation spectroscopy
UCP-1	Uncoupling protein 1
UPLC	Ultra-performance liquid chromatography
UPLC–ESI–QTOF/MS	Ultra-performance liquid chromatography–Electrospray ionization–Quadrupole-time-of-flight high definition–Mass spectrometry
UPLC–MS	Ultra-performance liquid chromatography–Mass spectrometry
UPLC–MS/MS–TIC	Ultra-performance liquid chromatography–tandem Mass spectrometry–Total ion chromatogram
VEGF	Vascular endothelial growth factor
VERO	Green monkey kidney cell line
VSMC	Human vascular smooth muscle cell line

References

1. Shannon, E.; Abu-Ghannam, N. Seaweeds as nutraceuticals for health and nutrition. *Phycologia* **2019**, *58*, 563–577. [CrossRef]
2. Dahlgren, E.; Enhus, C.; Lindqvist, D.; Eklund, B.; Asplund, L. Induced production of brominated aromatic compounds in the alga *Ceramium tenuicorne*. *Environ. Sci. Pollut. Res.* **2015**, *22*, 18107–18114. [CrossRef] [PubMed]
3. Maadane, A.; Merghoub, N.; Ainane, T.; El Arroussi, H.; Benhima, R.; Amzazi, S.; Bakri, Y.; Wahby, I. Antioxidant activity of some Moroccan marine microalgae: PUFA profiles, carotenoids and phenolic content. *J. Biotechnol.* **2015**, *215*, 13–19. [CrossRef] [PubMed]
4. Kirke, D.A.; Rai, D.K.; Smyth, T.J.; Stengel, D.B. An assessment of temporal variation in the low molecular weight phlorotannin profiles in four intertidal brown macroalgae. *Algal Res.* **2019**, *41*, 101550. [CrossRef]
5. Papazian, S.; Parrot, D.; Burýšková, B.; Weinberger, F.; Tasdemir, D. Surface chemical defence of the eelgrass *Zostera marina* against microbial foulers. *Sci. Rep.* **2019**, *9*, 1–12. [CrossRef]
6. Barbosa, M.; Lopes, G.; Ferreres, F.; Andrade, P.B.; Pereira, D.M.; Gil-Izquierdo, Á.; Valentão, P. Phlorotannin extracts from Fucales: Marine polyphenols as bioregulators engaged in inflammation-related mediators and enzymes. *Algal Res.* **2017**, *28*, 1–8. [CrossRef]
7. Lopes, G.; Barbosa, M.; Vallejo, F.; Gil-Izquierdo, Á.; Andrade, P.B.; Valentão, P.; Pereira, D.M.; Ferreres, F. Profiling phlorotannins from *Fucus* spp. of the Northern Portuguese coastline: Chemical approach by HPLC-DAD-ESI/MSn and UPLC-ESI-QTOF/MS. *Algal Res.* **2018**, *29*, 113–120. [CrossRef]
8. Lopes, G.; Barbosa, M.; Andrade, P.B.; Valentão, P. Phlorotannins from Fucales: Potential to control hyperglycemia and diabetes-related vascular complications. *J. Appl. Phycol.* **2019**, *31*, 3143–3152. [CrossRef]
9. Ravn, H.; Pedersen, M.F.; Borum, J.; Andary, C.; Anthoni, U.; Christophersen, C.; Nielsen, P.H. Seasonal variation and distribution of two phenolic compounds, rosmarinic acid and caffeic acid, in leaves and roots-rhizomes of eelgrass (*Zostera marina* L.). *Ophelia* **1994**, *40*, 51–61. [CrossRef]
10. Achamlale, S.; Rezzonico, B.; Grignon-Dubois, M. Rosmarinic acid from beach waste: Isolation and HPLC quantification in *Zostera detritus* from Arcachon lagoon. *Food Chem.* **2009**, *113*, 878–883. [CrossRef]
11. Achamlale, S.; Rezzonico, B.; Grignon-Dubois, M. Evaluation of *Zostera detritus* as a potential new source of zosteric acid. *J. Appl. Phycol.* **2009**, *21*, 347–352. [CrossRef]
12. Mishra, N.; Prasad, S.M.; Mishra, N. Influence of high light intensity and nitrate deprivation on growth and biochemical composition of the marine microalgae *Isochrysis galbana*. *Brazilian Arch. Biol. Technol.* **2019**, *62*, 19180398. [CrossRef]
13. Tierney, M.S.; Smyth, T.J.; Hayes, M.; Soler-Vila, A.; Croft, A.K.; Brunton, N. Influence of pressurised liquid extraction and solid-liquid extraction methods on the phenolic content and antioxidant activities of Irish macroalgae. *Int. J. Food Sci. Technol.* **2013**, *48*, 860–869. [CrossRef]
14. Hardouin, K.; Burlot, A.S.; Umami, A.; Tanniou, A.; Stiger-Pouvreau, V.; Widowati, I.; Bedoux, G.; Bourgougnon, N. Biochemical and antiviral activities of enzymatic hydrolysates from different invasive French seaweeds. *J. Appl. Phycol.* **2014**, *26*, 1029–1042. [CrossRef]
15. Casas, M.P.; Conde, E.; Domínguez, H.; Moure, A. Ecofriendly extraction of bioactive fractions from *Sargassum muticum*. *Process Biochem.* **2019**, *79*, 166–173. [CrossRef]
16. Kadam, S.U.; Tiwari, B.K.; Smyth, T.J.; O'Donnell, C.P. Optimization of ultrasound assisted extraction of bioactive components from brown seaweed *Ascophyllum nodosum* using response surface methodology. *Ultrason. Sonochem.* **2015**, *23*, 308–316. [CrossRef] [PubMed]
17. Parniakov, O.; Apicella, E.; Koubaa, M.; Barba, F.J.; Grimi, N.; Lebovka, N.; Pataro, G.; Ferrari, G.; Vorobiev, E. Ultrasound-assisted green solvent extraction of high-added value compounds from microalgae *Nannochloropsis* spp. *Bioresour. Technol.* **2015**, *198*, 262–267. [CrossRef]
18. Huang, J.J.H.; Xu, W.W.; Lin, S.L.; Cheung, P.C.K. Phytochemical profiles of marine phytoplanktons: An evaluation of their: In vitro antioxidant and anti-proliferative activities. *Food Funct.* **2016**, *7*, 5002–5017. [CrossRef]
19. Rajauria, G.; Foley, B.; Abu-Ghannam, N. Identification and characterization of phenolic antioxidant compounds from brown Irish seaweed *Himanthalia elongata* using LC-DAD–ESI-MS/MS. *Innov. Food Sci. Emerg. Technol.* **2016**, *37*, 261–268. [CrossRef]

20. Rajauria, G. Optimization and validation of reverse phase HPLC method for qualitative and quantitative assessment of polyphenols in seaweed. *J. Pharm. Biomed. Anal.* **2018**, *148*, 230–237. [CrossRef]

21. Heffernan, N.; Brunton, N.P.; FitzGerald, R.J.; Smyth, T.J. Profiling of the molecular weight and structural isomer abundance of macroalgae-derived phlorotannins. *Mar. Drugs* **2015**, *13*, 509–528. [CrossRef]

22. Bogolitsyn, K.; Druzhinina, A.; Kaplitsin, P.; Ovchinnikov, D.; Parshina, A.; Kuznetsova, M. Relationship between radical scavenging activity and polymolecular properties of brown algae polyphenols. *Chem. Pap.* **2019**, *73*, 2377–2385. [CrossRef]

23. Laabir, M.; Grignon-Dubois, M.; Masseret, E.; Rezzonico, B.; Soteras, G.; Rouquette, M.; Rieuvilleneuve, F.; Cecchi, P. Algicidal effects of *Zostera marina* L. and *Zostera noltii* Hornem. extracts on the neuro-toxic bloom-forming dinoflagellate *Alexandrium catenella*. *Aquat. Bot.* **2013**, *111*, 16–25. [CrossRef]

24. Zenthoefer, F.; Geisen, U.; Hofmann-Peiker, K.; Fuhrmann, M.; Kerber, J.; Kirchhöfer, R.; Hennig, S.; Peipp, M.; Geyer, R.; Piker, L.; et al. Isolation of polyphenols with anticancer activity from the Baltic Sea brown seaweed *Fucus vesiculosus* using bioassay-guided fractionation. *J. Appl. Phycol.* **2017**, *29*, 2021–2037. [CrossRef]

25. Tierney, M.S.; Soler-Vila, A.; Rai, D.K.; Croft, A.K.; Brunton, N.P.; Smyth, T.J. UPLC-MS profiling of low molecular weight phlorotannin polymers in *Ascophyllum nodosum*, *Pelvetia canaliculata* and *Fucus spiralis*. *Metabolomics* **2014**, *10*, 524–535. [CrossRef]

26. Zhong, B.; Robinson, N.A.; Warner, R.D.; Barrow, C.J.; Dunshea, F.R.; Suleria, H.A.R. LC-ESI-QTOF-MS/MS characterization of seaweed phenolics and their antioxidant potential. *Mar. Drugs* **2020**, *18*, 331. [CrossRef]

27. Scaglioni, P.T.; Pagnussatt, F.A.; Lemos, A.C.; Nicolli, C.P.; Del Ponte, E.M.; Badiale-Furlong, E. *Nannochloropsis* sp. and *Spirulina* sp. as a source of antifungal compounds to mitigate contamination by *Fusarium graminearum* sspecies complex. *Curr. Microbiol.* **2019**, *76*, 930–938. [CrossRef]

28. Klejdus, B.; Plaza, M.; Šnóblová, M.; Lojková, L. Development of new efficient method for isolation of phenolics from sea algae prior to their rapid resolution liquid chromatographic–tandem mass spectrometric determination. *J. Pharm. Biomed. Anal.* **2017**, *135*, 87–96. [CrossRef]

29. Pilavtepe, M.; Yucel, M.; Helvaci, S.S.; Demircioglu, M.; Yesil-Celiktas, O. Optimization and mathematical modeling of mass transfer between *Zostera marina* residues and supercritical CO_2 modified with ethanol. *J. Supercrit. Fluids* **2012**, *68*, 87–93. [CrossRef]

30. Baldrick, F.R.; McFadden, K.; Ibars, M.; Sung, C.; Moffatt, T.; Megarry, K.; Thomas, K.; Mitchell, P.; Wallace, J.M.W.; Pourshahidi, L.K.; et al. Impact of a (poly)phenol-rich extract from the brown algae Ascophyllum nodosum on DNA damage and antioxidant activity in an overweight or obese population: A randomized controlled trial. *Am. J. Clin. Nutr.* **2018**, *108*, 688–700. [CrossRef]

31. Akköz, C.; Arslan, D.; Ünver, A.; Özcan, M.M.; Yilmaz, B. Chemical composition, total phenolic and mineral contents of *Enteromorpha intestinalis* (l.) kütz. and cladophora glomerata (l.) kütz. seaweeds. *J. Food Biochem.* **2011**, *35*, 513–523. [CrossRef]

32. Wijesekara, I.; Kim, S.K.; Li, Y.; Li, Y.X. Phlorotannins as bioactive agents from brown algae. *Process Biochem.* **2011**, *46*, 2219–2224. [CrossRef]

33. Zidorn, C. Secondary metabolites of seagrasses (Alismatales and Potamogetonales; Alismatidae): Chemical diversity, bioactivity, and ecological function. *Phytochemistry* **2016**, *124*, 5–28. [CrossRef] [PubMed]

34. Catarino, M.D.; Silva, A.M.S.; Cardoso, S.M. Fucaceae: A source of bioactive phlorotannins. *Int. J. Mol. Sci.* **2017**, *18*, 1327. [CrossRef]

35. Montero, L.; del Pilar Sánchez-Camargo, A.; Ibáñez, E.; Gilbert-López, B. Phenolic compounds from edible algae: Bioactivity and health benefits. *Curr. Med. Chem.* **2018**, *25*, 4808–4826. [CrossRef]

36. Erpel, F.; Mateos, R.; Pérez-jiménez, J.; Pérez-correa, J.R. Phlorotannins: From isolation and structural characterization, to the evaluation of their antidiabetic and anticancer potential. *Food Res. Int.* **2020**, *137*, 109589. [CrossRef]

37. Bidleman, T.F.; Andersson, A.; Brugel, S.; Ericson, L.; Haglund, P.; Kupryianchyk, D.; Lau, D.C.P.; Liljelind, P.; Lundin, L.; Tysklind, A.; et al. Bromoanisoles and methoxylated bromodiphenyl ethers in macroalgae from Nordic coastal regions. *Environ. Sci. Process. Impacts* **2019**, *21*, 881–892. [CrossRef]

38. Cade, S.E.; Kuo, L.J.; Schultz, I.R. Polybrominated diphenyl ethers and their hydroxylated and methoxylated derivatives in seafood obtained from Puget Sound, WA. *Sci. Total Environ.* **2018**, *630*, 1149–1154. [CrossRef]

39. Malmvärn, A.; Marsh, G.; Kautsky, L.; Athanasiadou, M.; Bergman, Å.; Asplund, L. Hydroxylated and methoxylated brominated diphenyl ethers in the red algae *Ceramium tenuicorne* and blue mussels from the Baltic Sea. *Environ. Sci. Technol.* **2005**, *39*, 2990–2997. [CrossRef]
40. Koch, C.; Sures, B. Environmental concentrations and toxicology of 2,4,6-tribromophenol (TBP). *Environ. Pollut.* **2018**, *233*, 706–713. [CrossRef]
41. Ragupathi Raja Kannan, R.; Arumugam, R.; Iyapparaj, P.; Thangaradjou, T.; Anantharaman, P. In vitro antibacterial, cytotoxicity and haemolytic activities and phytochemical analysis of seagrasses from the Gulf of Mannar, South India. *Food Chem.* **2013**, *136*, 1484–1489. [CrossRef]
42. Zangrando, R.; Corami, F.; Barbaro, E.; Grosso, A.; Barbante, C.; Turetta, C.; Capodaglio, G.; Gambaro, A. Free phenolic compounds in waters of the Ross Sea. *Sci. Total Environ.* **2019**, *650*, 2117–2128. [CrossRef]
43. Glombitza, K.W.; Gerstberger, G. Phlorotannins with dibenzodioxin structural elements from the brown alga *Eisenia arborea*. *Phytochemistry* **1985**, *24*, 543–551. [CrossRef]
44. Pal Singh, I.; Bharate, S.B. Phloroglucinol compounds of natural origin. *Nat. Prod. Rep.* **2006**, *23*, 558–591. [CrossRef]
45. Koivikko, R.; Loponen, J.; Pihlaja, K.; Jormalainen, V. High-performance liquid chromatographic analysis of phlorotannins from the brown alga *Fucus vesiculosus*. *Phytochem. Anal.* **2007**, *18*, 326–332. [CrossRef] [PubMed]
46. Santana-Casiano, J.M.; González-Dávila, M.; González, A.G.; Millero, F.J. Fe(III) reduction in the presence of Catechol in seawater. *Aquat. Geochemistry* **2010**, *16*, 467–482. [CrossRef]
47. Sansone, C.; Brunet, C. Promises and challenges of microalgal antioxidant production. *Antioxidants* **2019**, *8*, 199. [CrossRef] [PubMed]
48. Tibbetts, S.M.; Milley, J.E.; Lall, S.P. Chemical composition and nutritional properties of freshwater and marine microalgal biomass cultured in photobioreactors. *J. Appl. Phycol.* **2015**, *27*, 1109–1119. [CrossRef]
49. Safafar, H.; van Wagenen, J.; Møller, P.; Jacobsen, C. Carotenoids, phenolic compounds and tocopherols contribute to the antioxidative properties of some microalgae species grown on industrial wastewater. *Mar. Drugs* **2015**, *13*, 7339–7356. [CrossRef] [PubMed]
50. Maadane, A.; Merghoub, N.; El Mernissi, N.; Ainane, T.; Amzazi, S.; Wahby, I.; Bakri, Y. Antimicrobial activity of marine microalgae isolated from Moroccan coastlines. *J. Microbiol. Biotechnol. Food Sci.* **2017**, *6*, 1257–1260. [CrossRef]
51. Bhuvana, P.; Sangeetha, P.; Anuradha, V.; Ali, M.S. Spectral characterization of bioactive compounds from microalgae: *N. oculata* and *C. vulgaris*. *Biocatal. Agric. Biotechnol.* **2019**, *19*, 101094. [CrossRef]
52. Gomez, A.L.; Lopez, J.A.; Rodriguez, A.; Fortiz, J.; Martinez, L.R.; Apolinar, A.; Enriquez, L.F. Produccion de compuestos fenolicos por cuatro especies de microalgas marinas sometidas a diferentes condiciones de iluminacion. *Lat. Am. J. Aquat. Res.* **2016**, *44*, 137–143. [CrossRef]
53. Fazelian, N.; Movafeghi, A.; Yousefzadi, M.; Rahimzadeh, M. Cytotoxic impacts of CuO nanoparticles on the marine microalga *Nannochloropsis oculata*. *Environ. Sci. Pollut. Res.* **2019**, *26*, 17499–17511. [CrossRef] [PubMed]
54. Rico, M.; López, A.; Santana-Casiano, J.M.; González, A.G.; González-Dávila, M. Variability of the phenolic profile in the diatom *Phaeodactylum tricornutum* growing under copper and iron stress. *Limnol. Oceanogr.* **2013**, *58*, 144–152. [CrossRef]
55. Santana-Casiano, J.M.; González-Dávila, M.; González, A.G.; Rico, M.; López, A.; Martel, A. Characterization of phenolic exudates from *Phaeodactylum tricornutum* and their effects on the chemistry of Fe(II)-Fe(III). *Mar. Chem.* **2014**, *158*, 10–16. [CrossRef]
56. Custódio, L.; Soares, F.; Pereira, H.; Barreira, L.; Vizetto-Duarte, C.; Rodrigues, M.J.; Rauter, A.P.; Alberício, F.; Varela, J. Fatty acid composition and biological activities of *Isochrysis galbana* T-ISO, *Tetraselmis* sp. and *Scenedesmus* sp.: Possible application in the pharmaceutical and functional food industries. *J. Appl. Phycol.* **2014**, *26*, 151–161. [CrossRef]
57. Mekdade, L.; Bey, M.; Hamed, B.; El-kebir, F.Z.; Mohamed, S.; Ayad, E.A.; June, M.; June, M. Evaluation of Antioxidant and Antiproliferative Activities of Nannochloropsis. *Res. Journal Pharm., Biol. Chem. Sci.* **2016**, *7*, 904–913.
58. Hemalatha, A.; Parthiban, C.; Saranya, C.; Girija, K.; Anantharaman, P. Evaluation of antioxidant activities and total phenolic contents of different solvent extracts of selected marine diatoms. *Indian J. Geo-Marine Sci.* **2015**, *44*, 1630–1636.

59. Sushanth, V.R.; Rajashekhar, M. Antioxidant and antimicrobial activities in the four species of marine microalgae isolated from Arabian Sea of Karnataka Coast. *Indian J. Geo-Marine Sci.* **2015**, *44*, 69–75.
60. Niwano, Y.; Sato, E.; Kohno, M.; Matsuyama, Y.; Kim, D.; Oda, T. Antioxidant properties of aqueous extracts from red tide plankton cultures. *Biosci. Biotechnol. Biochem.* **2007**, *71*, 1145–1153. [CrossRef]
61. Custódio, L.; Justo, T.; Silvestre, L.; Barradas, A.; Duarte, C.V.; Pereira, H.; Barreira, L.; Rauter, A.P.; Alberício, F.; Varela, J. Microalgae of different phyla display antioxidant, metal chelating and acetylcholinesterase inhibitory activities. *Food Chem.* **2012**, *131*, 134–140. [CrossRef]
62. Pang, X.; Lin, X.; Wang, P.; Zhou, X.; Yang, B.; Wang, J.; Liu, Y. Perylenequione derivatives with anticancer activities isolated from the marine sponge-derived fungus, *Alternaria* sp. SCSIO41014. *Mar. Drugs* **2018**, *16*, 280. [CrossRef]
63. Zhao, Z.; Ding, W.; Wang, P.-M.; Zheng, D.; Xu, J. Five polyketides isolated from the marine-derived fungus *Arthrinium* sp. *Nat. Prod. Res.* **2019**, 1–6. [CrossRef] [PubMed]
64. Liu, F.; Tian, L.; Chen, G.; Zhang, L.; Liu, B.; Zhang, W.; Bai, J.; Hua, H.; Wang, H.; Pei, Y.-H. Two new compounds from a marine-derived *Penicillium griseofulvum* T21-03. *J. Asian Nat. Prod. Res.* **2017**, *19*, 678–683. [CrossRef] [PubMed]
65. Sun, L.L.; Shao, C.L.; Chen, J.F.; Guo, Z.Y.; Fu, X.M.; Chen, M.; Chen, Y.Y.; Li, R.; De Voogd, N.J.; She, Z.G.; et al. New bisabolane sesquiterpenoids from a marine-derived fungus *Aspergillus* sp. isolated from the sponge *Xestospongia testudinaria*. *Bioorganic Med. Chem. Lett.* **2012**, *22*, 1326–1329. [CrossRef]
66. Liu, S.; Dai, H.; Konuklugil, B.; Orfali, R.S.; Lin, W.; Kalscheuer, R.; Liu, Z.; Proksch, P. Phenolic bisabolanes from the sponge-derived fungus *Aspergillus* sp. *Phytochem. Lett.* **2016**, *18*, 187–191. [CrossRef]
67. Wu, Z.; Wang, Y.; Liu, D.; Proksch, P.; Yu, S.; Lin, W. Antioxidative phenolic compounds from a marine-derived fungus *Aspergillus versicolor*. *Tetrahedron* **2016**, *72*, 50–57. [CrossRef]
68. Hulikere, M.M.; Joshi, C.G.; Ananda, D.; Poyya, J.; Nivya, T. Antiangiogenic, wound healing and antioxidant activity of *Cladosporium cladosporioides* (Endophytic Fungus) isolated from seaweed (*Sargassum wightii*). *Mycology* **2016**, *7*, 203–211. [CrossRef]
69. El-Hawary, S.S.; Sayed, A.M.; Mohammed, R.; Hassan, H.M.; Zaki, M.A.; Rateb, M.E.; Mohammed, T.A.; Amin, E.; Abdelmohsen, U.R. Epigenetic modifiers induce bioactive phenolic metabolites in the marine-derived fungus *Penicillium brevicompactum*. *Mar. Drugs* **2018**, *16*, 253. [CrossRef]
70. Zheng, Y.; Chen, X.; Chen, L.; Shen, L.; Fu, X.; Chen, Q.; Chen, M.; Wang, C. Isolation and neuroprotective activity of phenolic derivatives from the marine derived fungus *Penicillium janthinellum*. *J. Ocean Univ. China* **2020**, *19*, 700–706. [CrossRef]
71. Lu, Z.; Zhu, H.; Fu, P.; Wang, Y.; Zhang, Z.; Lin, H.; Liu, P.; Zhuang, Y.; Hong, K.; Zhu, W. Cytotoxic polyphenols from the marine-derived fungus *Penicillium expansum*. *J. Nat. Prod.* **2010**, *73*, 911–914. [CrossRef] [PubMed]
72. Elnaggar, M.S.; Ebada, S.S.; Ashour, M.L.; Ebrahim, W.; Müller, W.E.G.; Mándi, A.; Kurtán, T.; Singab, A.; Lin, W.; Liu, Z.; et al. Xanthones and sesquiterpene derivatives from a marine-derived fungus *Scopulariopsis* sp. *Tetrahedron* **2016**, *72*, 2411–2419. [CrossRef]
73. Wang, J.F.; Qin, X.; Xu, F.Q.; Zhang, T.; Liao, S.; Lin, X.; Yang, B.; Liu, J.; Wang, L.; Tu, Z.; et al. Tetramic acid derivatives and polyphenols from sponge-derived fungus and their biological evaluation. *Nat. Prod. Res.* **2015**, *29*, 1761–1765. [CrossRef] [PubMed]
74. Blagojević, D.; Babić, O.; Rašeta, M.; Šibul, F.; Janjušević, L.; Simeunović, J. Antioxidant activity and phenolic profile in filamentous cyanobacteria: The impact of nitrogen. *J. Appl. Phycol.* **2018**, *30*, 2337–2346. [CrossRef]
75. Ijaz, S.; Hasnain, S. Antioxidant potential of indigenous cyanobacterial strains in relation with their phenolic and flavonoid contents. *Nat. Prod. Res.* **2016**, *30*, 1297–1300. [CrossRef] [PubMed]
76. Jerez-Martel, I.; García-Poza, S.; Rodríguez-Martel, G.; Rico, M.; Afonso-Olivares, C.; Gómez-Pinchetti, J.L. Phenolic profile and antioxidant activity of crude extracts from microalgae and cyanobacteria strains. *J. Food Qual.* **2017**, *2017*. [CrossRef]
77. Milović, S.; Stanković, I.; Nikolić, D.; Radović, J.; Kolundžić, M.; Nikolić, V.; Stanojković, T.; Petović, S.; Kundaković-Vasović, T. Chemical analysis of selected seaweeds and seagrass from the Adriatic Coast of Montenegro. *Chem. Biodivers.* **2019**, *16*. [CrossRef] [PubMed]
78. Steele, L.T.; Valentine, J.F. Idiosyncratic responses of seagrass phenolic production following sea urchin grazing. *Mar. Ecol. Prog. Ser.* **2012**, *466*, 81–92. [CrossRef]

79. Mabrouk, S.B.; Reis, M.; Sousa, M.L.; Ribeiro, T.; Almeida, J.R.; Pereira, S.; Antunes, J.; Rosa, F.; Vasconcelos, V.; Achour, L.; et al. The marine seagrass *Halophila stipulacea* as a source of bioactive metabolites against obesity and biofouling. *Mar. Drugs* **2020**, *18*, 88. [CrossRef]

80. Cornara, L.; Pastorino, G.; Borghesi, B.; Salis, A.; Clericuzio, M.; Marchetti, C.; Damonte, G.; Burlando, B. *Posidonia oceanica* (L.) delile ethanolic extract modulates cell activities with skin health applications. *Mar. Drugs* **2018**, *16*, 21. [CrossRef]

81. Enerstvedt Hasle, K.; Lundberg, A.; Jordheim, M. Characterization of polyphenolic content in the aquatic plants *Ruppia cirrhosa* and *Ruppia maritima* -A source of nutritional natural products. *Molecules* **2018**, *23*, 16. [CrossRef] [PubMed]

82. Rengasamy, K.R.R.; Sadeer, N.B.; Zengin, G.; Mahomoodally, M.F.; Cziáky, Z.; Jekő, J.; Diuzheva, A.; Abdallah, H.H.; Kim, D.H. Biopharmaceutical potential, chemical profile and in silico study of the seagrass—*Syringodium isoetifolium* (Asch.) Dandy. *S. Afr. J. Bot.* **2019**, *127*, 167–175. [CrossRef]

83. Steele, L.T.; Caldwell, M.; Boettcher, A.; Arnold, T. Seagrass-pathogen interactions: "pseudo-induction" of turtlegrass phenolics near wasting disease lesions. *Mar. Ecol. Prog. Ser.* **2005**, *303*, 123–131. [CrossRef]

84. Trevathan-Tackett, S.M.; Lane, A.L.; Bishop, N.; Ross, C. Metabolites derived from the tropical seagrass Thalassia testudinum are bioactive against pathogenic *Labyrinthula* sp. *Aquat. Bot.* **2015**, *122*, 1–8. [CrossRef]

85. Styshova, O.N.; Popov, A.M.; Artyukov, A.A.; Klimovich, A.A. Main constituents of polyphenol complex from seagrasses of the genus *Zostera*, their antidiabetic properties and mechanisms of action. *Exp. Ther. Med.* **2017**, *13*, 1651–1659. [CrossRef]

86. Li, Y.; Mangoni, A.; Shulha, O.; Çiçek, S.S.; Zidorn, C. Cyclic diarylheptanoids deoxycymodienol and isotedarene A from *Zostera marina* (Zosteraceae). *Tetrahedron Lett.* **2019**, *60*, 150930. [CrossRef]

87. Grignon-Dubois, M.; Rezzonico, B.; Alcoverro, T. Regional scale patterns in seagrass defences: Phenolic acid content in *Zostera noltii*. *Estuar. Coast. Shelf Sci.* **2012**, *114*, 18–22. [CrossRef]

88. Manck, L.; Quintana, E.; Suárez, R.; Brun, F.G.; Hernández, I.; Ortega, M.J.; Zubía, E. Profiling of phenolic natural products in the seagrass *Zostera noltei* by UPLC-MS. *Nat. Prod. Commun.* **2017**, *12*, 687–690. [CrossRef]

89. Grignon-Dubois, M.; Rezzonico, B. Phenolic chemistry of the seagrass Zostera noltei Hornem. Part 1: First evidence of three infraspecific flavonoid chemotypes in three distinctive geographical regions. *Phytochemistry* **2018**, *146*, 91–101. [CrossRef]

90. Arnold, T.; Freundlich, G.; Weilnau, T.; Verdi, A.; Tibbetts, I.R. Impacts of groundwater discharge at Myora Springs (North Stradbroke Island, Australia) on the phenolic metabolism of eelgrass, *Zostera muelleri*, and grazing by the juvenile rabbitfish, *Siganus fuscescens*. *PLoS ONE* **2014**, *9*, e104738. [CrossRef]

91. Cichewicz, R.H.; Clifford, L.J.; Lassen, P.R.; Cao, X.; Freedman, T.B.; Nafie, L.A.; Deschamps, J.D.; Kenyon, V.A.; Flanary, J.R.; Holman, T.R.; et al. Stereochemical determination and bioactivity assessment of (S)-(+)-curcuphenol dimers isolated from the marine sponge *Didiscus aceratus* and synthesized through laccase biocatalysis. *Bioorganic Med. Chem.* **2005**, *13*, 5600–5612. [CrossRef] [PubMed]

92. Kaweetripob, W.; Mahidol, C.; Wongbundit, S.; Tuntiwachwuttikul, P.; Ruchirawat, S.; Prawat, H. Sesterterpenes and phenolic alkenes from the Thai sponge *Hyrtios erectus*. *Tetrahedron* **2018**, *74*, 316–323. [CrossRef]

93. Costa, M.; Coello, L.; Urbatzka, R.; Pérez, M.; Thorsteinsdottir, M. New aromatic bisabolane derivatives with lipid-reducing activity from the marine sponge *Myrmekioderma* sp. *Mar. Drugs* **2019**, *17*, 375. [CrossRef] [PubMed]

94. Wongbundit, S.; Mahidol, C.; Kaweetripob, W.; Prachyawarakorn, V.; Eurtivong, C.; Sahakitpichan, P.; Tuntiwachwuttikul, P.; Ruchirawat, S.; Prawat, H. Biscurcudiols, myrmekioperoxides, and myrmekiodermaral from the Thai marine sponge *Myrmekioderma* sp. *Tetrahedron* **2020**, *76*, 131162. [CrossRef]

95. Olsen, E.K.; Hansen, E.; Isaksson, J.; Andersen, J.H. Cellular antioxidant effect of four bromophenols from the red algae, *Vertebrata lanosa*. *Mar. Drugs* **2013**, *11*, 2769–2784. [CrossRef]

96. Ferreres, F.; Lopes, G.; Gil-Izquierdo, A.; Andrade, P.B.; Sousa, C.; Mouga, T.; Valentão, P. Phlorotannin extracts from fucales characterized by HPLC-DAD-ESI-MS n: Approaches to hyaluronidase inhibitory capacity and antioxidant properties. *Mar. Drugs* **2012**, *10*, 2766–2781. [CrossRef]

97. Prieto, P.; Pineda, M.; Aguilar, M. Spectrophotometric quantitation of antioxidant capacity through the formation of a phosphomolybdenum complex: Specific application to the determination of vitamin E. *Anal. Biochem.* **1999**, *269*, 337–341. [CrossRef]

98. Cabrita, M.T.; Vale, C.; Rauter, A.P. Halogenated compounds from marine algae. *Mar. Drugs* **2010**, *8*, 2301–2317. [CrossRef]

99. Topcu, G.; Aydogmus, Z.; Imre, S.; Gören, A.C.; Pezzuto, J.M.; Clement, J.A.; Kingston, D.G.I. Brominated sesquiterpenes from the red alga *Laurencia obtusa*. *J. Nat. Prod.* **2003**, *66*, 1505–1508. [CrossRef]

100. Machado, L.; Magnusson, M.; Paul, N.A.; Kinley, R.; de Nys, R.; Tomkins, N. Identification of bioactives from the red seaweed *Asparagopsis taxiformis* that promote antimethanogenic activity in vitro. *J. Appl. Phycol.* **2016**, *28*, 3117–3126. [CrossRef]

101. Whitfield, F.B.; Last, J.H.; Shaw, K.J.; Tindale, C.R. 2,6-Dibromophenol: The cause of an iodoform-like off-flavour in some Australian crustacea. *J. Sci. Food Agric.* **1988**, *46*, 29–42. [CrossRef]

102. Kim, K.C.; Hyun, Y.J.; Hewage, S.R.K.M.; Piao, M.J.; Kang, K.A.; Kang, H.K.; Koh, Y.S.; Ahn, M.J.; Hyun, J.W. 3-Bromo-4,5-dihydroxybenzaldehyde enhances the level of reduced glutathione via the Nrf2-mediated pathway in human keratinocytes. *Mar. Drugs* **2017**, *15*, 291. [CrossRef] [PubMed]

103. Yuan, Y.; Zheng, Y.; Zhou, J.; Geng, Y.; Zou, P.; Li, Y.; Zhang, C. Polyphenol-rich extracts from brown macroalgae *Lessonia trabeculate* attenuate hyperglycemia and modulate gut microbiota in high-rfat diet and Streptozotocin-Induced diabetic rats. *J. Agric. Food Chem.* **2019**, *67*, 12472–12480. [CrossRef] [PubMed]

104. Rodrigues, D.; Freitas, A.C.; Pereira, L.; Rocha-Santos, T.A.P.; Vasconcelos, M.W.; Roriz, M.; Rodríguez-Alcalá, L.M.; Gomes, A.M.P.; Duarte, A.C. Chemical composition of red, brown and green macroalgae from Buarcos bay in Central West Coast of Portugal. *Food Chem.* **2015**, *183*, 197–207. [CrossRef]

105. Uribe, E.; Vega-Gálvez, A.; Heredia, V.; Pastén, A.; Di Scala, K. An edible red seaweed (*Pyropia orbicularis*): Influence of vacuum drying on physicochemical composition, bioactive compounds, antioxidant capacity, and pigments. *J. Appl. Phycol.* **2018**, *30*, 673–683. [CrossRef]

106. Tanna, B.; Brahmbhatt, H.R.; Mishra, A. Phenolic, flavonoid, and amino acid compositions reveal that selected tropical seaweeds have the potential to be functional food ingredients. *J. Food Process. Preserv.* **2019**, *43*, 1–10. [CrossRef]

107. Ford, L.; Theodoridou, K.; Sheldrake, G.N.; Walsh, P.J. A critical review of analytical methods used for the chemical characterisation and quantification of phlorotannin compounds in brown seaweeds. *Phytochem. Anal.* **2019**, *30*, 587–599. [CrossRef]

108. Steevensz, A.J.; MacKinnon, S.L.; Hankinson, R.; Craft, C.; Connan, S.; Stengel, D.B.; Melanson, J.E. Profiling phlorotannins in brown macroalgae by liquid chromatography-high resolution mass spectrometry. *Phytochem. Anal.* **2012**, *23*, 547–553. [CrossRef]

109. Zhang, R.; Yuen, A.K.L.; Magnusson, M.; Wright, J.T.; de Nys, R.; Masters, A.F.; Maschmeyer, T. A comparative assessment of the activity and structure of phlorotannins from the brown seaweed *Carpophyllum flexuosum*. *Algal Res.* **2018**, *29*, 130–141. [CrossRef]

110. Sellimi, S.; Benslima, A.; Barragan-Montero, V.; Hajji, M.; Nasri, M. Polyphenolic-protein-polysaccharide ternary conjugates from *Cystoseira barbata* Tunisian seaweed as potential biopreservatives: Chemical, antioxidant and antimicrobial properties. *Int. J. Biol. Macromol.* **2017**, *105*, 1375–1383. [CrossRef]

111. Trifan, A.; Vasincu, A.; Luca, S.V.; Neophytou, C.; Wolfram, E.; Opitz, S.E.W.; Sava, D.; Bucur, L.; Cioroiu, B.I.; Miron, A.; et al. Unravelling the potential of seaweeds from the Black Sea coast of Romania as bioactive compounds sources. Part I: *Cystoseira barbata* (Stackhouse) C. Agardh. *Food Chem. Toxicol.* **2019**, *134*, 110820. [CrossRef]

112. Olate-Gallegos, C.; Barriga, A.; Vergara, C.; Fredes, C.; García, P.; Giménez, B.; Robert, P. Identification of polyphenols from Chilean brown seaweeds extracts by LC-DAD-ESI-MS/MS. *J. Aquat. Food Prod. Technol.* **2019**, *28*, 375–391. [CrossRef]

113. Kim, S.M.; Kang, S.W.; Jeon, J.S.; Jung, Y.J.; Kim, W.R.; Kim, C.Y.; Um, B.H. Determination of major phlorotannins in *Eisenia bicyclis* using hydrophilic interaction chromatography: Seasonal variation and extraction characteristics. *Food Chem.* **2013**, *138*, 2399–2406. [CrossRef]

114. Cho, S.; Yang, H.; Jeon, Y.J.; Lee, C.J.; Jin, Y.H.; Baek, N.I.; Kim, D.; Kang, S.M.; Yoon, M.; Yong, H.; et al. Phlorotannins of the edible brown seaweed *Ecklonia cava* Kjellman induce sleep via positive allosteric modulation of gamma-aminobutyric acid type A-benzodiazepine receptor: A novel neurological activity of seaweed polyphenols. *Food Chem.* **2012**, *132*, 1133–1142. [CrossRef] [PubMed]

115. Lee, J.H.; Ko, J.Y.; Oh, J.Y.; Kim, C.Y.; Lee, H.J.; Kim, J.; Jeon, Y.J. Preparative isolation and purification of phlorotannins from *Ecklonia cava* using centrifugal partition chromatography by one-step. *Food Chem.* **2014**, *158*, 433–437. [CrossRef] [PubMed]

116. Nho, J.A.; Shin, Y.S.; Jeong, H.-R.; Cho, S.; Heo, H.J.; Kim, G.H.; Kim, D.-O. Neuroprotective effects of phlorotannin-rich extract from brown seaweed *Ecklonia cava* on neuronal PC-12 and SH-SY5Y Cells with Oxidative Stress. *J. Microbiol. Biotechnol.* **2020**, *30*, 359–367. [CrossRef] [PubMed]

117. Wei, R.; Lee, M.S.; Lee, B.; Oh, C.W.; Choi, C.G.; Kim, H.R. Isolation and identification of anti-inflammatory compounds from ethyl acetate fraction of *Ecklonia stolonifera* and their anti-inflammatory action. *J. Appl. Phycol.* **2016**, *28*, 3535–3545. [CrossRef]

118. Kirke, D.A.; Smyth, T.J.; Rai, D.K.; Kenny, O.; Stengel, D.B. The chemical and antioxidant stability of isolated low molecular weight phlorotannins. *Food Chem.* **2017**, *221*, 1104–1112. [CrossRef]

119. Wang, T.; Jónsdóttir, R.; Liu, H.; Gu, L.; Kristinsson, H.G.; Raghavan, S.; Ólafsdóttir, G. Antioxidant capacities of phlorotannins extracted from the brown algae *Fucus vesiculosus*. *J. Agric. Food Chem.* **2012**, *60*, 5874–5883. [CrossRef]

120. Catarino, M.D.; Silva, A.M.S.; Mateus, N.; Cardoso, S.M. Optimization of phlorotannins extraction from *Fucus vesiculosus* and evaluation of their potential to prevent metabolic disorders. *Mar. Drugs* **2019**, *17*, 162. [CrossRef]

121. Le Lann, K.; Surget, G.; Couteau, C.; Coiffard, L.; Cérantola, S.; Gaillard, F.; Larnicol, M.; Zubia, M.; Guérard, F.; Poupart, N.; et al. Sunscreen, antioxidant, and bactericide capacities of phlorotannins from the brown macroalga *Halidrys siliquosa*. *J. Appl. Phycol.* **2016**, *28*, 3547–3559. [CrossRef]

122. Haraguchi, K.; Kotaki, Y.; Relox, J.R.; Romero, M.L.J.; Terada, R. Monitoring of naturally produced brominated phenoxyphenols and phenoxyanisoles in aquatic plants from the Philippines. *J. Agric. Food Chem.* **2010**, *58*, 12385–12391. [CrossRef] [PubMed]

123. Vissers, A.M.; Caligiani, A.; Sforza, S.; Vincken, J.P.; Gruppen, H. Phlorotannin Composition of *Laminaria digitata*. *Phytochem. Anal.* **2017**, *28*, 487–495. [CrossRef] [PubMed]

124. Xu, K.; Guo, S.; Jia, X.; Li, X.; Shi, D. Phytochemical and chemotaxonomic study on *Leathesia nana* (Chordariaceae). *Biochem. Syst. Ecol.* **2018**, *81*, 42–44. [CrossRef]

125. Nair, D.; Vanuopadath, M.; Balasubramanian, A.; Iyer, A.; Ganesh, S.; Anil, A.N.; Vikraman, V.; Pillai, P.; Bose, C.; Nair, B.G.; et al. Phlorotannins from *Padina tetrastromatica*: Structural characterisation and functional studies. *J. Appl. Phycol.* **2019**, *31*, 3131–3141. [CrossRef]

126. Li, Y.; Fu, X.; Duan, D.; Liu, X.; Xu, J.; Gao, X. Extraction and identification of phlorotannins from the brown aqlga, *Sargassum fusiforme* (Harvey) Setchell. *Mar. Drugs* **2017**, *15*, 49. [CrossRef]

127. Montero, L.; Sánchez-Camargo, A.P.; García-Cañas, V.; Tanniou, A.; Stiger-Pouvreau, V.; Russo, M.; Rastrelli, L.; Cifuentes, A.; Herrero, M.; Ibáñez, E. Anti-proliferative activity and chemical characterization by comprehensive two-dimensional liquid chromatography coupled to mass spectrometry of phlorotannins from the brown macroalga *Sargassum muticum* collected on North-Atlantic coasts. *J. Chromatogr. A* **2016**, *1428*, 115–125. [CrossRef]

128. Vázquez-Rodríguez, B.; Gutiérrez-Uribe, J.A.; Antunes-Ricardo, M.; Santos-Zea, L.; Cruz-Suárez, L.E. Ultrasound-assisted extraction of phlorotannins and polysaccharides from *Silvetia compressa* (Phaeophyceae). *J. Appl. Phycol.* **2020**, *32*, 1441–1453. [CrossRef]

129. Murugan, K.; Iyer, V.V. Differential growth inhibition of cancer cell lines and antioxidant activity of extracts of red, brown, and green marine algae. *Vitr. Cell. Dev. Biol. - Anim.* **2013**, *49*, 324–334. [CrossRef]

130. Aravindan, S.; Delma, C.R.; Thirugnanasambandan, S.S.; Herman, T.S.; Aravindan, N. Anti-pancreatic cancer deliverables from sea: First-hand evidence on the efficacy, molecular targets and mode of action for multifarious polyphenols from five different brown-algae. *PLoS ONE* **2013**, *8*, e61977. [CrossRef]

131. Lopes, G.; Sousa, C.; Silva, L.R.; Pinto, E.; Andrade, P.B.; Bernardo, J.; Mouga, T.; Valentão, P. Can phlorotannins purified extracts constitute a novel pharmacological alternative for microbial infections with associated inflammatory conditions? *PLoS ONE* **2012**, *7*, e31145. [CrossRef] [PubMed]

132. Audibert, L.; Fauchon, M.; Blanc, N.; Hauchard, D.; Ar Gall, E. Phenolic compounds in the brown seaweed *Ascophyllum nodosum*: Distribution and radical-scavenging activities. *Phytochem. Anal.* **2010**, *21*, 399–405. [CrossRef] [PubMed]

133. Kurth, C.; Welling, M.; Pohnert, G. Sulfated phenolic acids from *Dasycladales siphonous* green algae. *Phytochemistry* **2015**, *117*, 417–423. [CrossRef] [PubMed]

134. De Oliveira, A.L.L.; Da Silva, D.B.; Lopes, N.P.; Debonsi, H.M.; Yokoya, N.S. Chemical constituents from red algae *Bostrychia radicans* (Rhodomelaceae): New amides and phenolic compounds. *Quim. Nova* **2012**, *35*, 2186–2188. [CrossRef]

135. Dahlgren, E.; Lindqvist, D.; Dahlgren, H.; Asplund, L.; Lehtilä, K. Trophic transfer of naturally produced brominated aromatic compounds in a Baltic Sea food chain. *Chemosphere* **2016**, *144*, 1597–1604. [CrossRef] [PubMed]

136. Lindqvist, D.; Dahlgren, E.; Asplund, L. Biosynthesis of hydroxylated polybrominated diphenyl ethers and the correlation with photosynthetic pigments in the red alga *Ceramium tenuicorne*. *Phytochemistry* **2017**, *133*, 51–58. [CrossRef]

137. Mikami, D.; Kurihara, H.; Kim, S.M.; Takahashi, K. Red algal bromophenols as glucose 6-phosphate dehydrogenase inhibitors. *Mar. Drugs* **2013**, *11*, 4050–4057. [CrossRef]

138. Mikami, D.; Kurihara, H.; Ono, M.; Kim, S.M.; Takahashi, K. Inhibition of algal bromophenols and their related phenols against glucose 6-phosphate dehydrogenase. *Fitoterapia* **2016**, *108*, 20–25. [CrossRef]

139. Islam, M.R.; Mikami, D.; Kurihara, H. Two new algal bromophenols from *Odonthalia corymbifera*. *Tetrahedron Lett.* **2017**, *58*, 4119–4121. [CrossRef]

140. Lever, J.; Curtis, G.; Brkljača, R.; Urban, S. Bromophenolics from the reed alga *Polysiphonia decipiens*. *Mar. Drugs* **2019**, *17*, 497. [CrossRef]

141. Kim, S.Y.; Kim, S.R.; Oh, M.J.; Jung, S.J.; Kang, S.Y. In Vitro antiviral activity of red alga, Polysiphonia morrowii extract and its bromophenols against fish pathogenic infectious hematopoietic necrosis virus and infectious pancreatic necrosis virus. *J. Microbiol.* **2011**, *49*, 102–106. [CrossRef] [PubMed]

142. Choi, Y.K.; Ye, B.R.; Kim, E.A.; Kim, J.; Kim, M.S.; Lee, W.W.; Ahn, G.N.; Kang, N.; Jung, W.K.; Heo, S.J. Bis (3-bromo-4,5-dihydroxybenzyl) ether, a novel bromophenol from the marine red alga *Polysiphonia morrowii* that suppresses LPS-induced inflammatory response by inhibiting ROS-mediated ERK signaling pathway in RAW 264.7 macrophages. *Biomed. Pharmacother.* **2018**, *103*, 1170–1177. [CrossRef] [PubMed]

143. Li, K.; Li, X.M.; Gloer, J.B.; Wang, B.G. New nitrogen-containing bromophenols from the marine red alga *Rhodomela confervoides* and their radical scavenging activity. *Food Chem.* **2012**, *135*, 868–872. [CrossRef] [PubMed]

144. Liu, X.; Li, X.; Gao, L.; Cui, C.; Li, C.; Li, J.; Wang, B. Extraction and PTP1B inhibitory activity of bromophenols from the marine red alga *Symphyocladia latiuscula*. *Chinese J. Oceanol. Limnol.* **2011**, *29*, 686–690. [CrossRef]

145. Xu, X.; Yin, L.; Gao, L.; Gao, J.; Chen, J.; Li, J.; Song, F. Two new bromophenols with radical scavenging activity from marine red alga *Symphyocladia latiuscula*. *Mar. Drugs* **2013**, *11*, 842–847. [CrossRef]

146. Xu, X.; Yin, L.; Gao, J.; Gao, L.; Song, F. Antifungal bromophenols from marine red alga *Symphyocladia latiuscula*. *Chem. Biodivers.* **2014**, *11*, 807–811. [CrossRef]

147. Xu, X.; Yang, H.; Khalil, Z.G.; Yin, L.; Xiao, X.; Neupane, P.; Bernhardt, P.V.; Salim, A.A.; Song, F.; Capon, R.J. Chemical diversity from a Chinese marine red alga, *Symphyocladia latiuscula*. *Mar. Drugs* **2017**, *15*, 374. [CrossRef]

148. Xu, X.; Yang, H.; Khalil, Z.G.; Yin, L.; Xiao, X.; Salim, A.A.; Song, F.; Capon, R.J. Bromocatechol conjugates from a Chinese marine red alga, *Symphyocladia latiuscula*. *Phytochemistry* **2019**, *158*, 20–25. [CrossRef]

149. Paudel, P.; Seong, S.H.; Zhou, Y.; Park, H.J.; Jung, H.A.; Choi, J.S. Anti-Alzheimer's Disease Activity of Bromophenols from a Red Alga, *Symphyocladia latiuscula* (Harvey) Yamada. *ACS Omega* **2019**, *4*, 12259–12270. [CrossRef]

150. Hofer, S.; Hartmann, A.; Orfanoudaki, M.; Ngoc, H.N.; Nagl, M.; Karsten, U.; Heesch, S.; Ganzera, M. Development and validation of an HPLC method for the quantitative analysis of bromophenolic compounds in the red alga *Vertebrata lanosa*. *Mar. Drugs* **2019**, *17*, 675. [CrossRef]

151. Huang, D.; Boxin, O.U.; Prior, R.L. The chemistry behind antioxidant capacity assays. *J. Agric. Food Chem.* **2005**, *53*, 1841–1856. [CrossRef] [PubMed]

152. Tierney, M.S.; Smyth, T.J.; Rai, D.K.; Soler-Vila, A.; Croft, A.K.; Brunton, N. Enrichment of polyphenol contents and antioxidant activities of Irish brown macroalgae using food-friendly techniques based on polarity and molecular size. *Food Chem.* **2013**, *139*, 753–761. [CrossRef] [PubMed]

153. Lim, S.N.; Cheung, P.C.K.; Ooi, V.E.C.; Ang, P.O. Evaluation of antioxidative activity of extracts from a brown seaweed, *Sargassum siliquastrum*. *J. Agric. Food Chem.* **2002**, *50*, 3862–3866. [CrossRef] [PubMed]

154. Martins, C.D.L.; Ramlov, F.; Nocchi Carneiro, N.P.; Gestinari, L.M.; dos Santos, B.F.; Bento, L.M.; Lhullier, C.; Gouvea, L.; Bastos, E.; Horta, P.A.; et al. Antioxidant properties and total phenolic contents of some tropical seaweeds of the Brazilian coast. *J. Appl. Phycol.* **2013**, *25*, 1179–1187. [CrossRef]

155. Sevimli-Gur, C.; Yesil-Celiktas, O. Cytotoxicity screening of supercritical fluid extracted seaweeds and phenylpropanoids. *Mol. Biol. Rep.* **2019**, *46*, 3691–3699. [CrossRef] [PubMed]

156. Fraga, C.G.; Croft, K.D.; Kennedy, D.O.; Tomás-Barberán, F.A. The effects of polyphenols and other bioactives on human health. *Food Funct.* **2019**, *10*, 514–528. [CrossRef]
157. Williams, R.; Colagiuri, S.; Chan, J.; Gregg, E.W.; Ke, C.; Lim, L.-L.; Yang, X. *IDF Atlas*, 9th ed.; International Diabetes Federation: Brussels, Belgium, 2019; ISBN 978-2-930229-87-4.
158. World Health Organization. Global Report on Diabetes. Available online: https://apps.who.int/iris/handle/10665/204871 (accessed on 7 August 2020).
159. Hermans, M.P.; Dath, N. Prevalence and co-prevalence of comorbidities in Belgian patients with type 2 diabetes mellitus: A transversal, descriptive study. *Acta Clin. Belgica Int. J. Clin. Lab. Med.* **2018**, *73*, 68–74. [CrossRef]
160. Lee, S.H.; Jeon, Y.J. Anti-diabetic effects of brown algae derived phlorotannins, marine polyphenols through diverse mechanisms. *Fitoterapia* **2013**, *86*, 129–136. [CrossRef]
161. Gunathilaka, T.L.; Samarakoon, K.; Ranasinghe, P.; Peiris, L.D.C. Antidiabetic potential of marine brown algae - A Mini Review. *J. Diabetes Res.* **2020**, *2020*. [CrossRef]
162. Murugan, A.C.; Karim, M.R.; Yusoff, M.B.M.; Tan, S.H.; Asras, M.F.B.F.; Rashid, S.S. New insights into seaweed polyphenols on glucose homeostasis. *Pharm. Biol.* **2015**, *53*, 1087–1097. [CrossRef]
163. Zhao, C.; Yang, C.; Liu, B.; Lin, L.; Sarker, S.D.; Nahar, L.; Yu, H.; Cao, H.; Xiao, J. Bioactive compounds from marine macroalgae and their hypoglycemic benefits. *Trends Food Sci. Technol.* **2018**, *72*, 1–12. [CrossRef]
164. Benalla, W.; Bellahcen, S.; Bnouham, M. Antidiabetic medicinal plants as a source of alpha glucosidase Inhibitors. *Curr. Diabetes Rev.* **2010**, *6*, 247–254. [CrossRef] [PubMed]
165. Lee, S.H.; Yong-Li; Karadeniz, F.; Kim, M.M.; Kim, S.K. α-Glucosidase and α-amylase inhibitory activities of phloroglucinal derivatives from edible marine brown alga, *Ecklonia cava*. *J. Sci. Food Agric.* **2009**, *89*, 1552–1558. [CrossRef]
166. Nwosu, F.; Morris, J.; Lund, V.A.; Stewart, D.; Ross, H.A.; McDougall, G.J. Anti-proliferative and potential anti-diabetic effects of phenolic-rich extracts from edible marine algae. *Food Chem.* **2011**, *126*, 1006–1012. [CrossRef]
167. Lordan, S.; Smyth, T.J.; Soler-Vila, A.; Stanton, C.; Paul Ross, R. The α-amylase and α-glucosidase inhibitory effects of Irish seaweed extracts. *Food Chem.* **2013**, *141*, 2170–2176. [CrossRef]
168. Kellogg, J.; Grace, M.H.; Lila, M.A. Phlorotannins from alaskan seaweed inhibit carbolytic enzyme activity. *Mar. Drugs* **2014**, *12*, 5277–5294. [CrossRef]
169. Sharifuddin, Y.; Chin, Y.X.; Lim, P.E.; Phang, S.M. Potential bioactive compounds from seaweed for diabetes management. *Mar. Drugs* **2015**, *13*, 5447–5491. [CrossRef]
170. Park, S.R.; Kim, J.H.; Jang, H.D.; Yang, S.Y.; Kim, Y.H. Inhibitory activity of minor phlorotannins from Ecklonia cava on α-glucosidase. *Food Chem.* **2018**, *257*, 128–134. [CrossRef]
171. Ezzat, S.M.; El Bishbishy, M.H.; Habtemariam, S.; Salehi, B.; Sharifi-Rad, M.; Martins, N.; Sharifi-Rad, J. Looking at marine-derived bioactive molecules as upcoming anti-diabetic agents: A special emphasis on PTP1B inhibitors. *Molecules* **2018**, *23*, 3334. [CrossRef]
172. Xu, Q.; Luo, J.; Wu, N.; Zhang, R.; Shi, D. BPN, a marine-derived PTP1B inhibitor, activates insulin signaling and improves insulin resistance in C2C12 myotubes. *Int. J. Biol. Macromol.* **2018**, *106*, 379–386. [CrossRef]
173. Luo, J.; Hou, Y.; Xie, M.; Ma, W.; Shi, D.; Jiang, B. CYC31, A Natural bromophenol PTP1B inhibitor, activates insulin signaling and improves long chain-fatty acid oxidation in C2C12 Myotubes. *Mar. Drugs* **2020**, *18*, 267. [CrossRef]
174. Boyd, A.C.; Abdel-Wahab, Y.H.A.; McKillop, A.M.; McNulty, H.; Barnett, C.R.; O'Harte, F.P.M.; Flatt, P.R. Impaired ability of glycated insulin to regulate plasma glucose and stimulate glucose transport and metabolism in mouse abdominal muscle. *Biochim. Biophys. Acta Gen. Subj.* **2000**, *1523*, 128–134. [CrossRef]
175. Iannuzzi, C.; Borriello, M.; Carafa, V.; Altucci, L.; Vitiello, M.; Balestrieri, M.L.; Ricci, G.; Irace, G.; Sirangelo, I. D-ribose-glycation of insulin prevents amyloid aggregation and produces cytotoxic adducts. *Biochim. Biophys. Acta Mol. Basis Dis.* **2016**, *1862*, 93–104. [CrossRef] [PubMed]
176. Singh, V.P.; Bali, A.; Singh, N.; Jaggi, A.S. Advanced glycation end products and diabetic complications. *Korean J. Physiol. Pharmacol.* **2014**, *18*, 1–14. [CrossRef] [PubMed]
177. Sugiura, S.; Minami, Y.; Taniguchi, R.; Tanaka, R.; Miyake, H.; Mori, T.; Ueda, M.; Shibata, T. Evaluation of anti-glycation activities of phlorotannins in human and bovine serum albumin-methylglyoxal models. *Nat. Prod. Commun.* **2017**, *12*, 1793–1796. [CrossRef]

178. Shakambari, G.; Ashokkumar, B.; Varalakshmi, P. Phlorotannins from Brown Algae: Inhibition of advanced glycation end products formation in high glucose induced *Caenorhabditis elegans*. *Indian J. Exp. Biol.* **2015**, *53*, 371–379.

179. Seong, S.H.; Paudel, P.; Jung, H.A.; Choi, J.S. Identifying phlorofucofuroeckol-A as a dual inhibitor of amyloid-β25-35 self-aggregation and insulin glycation: Elucidation of the molecular mechanism of action. *Mar. Drugs* **2019**, *17*, 600. [CrossRef]

180. Malin, S.K.; Finnegan, S.; Fealy, C.E.; Filion, J.; Rocco, M.B.; Kirwan, J.P. B-Cell dysfunction is associated with metabolic syndrome severity in adults. *Metab. Syndr. Relat. Disord.* **2014**, *12*, 79–85. [CrossRef]

181. Lee, S.H.; Kang, S.M.; Ko, S.C.; Kang, M.C.; Jeon, Y.J. Octaphlorethol A, a novel phenolic compound isolated from *Ishige foliacea*, protects against streptozotocin-induced pancreatic β cell damage by reducing oxidative stress and apoptosis. *Food Chem. Toxicol.* **2013**, *59*, 643–649. [CrossRef]

182. Kim, E.A.; Kang, M.C.; Lee, J.H.; Kang, N.; Lee, W.; Oh, J.Y.; Yang, H.W.; Lee, J.S.; Jeon, Y.J. Protective effect of marine brown algal polyphenols against oxidative stressed zebrafish with high glucose. *RSC Adv.* **2015**, *5*, 25738–25746. [CrossRef]

183. Cha, S.H.; Kim, H.S.; Hwang, Y.; Jeon, Y.J.; Jun, H.S. Polysiphonia japonica extract attenuates palmitate-induced toxicity and enhances insulin secretion in pancreatic beta-cells. *Oxid. Med. Cell. Longev.* **2018**, *2018*. [CrossRef] [PubMed]

184. Sheetz, M.J.; King, G.L. Molecular understanding of hyperglycemia's adverse effects for diabetic complications. *J. Am. Med. Assoc.* **2002**, *288*, 2579–2588. [CrossRef] [PubMed]

185. Lee, S.H.; Kang, S.M.; Ko, S.C.; Lee, D.H.; Jeon, Y.J. Octaphlorethol A, a novel phenolic compound isolated from a brown alga, Ishige foliacea, increases glucose transporter 4-mediated glucose uptake in skeletal muscle cells. *Biochem. Biophys. Res. Commun.* **2012**, *420*, 576–581. [CrossRef] [PubMed]

186. Cai, Y.; Wang, Q.; Ling, Z.; Pipeleers, D.; McDermott, P.; Pende, M.; Heimberg, H.; Van de Casteele, M. Akt activation protects pancreatic beta cells from AMPK-mediated death through stimulation of mTOR. *Biochem. Pharmacol.* **2008**, *75*, 1981–1993. [CrossRef] [PubMed]

187. Lee, S.H.; Park, M.H.; Heo, S.J.; Kang, S.M.; Ko, S.C.; Han, J.S.; Jeon, Y.J. Dieckol isolated from *Ecklonia cava* inhibits α-glucosidase and α-amylase in vitro and alleviates postprandial hyperglycemia in streptozotocin-induced diabetic mice. *Food Chem. Toxicol.* **2010**, *48*, 2633–2637. [CrossRef] [PubMed]

188. Lee, H.A.; Lee, J.H.; Han, J.S. A phlorotannin constituent of *Ecklonia cava* alleviates postprandial hyperglycemia in diabetic mice. *Pharm. Biol.* **2017**, *55*, 1149–1154. [CrossRef]

189. Murray, M.; Dordevic, A.L.; Ryan, L.; Bonham, M.P. The impact of a single dose of a polyphenol-rich seaweed extract on postprandial glycaemic control in healthy adults: A randomised cross-over trial. *Nutrients* **2018**, *10*, 270. [CrossRef]

190. Kang, M.C.; Wijesinghe, W.A.J.P.; Lee, S.H.; Kang, S.M.; Ko, S.C.; Yang, X.; Kang, N.; Jeon, B.T.; Kim, J.; Lee, D.H.; et al. Dieckol isolated from brown seaweed *Ecklonia cava* attenuates type II diabetes in db/db mouse model. *Food Chem. Toxicol.* **2013**, *53*, 294–298. [CrossRef]

191. Yan, X.; Yang, C.; Lin, G.; Chen, Y.; Miao, S.; Liu, B.; Zhao, C. Antidiabetic potential of green seaweed *Enteromorpha prolifera* flavonoids regulating insulin signaling pathway and gut microbiota in Type 2 Diabetic Mice. *J. Food Sci.* **2019**, *84*, 165–173. [CrossRef]

192. Lee, S.H.; Jeon, Y.J. Efficacy and safety of a dieckol-rich extract (AG-dieckol) of brown algae, *Ecklonia cava*, in pre-diabetic individuals: A double-blind, randomized, placebo-controlled clinical trial. *Food Funct.* **2015**, *6*, 853–858. [CrossRef]

193. Cavalot, F.; Petrelli, A.; Traversa, M.; Bonomo, K.; Fiora, E.; Conti, M.; Anfossi, G.; Costa, G.; Trovati, M. Postprandial blood glucose is a stronger predictor of cardiovascular events than fasting blood glucose in type 2 diabetes mellitus, particularly in women: Lessons from the San Luigi Gonzaga diabetes study. *J. Clin. Endocrinol. Metab.* **2006**, *91*, 813–819. [CrossRef] [PubMed]

194. Abdali, D.; Samson, S.E.; Grover, A.K. How effective are antioxidant supplements in obesity and diabetes? *Med. Princ. Pract.* **2015**, *24*, 201–215. [CrossRef] [PubMed]

195. Gregor, M.F.; Hotamisligil, G.S. Inflammatory mechanisms in obesity. *Annu. Rev. Immunol.* **2011**, *29*, 415–445. [CrossRef] [PubMed]

196. Jung, U.J.; Choi, M.S. Obesity and its metabolic complications: The role of adipokines and the relationship between obesity, inflammation, insulin resistance, dyslipidemia and nonalcoholic fatty liver disease. *Int. J. Mol. Sci.* **2014**, *15*, 6184–6223. [CrossRef]

197. Marseglia, L.; Manti, S.; D'Angelo, G.; Nicotera, A.; Parisi, E.; Di Rosa, G.; Gitto, E.; Arrigo, T. Oxidative stress in obesity: A critical component in human diseases. *Int. J. Mol. Sci.* **2015**, *16*, 378–400. [CrossRef]
198. Eom, S.H.; Lee, M.S.; Lee, E.W.; Kim, Y.M.; Kim, T.H. Pancreatic lipase inhibitory activity of phlorotannins isolated from *Eisenia bicyclis*. *Phyther. Res.* **2013**, *27*, 148–151. [CrossRef]
199. Austin, C.; Stewart, D.; Allwood, J.W.; McDougall, G.J. Extracts from the edible seaweed *Ascophyllum nodosum*, inhibit lipase activity in vitro: Contributions of phenolic and polysaccharide components. *Food Funct.* **2018**, *9*, 502–510. [CrossRef]
200. Franssen, R.; Monajemi, H.; Stroes, E.S.G.; Kastelein, J.J.P. Obesity and dyslipidemia. *Med. Clin. N. Am.* **2011**, *95*, 893–902. [CrossRef]
201. Ko, S.C.; Lee, M.; Lee, J.H.; Lee, S.H.; Lim, Y.; Jeon, Y.J. Dieckol, a phlorotannin isolated from a brown seaweed, *Ecklonia cava*, inhibits adipogenesis through AMP-activated protein kinase (AMPK) activation in 3T3-L1 preadipocytes. *Environ. Toxicol. Pharmacol.* **2013**, *36*, 1253–1260. [CrossRef]
202. Jung, H.A.; Jung, H.J.; Jeong, H.Y.; Kwon, H.J.; Ali, M.Y.; Choi, J.S. Phlorotannins isolated from the edible brown alga *Ecklonia stolonifera* exert anti-adipogenic activity on 3T3-L1 adipocytes by downregulating C/EBPα and PPARγ. *Fitoterapia* **2014**, *92*, 260–269. [CrossRef]
203. Kwon, T.H.; Wu, Y.X.; Kim, J.S.; Woo, J.H.; Park, K.T.; Kwon, O.J.; Seo, H.J.; Kim, T.; Park, N.H. 6,6'-Bieckol inhibits adipocyte differentiation through downregulation of adipogenesis and lipogenesis in 3T3-L1 cells. *J. Sci. Food Agric.* **2015**, *95*, 1830–1837. [CrossRef] [PubMed]
204. Kong, C.S.; Kim, H.; Seo, Y. Edible brown alga *Ecklonia cava* derived phlorotannin-induced anti-adipogenic activity in vitro. *J. Food Biochem.* **2015**, *39*, 1–10. [CrossRef]
205. Salas, A.; Noé, V.; Ciudad, C.J.; Romero, M.M.; Remesar, X.; Esteve, M. Short-term oleoyl-estrone treatment affects capacity to manage lipids in rat adipose tissue. *BMC Genom.* **2007**, *8*, 292. [CrossRef] [PubMed]
206. Izquierdo, A.G.; Crujeiras, A.B.; Casanueva, F.F.; Carreira, M.C. Leptin, obesity, and leptin resistance: Where are we 25 years later? *Nutrients* **2019**, *11*, 2704. [CrossRef]
207. Kim, I.H.; Nam, T.J. Enzyme-treated *Ecklonia cava* extract inhibits adipogenesis through the downregulation of C/EBPα in 3T3-L1 adipocytes. *Int. J. Mol. Med.* **2017**, *39*, 636–644. [CrossRef]
208. Karadeniz, F.; Ahn, B.N.; Kim, J.A.; Seo, Y.; Jang, M.S.; Nam, K.H.; Kim, M.; Lee, S.H.; Kong, C.S. Phlorotannins suppress adipogenesis in pre-adipocytes while enhancing osteoblastogenesis in pre-osteoblasts. *Arch. Pharm. Res.* **2015**, *38*, 2172–2182. [CrossRef]
209. Choi, H.S.; Jeon, H.J.; Lee, O.H.; Lee, B.Y. Dieckol, a major phlorotannin in Ecklonia cava, suppresses lipid accumulation in the adipocytes of high-fat diet-fed zebrafish and mice: Inhibition of early adipogenesis via cell-cycle arrest and AMPKα activation. *Mol. Nutr. Food Res.* **2015**, *59*, 1458–1471. [CrossRef]
210. Ko, S.C.; Ding, Y.; Kim, J.; Ye, B.R.; Kim, E.A.; Jung, W.K.; Heo, S.J.; Lee, S.H. Bromophenol (5-bromo-3,4-dihydroxybenzaldehyde) isolated from red alga *Polysiphonia morrowii* inhibits adipogenesis by regulating expression of adipogenic transcription factors and AMP-activated protein kinase activation in 3T3-L1 adipocytes. *Phyther. Res.* **2019**, *33*, 737–744. [CrossRef]
211. Kellogg, J.; Esposito, D.; Grace, M.H.; Komarnytsky, S.; Lila, M.A. Alaskan seaweeds lower inflammation in RAW 264.7 macrophages and decrease lipid accumulation in 3T3-L1 adipocytes. *J. Funct. Foods* **2015**, *15*, 396–407. [CrossRef]
212. Song, F.; Del Pozo, C.H.; Rosario, R.; Zou, Y.S.; Ananthakrishnan, R.; Xu, X.; Patel, P.R.; Benoit, V.M.; Yan, S.F.; Li, H.; et al. RAGE regulates the metabolic and inflammatory response to high-fat feeding in mice. *Diabetes* **2014**, *63*, 1948–1965. [CrossRef]
213. Choi, J.; Oh, S.; Son, M.; Byun, K. Pyrogallol-phloroglucinol-6,6-bieckol alleviates obesity and systemic inflammation in a mouse model by reducing expression of RAGE and RAGE ligands. *Mar. Drugs* **2019**, *17*, 612. [CrossRef] [PubMed]
214. Yeo, A.R.; Lee, J.; Tae, I.H.; Park, S.R.; Cho, Y.H.; Lee, B.H.; Cheol Shin, H.; Kim, S.H.; Yoo, Y.C. Anti-hyperlipidemic effect of polyphenol extract (Seapolynol™) and dieckol isolated from *Ecklonia cava* in in vivo and in vitro models. *Prev. Nutr. Food Sci.* **2012**, *17*, 1–7. [CrossRef] [PubMed]
215. Park, E.Y.; Kim, E.H.; Kim, M.H.; Seo, Y.W.; Lee, J.I.; Jun, H.S. Polyphenol-rich fraction of brown alga *Ecklonia cava* collected from Gijang, Korea, reduces obesity and glucose levels in high-fat diet-induced obese mice. *Evidence-based Complement. Altern. Med.* **2012**, *2012*. [CrossRef] [PubMed]

216. Eo, H.; Jeon, Y.J.; Lee, M.; Lim, Y. Brown alga *Ecklonia cava* polyphenol extract ameliorates hepatic lipogenesis, oxidative stress, and inflammation by activation of AMPK and SIRT1 in high-fat diet-induced obese mice. *J. Agric. Food Chem.* **2015**, *63*, 349–359. [CrossRef]

217. Ding, Y.; Wang, L.; Im, S.; Hwang, O.; Kim, H.S.; Kang, M.C.; Lee, S.H. Anti-obesity effect of diphlorethohydroxycarmalol isolated from brown alga *Ishige okamurae* in high-fat diet-induced obese mice. *Mar. Drugs* **2019**, *17*, 637. [CrossRef]

218. Shin, H.C.; Kim, S.H.; Park, Y.; Lee, B.H.; Hwang, H.J. Effects of 12-week oral supplementation of *Ecklonia cava* polyphenols on anthropometric and blood lipid parameters in overweight Korean individuals: A double-blind randomized clinical trial. *Phyther. Res.* **2012**, *26*, 363–368. [CrossRef]

219. Samson, S.L.; Garber, A.J. Metabolic syndrome. *Endocrinol. Metab. Clin. N. Am.* **2014**, *43*, 1–23. [CrossRef]

220. Alberti, K.G.M.M.; Eckel, R.H.; Grundy, S.M.; Zimmet, P.Z.; Cleeman, J.I.; Donato, K.A.; Fruchart, J.C.; James, W.P.T.; Loria, C.M.; Smith, S.C. Harmonizing the metabolic syndrome: A joint interim statement of the international diabetes federation task force on epidemiology and prevention; National Heart, Lung, and Blood Institute; American Heart Association; World Heart Federation; International. *Circulation* **2009**, *120*, 1640–1645. [CrossRef]

221. Stefanska, A.; Bergmann, K.; Sypniewska, G. Metabolic Syndrome and Menopause. In *Advances in Clinical Chemistry*; Elsevier Inc.: Amsterdam, The Netherlands, 2015; Volume 72, pp. 1–75. ISBN 9780128033142.

222. Park, J.K.; Woo, H.W.; Kim, M.K.; Shin, J.; Lee, Y.H.; Shin, D.H.; Shin, M.H.; Choi, B.Y. Dietary iodine, seaweed consumption, and incidence risk of metabolic syndrome among postmenopausal women: A prospective analysis of the Korean Multi-Rural Communities Cohort Study (MRCohort). *Eur. J. Nutr.* **2020**. [CrossRef]

223. Vijayan, R.; Chitra, L.; Penislusshiyan, S.; Palvannan, T. Exploring bioactive fraction of *Sargassum wightii*: In vitro elucidation of angiotensin-i-converting enzyme inhibition and antioxidant potential. *Int. J. Food Prop.* **2018**, *21*, 674–684. [CrossRef]

224. Son, M.; Oh, S.; Lee, H.S.; Ryu, B.M.; Jiang, Y.; Jang, J.T.; Jeon, Y.J.; Byun, K. Pyrogallol-phloroglucinol-6,6'-bieckol from *Ecklonia cava* improved blood circulation in diet-induced obese and diet-induced hypertension mouse models. *Mar. Drugs* **2019**, *17*, 272. [CrossRef]

225. Kammoun, I.; Ben Salah, H.; Ben Saad, H.; Cherif, B.; Droguet, M.; Magné, C.; Kallel, C.; Boudawara, O.; Hakim, A.; Gharsallah, N.; et al. Hypolipidemic and cardioprotective effects of *Ulva lactuca* ethanolic extract in hypercholesterolemic mice. *Arch. Physiol. Biochem.* **2018**, *124*, 313–325. [CrossRef] [PubMed]

226. Wanyonyi, S.; Du Preez, R.; Brown, L.; Paul, N.A.; Panchal, S.K. *Kappaphycus alvarezii* as a food supplement prevents diet-induced metabolic syndrome in rats. *Nutrients* **2017**, *9*, 1261. [CrossRef]

227. Gómez-Guzmán, M.; Rodríguez-Nogales, A.; Algieri, F.; Gálvez, J. Potential role of seaweed polyphenols in cardiovascular-associated disorders. *Mar. Drugs* **2018**, *16*, 250. [CrossRef]

228. Alam, T.; Khan, S.; Gaba, B.; Haider, M.F.; Baboota, S.; Ali, J. Nanocarriers as treatment modalities for hypertension. *Drug Deliv.* **2017**, *24*, 358–369. [CrossRef]

229. Seca, A.M.L.; Pinto, D.C.G.A. Overview on the antihypertensive and anti-obesity effects of secondary metabolites from seaweeds. *Mar. Drugs* **2018**, *16*, 237. [CrossRef] [PubMed]

230. Olasehinde, T.A.; Olaniran, A.O.; Okoh, A.I. Macroalgae as a valuable source of naturally occurring bioactive compounds for the treatment of Alzheimer's disease. *Mar. Drugs* **2019**, *17*, 609. [CrossRef] [PubMed]

231. Gouras, G.K.; Olsson, T.T.; Hansson, O. β-amyloid peptides and amyloid plaques in Alzheimer's disease. *Neurotherapeutics* **2015**, *12*, 3–11. [CrossRef]

232. Frozza, R.L.; Lourenco, M.V.; de Felice, F.G. Challenges for Alzheimer's disease therapy: Insights from novel mechanisms beyond memory defects. *Front. Neurosci.* **2018**, *12*, 1–13. [CrossRef]

233. Cai, H.; Wang, Y.; McCarthy, D.; Wen, H.; Borchelt, D.R.; Price, D.L.; Wong, P.C. BACE1 is the major β-secretase for generation of Aβ peptides by neurons. *Nat. Neurosci.* **2001**, *4*, 233–234. [CrossRef]

234. Choi, B.W.; Lee, H.S.; Shin, H.C.; Lee, B.H. Multifunctional activity of polyphenolic compounds associated with a potential for Alzheimer's disease therapy from ecklonia cava. *Phyther. Res.* **2015**, *29*, 549–553. [CrossRef] [PubMed]

235. Choi, J.S.; Haulader, S.; Karki, S.; Jung, H.J.; Kim, H.R.; Jung, H.A. Acetyl- and butyryl-cholinesterase inhibitory activities of the edible brown alga *Eisenia bicyclis*. *Arch. Pharm. Res.* **2015**, *38*, 1477–1487. [CrossRef] [PubMed]

236. Olasehinde, T.A.; Olaniran, A.O.; Okoh, A.I. Phenolic composition, antioxidant activity, anticholinesterase potential and modulatory effects of aqueous extracts of some seaweeds on β-amyloid aggregation and disaggregation. *Pharm. Biol.* **2019**, *57*, 460–469. [CrossRef] [PubMed]

237. Olasehinde, T.A.; Olaniran, A.O.; Okoh, A.I. Aqueous–ethanol extracts of some South African seaweeds inhibit beta-amyloid aggregation, cholinesterases, and beta-secretase activities in vitro. *J. Food Biochem.* **2019**, *43*, 1–10. [CrossRef]

238. Kim, J.J.; Kang, Y.J.; Shin, S.A.; Bak, D.H.; Lee, J.W.; Lee, K.B.; Yoo, Y.C.; Kim, D.K.; Lee, B.H.; Kim, D.W.; et al. Phlorofucofuroeckol improves glutamate- induced neurotoxicity through modulation of oxidative stress-mediated mitochondrial dysfunction in PC12 cells. *PLoS ONE* **2016**, *11*, e0163433. [CrossRef]

239. Yang, E.J.; Ahn, S.; Ryu, J.; Choi, M.S.; Choi, S.; Chong, Y.H.; Hyun, J.W.; Chang, M.J.; Kim, H.S. Phloroglucinol attenuates the cognitive deficits of the 5XFAD mouse model of Alzheimer's disease. *PLoS ONE* **2015**, *10*, e0135686. [CrossRef]

240. Wang, J.; Zheng, J.; Huang, C.; Zhao, J.; Lin, J.; Zhou, X.; Naman, C.B.; Wang, N.; Gerwick, W.H.; Wang, Q.; et al. Eckmaxol, a phlorotannin extracted from *Ecklonia maxima*, produces anti-β-amyloid oligomer neuroprotective effects possibly via directly acting on glycogen synthase kinase 3β. *ACS Chem. Neurosci.* **2018**, *9*, 1349–1356. [CrossRef]

241. Lee, S.; Youn, K.; Kim, D.H.; Ahn, M.R.; Yoon, E.; Kim, O.Y.; Jun, M. Anti-neuroinflammatory property of phlorotannins from *Ecklonia cava* on Aβ 25-35 -induced damage in PC12 cells. *Mar. Drugs* **2019**, *17*, 7. [CrossRef]

242. Um, M.Y.; Lim, D.W.; Son, H.J.; Cho, S.; Lee, C. Phlorotannin-rich fraction from Ishige foliacea brown seaweed prevents the scopolamine-induced memory impairment via regulation of ERK-CREB-BDNF pathway. *J. Funct. Foods* **2018**, *40*, 110–116. [CrossRef]

243. Lesné, S.; Ali, C.; Gabriel, C.; Croci, N.; MacKenzie, E.T.; Glabe, C.G.; Plotkine, M.; Marchand-Verrecchia, C.; Vivien, D.; Buisson, A. NMDA receptor activation inhibits α-secretase and promotes neuronal amyloid-β production. *J. Neurosci.* **2005**, *25*, 9367–9377. [CrossRef]

244. Figueira, I.; Garcia, G.; Pimpão, R.C.; Terrasso, A.P.; Costa, I.; Almeida, A.F.; Tavares, L.; Pais, T.F.; Pinto, P.; Ventura, M.R.; et al. Polyphenols journey through blood-brain barrier towards neuronal protection. *Sci. Rep.* **2017**, *7*, 1–16. [CrossRef] [PubMed]

245. Schram, F.R.; Ng, P.K.L. What is cancer? *J. Crustac. Biol.* **2012**, *32*, 665–672. [CrossRef]

246. Gutiérrez-Rodríguez, A.G.; Juárez-Portilla, C.; Olivares-Bañuelos, T.; Zepeda, R.C. Anticancer activity of seaweeds. *Drug Discov. Today* **2018**, *23*, 434–447. [CrossRef]

247. Matulja, D.; Wittine, K.; Malatesti, N.; Laclef, S.; Turks, M.; Markovic, M.K.; Ambrožić, G.; Marković, D. marine natural products with high anticancer activities. *Curr. Med. Chem.* **2020**, *27*, 1243–1307. [CrossRef] [PubMed]

248. Kim, J.; Lee, J.; Oh, J.H.; Chang, H.J.; Sohn, D.K.; Shin, A.; Kim, J. Associations among dietary seaweed intake, c-MYC rs6983267 polymorphism, and risk of colorectal cancer in a Korean population: A case–control study. *Eur. J. Nutr.* **2020**, *59*, 1963–1974. [CrossRef] [PubMed]

249. Mhadhebi, L.; Mhadhebi, A.; Robert, J.; Bouraoui, A. Antioxidant, anti-inflammatory and antiproliferative effects of aqueous extracts of three mediterranean brown seaweeds of the Genus *Cystoseira*. *Iran. J. Pharm. Res.* **2014**, *13*, 207–220. [CrossRef]

250. Corona, G.; Coman, M.M.; Spencer, J.P.E.; Rowland, I. Digested and fermented seaweed phlorotannins reduce DNA damage and inhibit growth of HT-29 colon cancer cells. *Proc. Nutr. Soc.* **2014**, *73*, 262519. [CrossRef]

251. Mwangi, H.M.; Njue, W.M.; Onani, M.O.; Thovhoghi, N.; Mabusela, W.T. Phlorotannins and a sterol isolated from a brown alga *Ecklonia maxima*, and their cytotoxic activity against selected cancer cell lines HeLa, H157 and MCF7. *Interdiscip. J. Chem.* **2017**, *2*, 1–6. [CrossRef]

252. Namvar, F.; Baharara, J.; Mahdi, A.A. Antioxidant and anticancer activities of selected Persian Gulf algae. *Indian J. Clin. Biochem.* **2014**, *29*, 13–20. [CrossRef]

253. Lopes-Costa, E.; Abreu, M.; Gargiulo, D.; Rocha, E.; Ramos, A.A. Anticancer effects of seaweed compounds fucoxanthin and phloroglucinol, alone and in combination with 5-fluorouracil in colon cells. *J. Toxicol. Environ. Heal. Part A Curr. Issues* **2017**, *80*, 776–787. [CrossRef]

254. Arai, M.; Shin, D.; Kamiya, K.; Ishida, R.; Setiawan, A.; Kotoku, N.; Kobayashi, M. Marine spongean polybrominated diphenyl ethers, selective growth inhibitors against the cancer cells adapted to glucose starvation, inhibits mitochondrial complex II. *J. Nat. Med.* **2017**, *71*, 44–49. [CrossRef] [PubMed]

255. Bernardini, G.; Minetti, M.; Polizzotto, G.; Biazzo, M.; Santucci, A. Pro-Apoptotic activity of French polynesian *Padina pavonica* extract on human osteosarcoma cells. *Mar. Drugs* **2018**, *16*, 504. [CrossRef] [PubMed]

256. Kosanić, M.; Ranković, B.; Stanojković, T. Brown macroalgae from the Adriatic Sea as a promising source of bioactive nutrients. *J. Food Meas. Charact.* **2019**, *13*, 330–338. [CrossRef]

257. Abdelhamid, A.; Lajili, S.; Elkaibi, M.A.; Ben Salem, Y.; Abdelhamid, A.; Muller, C.D.; Majdoub, H.; Kraiem, J.; Bouraoui, A. Optimized extraction, preliminary characterization and evaluation of the in vitro anticancer activity of phlorotannin-rich fraction from the brown seaweed, *Cystoseira sedoides*. *J. Aquat. Food Prod. Technol.* **2019**, *28*, 892–909. [CrossRef]

258. Abu-Khudir, R.; Ismail, G.A.; Diab, T. Antimicrobial, antioxidant, and anti-tumor activities of *Sargassum linearifolium* and *Cystoseira crinita* from Egyptian Mediterranean Coast. *Nutr. Cancer* **2020**, *0*, 1–16. [CrossRef]

259. Premarathna, A.D.; Ranahewa, T.H.; Wijesekera, S.K.; Harishchandra, D.L.; Karunathilake, K.J.K.; Waduge, R.N.; Wijesundara, R.R.M.K.K.; Jayasooriya, A.P.; Wijewardana, V.; Rajapakse, R.P.V.J. Preliminary screening of the aqueous extracts of twenty-three different seaweed species in Sri Lanka with in-vitro and in-vivo assays. *Heliyon* **2020**, *6*, e03918. [CrossRef]

260. Ahn, J.H.; Yang, Y.I.; Lee, K.T.; Choi, J.H. Dieckol, isolated from the edible brown algae Ecklonia cava, induces apoptosis of ovarian cancer cells and inhibits tumor xenograft growth. *J. Cancer Res. Clin. Oncol.* **2015**, *141*, 255–268. [CrossRef]

261. Yang, Y.I.; Ahn, J.H.; Choi, Y.S.; Choi, J.H. Brown algae phlorotannins enhance the tumoricidal effect of cisplatin and ameliorate cisplatin nephrotoxicity. *Gynecol. Oncol.* **2015**, *136*, 355–364. [CrossRef]

262. Eo, H.J.; Kwon, T.H.; Park, G.H.; Song, H.M.; Lee, S.J.; Park, N.H.; Jeong, J.B. In vitro anticancer activity of phlorofucofuroeckol a via upregulation of activating transcription factor 3 against human colorectal cancer cells. *Mar. Drugs* **2016**, *14*, 69. [CrossRef]

263. Park, C.; Lee, H.; Hwangbo, H.; Ji, S.Y.; Kim, M.Y.; Kim, S.Y.; Hong, S.H.; Kim, G.Y.; Choi, Y.H. Ethanol extract of *Hizikia fusiforme* induces apoptosis in B16F10 mouse melanoma cells through ROS-dependent inhibition of the PI3K/Akt signaling pathway. *Asian Pacific J. Cancer Prev.* **2020**, *21*, 1275–1282. [CrossRef]

264. Lee, H.; Kang, C.; Jung, E.S.; Kim, J.S.; Kim, E. Antimetastatic activity of polyphenol-rich extract of Ecklonia cava through the inhibition of the Akt pathway in A549 human lung cancer cells. *Food Chem.* **2011**, *127*, 1229–1236. [CrossRef] [PubMed]

265. Kim, R.K.; Uddin, N.; Hyun, J.W.; Kim, C.; Suh, Y.; Lee, S.J. Novel anticancer activity of phloroglucinol against breast cancer stem-like cells. *Toxicol. Appl. Pharmacol.* **2015**, *286*, 143–150. [CrossRef]

266. Kim, R.K.; Suh, Y.; Yoo, K.C.; Cui, Y.H.; Hwang, E.; Kim, H.J.; Kang, J.S.; Kim, M.J.; Lee, Y.Y.; Lee, S.J. Phloroglucinol suppresses metastatic ability of breast cancer cells by inhibition of epithelial-mesenchymal cell transition. *Cancer Sci.* **2015**, *106*, 94–101. [CrossRef] [PubMed]

267. Qi, X.; Liu, G.; Qiu, L.; Lin, X.; Liu, M. Marine bromophenol bis(2,3-dibromo-4,5-dihydroxybenzyl) ether, represses angiogenesis in HUVEC cells and in zebrafish embryos via inhibiting the VEGF signal systems. *Biomed. Pharmacother.* **2015**, *75*, 58–66. [CrossRef] [PubMed]

268. Sadeeshkumar, V.; Duraikannu, A.; Ravichandran, S.; Kodisundaram, P.; Fredrick, W.S.; Gobalakrishnan, R. Modulatory efficacy of dieckol on xenobiotic-metabolizing enzymes, cell proliferation, apoptosis, invasion and angiogenesis during NDEA-induced rat hepatocarcinogenesis. *Mol. Cell. Biochem.* **2017**, *433*, 195–204. [CrossRef] [PubMed]

269. Li, Y.X.; Li, Y.; Je, J.Y.; Kim, S.K. Dieckol as a novel anti-proliferative and anti-angiogenic agent and computational anti-angiogenic activity evaluation. *Environ. Toxicol. Pharmacol.* **2015**, *39*, 259–270. [CrossRef]

270. Zhen, A.X.; Hyun, Y.J.; Piao, M.J.; Devage, P.; Madushan, S. Eckol inhibits particulate matter 2.5-induced skin damage via MAPK signaling pathway. *Mar. Drugs* **2019**, *17*, 444. [CrossRef]

271. Zhang, M.Y.; Guo, J.; Hu, X.M.; Zhao, S.Q.; Li, S.L.; Wang, J. An in vivo anti-tumor effect of eckol from marine brown algae by improving the immune response. *Food Funct.* **2019**, *10*, 4361–4371. [CrossRef] [PubMed]

272. Ford, J.M.; Kastan, M.B. DNA Damage response pathways and fcancer. In *Abeloff's Clinical Oncology: Fifth Edition*; Saunders: Philadelphia, PA, USA, 2013; pp. 142–153.e4, ISBN 9780323222112.

273. Singh, A.K.; Cabral, C.; Kumar, R.; Ganguly, R.; Rana, H.K.; Gupta, A.; Lauro, M.R.; Carbone, C.; Reis, F.; Pandey, A.K. Beneficial effects of dietary polyphenols on gut microbiota and strategies to improve delivery efficiency. *Nutrients* **2019**, *11*, 2216. [CrossRef]

274. Guo, B.; Liu, B.; Wei, H.; Cheng, K.W.; Chen, F. Extract of the microalga *Nitzschia laevis* prevents high-fat-diet-induced obesity in mice by modulating the composition of gut microbiota. *Mol. Nutr. Food Res.* **2019**, *63*, 1–14. [CrossRef]

275. Newman, D.J.; Cragg, G.M. Natural products as sources of new drugs from 1981 to 2014. *J. Nat. Prod.* **2016**, *79*, 629–661. [CrossRef] [PubMed]

276. Rajauria, G.; Jaiswal, A.K.; Abu-Gannam, N.; Gupta, S. Antimicrobial, antioxidant and free radical-scavenging capacity of brown seaweed *Himanthalia elongata* from western coast of Ireland. *J. Food Biochem.* **2013**, *37*, 322–335. [CrossRef]

277. Besednova, N.N.; Zvyagintseva, T.N.; Kuznetsova, T.A.; Makarenkova, I.D.; Smolina, T.P.; Fedyanina, L.N.; Kryzhanovsky, S.P.; Zaporozhets, T.S. Marine algae metabolites as promising therapeutics for the prevention and treatment of HIV/AIDS. *Metabolites* **2019**, *9*, 87. [CrossRef] [PubMed]

278. Ahn, M.-J.; Yoon, K.-D.; Kim, C.Y.; Kim, J.H.; Shin, C.-G.; Kim, J.M. Inhibitory activity on HIV-1 reverse transcriptase and integrase of a carmalol derivative from a brown alga, *Ishige okamurae*. *Phyther. Res.* **2006**, *20*, 711–713. [CrossRef] [PubMed]

279. Artan, M.; Li, Y.; Karadeniz, F.; Lee, S.H.; Kim, M.M.; Kim, S.K. Anti-HIV-1 activity of phloroglucinol derivative, 6,6′-bieckol, from *Ecklonia cava*. *Bioorganic Med. Chem.* **2008**, *16*, 7921–7926. [CrossRef]

280. Karadeniz, F.; Kang, K.H.; Park, J.W.; Park, S.J.; Kim, S.K. Anti-HIV-1 activity of phlorotannin derivative 8,4‴-dieckol from Korean brown alga *Ecklonia cava*. *Biosci. Biotechnol. Biochem.* **2014**, *78*, 1151–1158. [CrossRef]

Review

Seaweed Phenolics: From Extraction to Applications

João Cotas [1], Adriana Leandro [1], Pedro Monteiro [2], Diana Pacheco [1], Artur Figueirinha [3,4], Ana M. M. Gonçalves [1,5], Gabriela Jorge da Silva [2] and Leonel Pereira [1,*]

[1]   MARE-Marine and Environmental Sciences Centre, Department of Life Sciences, University of Coimbra, 3001-456 Coimbra, Portugal; jcotas@gmail.com (J.C.); adrianaleandro94@hotmail.com (A.L.); dianampacheco96@gmail.com (D.P.); amgoncalves@uc.pt (A.M.M.G.)
[2]   Faculty of Pharmacy and Center for Neurosciences and Cell Biology, Health Sciences Campus, University of Coimbra, Azinhaga de Santa Comba, 3000-548 Coimbra, Portugal; pmdmonteiro1992@gmail.com (P.M.); gjsilva@ci.uc.pt (G.J.d.S.)
[3]   LAQV, REQUIMTE, Faculty of Pharmacy of the University of Coimbra, University of Coimbra, Azinhaga de Santa Comba, 3000-548 Coimbra, Portugal; amfigueirinha@gmail.com
[4]   Faculty of Pharmacy of University of Coimbra, University of Coimbra, 3000-548 Coimbra, Portugal
[5]   Department of Biology and CESAM, University of Aveiro, 3810-193 Aveiro, Portugal
*   Correspondence: leonel.pereira@uc.pt; Tel.: +351-239-855-229

Received: 27 June 2020; Accepted: 20 July 2020; Published: 24 July 2020

Abstract: Seaweeds have attracted high interest in recent years due to their chemical and bioactive properties to find new molecules with valuable applications for humankind. Phenolic compounds are the group of metabolites with the most structural variation and the highest content in seaweeds. The most researched seaweed polyphenol class is the phlorotannins, which are specifically synthesized by brown seaweeds, but there are other polyphenolic compounds, such as bromophenols, flavonoids, phenolic terpenoids, and mycosporine-like amino acids. The compounds already discovered and characterized demonstrate a full range of bioactivities and potential future applications in various industrial sectors. This review focuses on the extraction, purification, and future applications of seaweed phenolic compounds based on the bioactive properties described in the literature. It also intends to provide a comprehensive insight into the phenolic compounds in seaweed.

Keywords: seaweed polyphenolics; polyphenolics extractions; phlorotannins; bromophenols; flavonoids; phenolic terpenoids; polyphenolics bioactivities

1. Introduction

Recently, organisms living in aquatic habitats have been gaining more interest as the target of studies by numerous scientific groups that have mainly studied their pharmaceutical and biomedical properties, such as the antioxidant, anti-inflammatory, anti-fungal, anti-bacterial, and neuroprotective activity of their large diversity of bioactive compounds [1–5].

Seaweeds are considered the sea vegetables and the basis of life in the aquatic habitats, and they have been employed as fertilizer, human food, and animal feed from ancient to modern times [6,7]. Seaweeds are classified into three main groups, Chlorophyta (green seaweeds), Rhodophyta (red seaweeds), and Phaeophyceae (brown seaweeds), according to their composition of pigments, and their composition of metabolites differs vastly [8].

Seaweeds are commonly exposed to harsh environmental conditions, and the damage effects on them are not visible; consequently, the seaweed produces a large range of metabolites (xanthophylls, tocopherols, and polysaccharides) to protect from abiotic and biotic factors, such as herbivory and sea mechanical aggression [9–15]. Note that seaweed metabolites content and diversity is subject to abiotic and biotic factors, such as species, life stage, size, age, reproductive status, location, depth, nutrient enrichment, salinity, light intensity exposure, ultraviolet radiation, intensity of herbivory, and time of collection; thus, the full exploitation of algal diversity and complexity requires knowledge of environmental impacts and an understanding of biochemical and biological variability [16,17]. Likewise, seaweeds have nutraceutical and pharmaceutical compounds, such as phenols and chlorophylls [6].

Phenolic compounds are found in land plants and in seaweeds [18,19]. Polyphenols synthesized by seaweeds, as one of the largest and most widely distributed groups of seaweed phytochemicals, have gained special attention due to their pharmacological activity and array of health-promoting benefits, as polyphenols play a significant role in the high variety of seaweed biological activities [2,8,14].

Seaweed phenolic compounds are metabolites that are chemically characterized as molecules containing hydroxylated aromatic rings [14,15,20]. These phytochemicals show a wide variety of chemical structures, from simple moieties to high molecular polymers. The biogenetically primary synthetic pathways that produce these phytochemicals are the shikimate or the acetate pathways [21–23].

Seaweeds are a valuable source of polyphenolic compounds such as phlorotannins, bromophenols, flavonoids, phenolic terpenoids, and mycosporine-like amino acids. In the brown seaweeds, the phlorotannins are the major polyphenolic class found only in the marine brown seaweeds. On the other hand, the largest proportion of phenolic compounds present in green and red seaweeds are of bromophenols, flavonoids, phenolics acids, phenolic terpenoids, and mycosporine-like amino acids [24–27]. These molecules are considered secondary metabolites, as they are protective agents that are produced in response to different stimuli and are defense mechanisms of the seaweeds against herbivory and UV radiation [4].

Phenolic compounds are very difficult to isolate quantitatively at an industrial scale due to their structural similarity and tendency to react with other compounds [14]. However, these compounds possess chemical properties that enable their extraction and purification, allowing highly purified extracts to be obtained at a lab scale [28].

Most phenolic compounds possess a broad variety of biological activities, as anti-diabetic, anti-inflammatory, anti-microbial, anti-viral, anti-allergic, anti-diabetic, antioxidant, anti-photoaging, anti-pruritic, hepatoprotective, hypotension, neuroprotective and anticancer properties [14,27,29–40]. Most of the bioactivities are related to the interaction of phenolic compounds with proteins (enzymes or cellular receptors) [41].

These wide-range bioactivities make seaweeds candidates for the development of products or ingredients used in industrial applications as pharmaceuticals, cosmetics, functional foods, and even in bioactive food packaging films to maintain the quality of food products [37,39,42–44].

Simple phenolic compounds can act as an intermediate in the biosynthesis of many polyphenolic secondary metabolites, being also an essential precursor for the industrial synthesis of many other organic substances. Inclusively, salts of benzoic acid are used as industrial food preservers [15,35].

This review aims to present a comprehensive insight into the seaweed phenolic compounds, providing important information about the current and potential status of these compounds from their origin to extraction and isolation methodology, highlighting the potential activities and commercial applications of these compounds in various industries or the potential to become new products.

2. Phenols Found in Seaweeds

Structurally, phenolic molecules are characterized by the presence of an aromatic ring with one or more hydroxyl groups, with structures ranging from simple molecules, such as hidroxycinnamic acids or flavonoids, to more complex polymers, which are characterized by a wide range of molecular sizes (126–650 kDa) [45–51]. The name "phenol" refers to a substructure with one phenolic hydroxyl group; cathecol and resorcinol (benzenediols) are characterized by two phenolic hydroxyl groups; and pyrogallol and phloroglucinol are characterized by three hydroxyl groups (benzenetriols) [48].

2.1. Phenolic Acids

Phenolic acids (PAs) are bioactive compounds involved in several functions, including nutrient absorption, protein synthesis, enzymatic activity, photosynthesis, and allelopathy. These are regularly bound to other molecules, such as simple and/or complex carbohydrates, organic acids, and other bioactive molecules such as flavonoids or terpenoids [52–54]. These PAs are formed by a single phenol ring and at least a functional carboxylic acid group, and they are usually classified depending on the number or carbons in the chain, which is bound to the phenolic ring. Accordingly, these phenolic acids are classified as C6-C1 for hydroxybenzoic acid (HBA; with one carbon chain attached to the phenolic ring), C6-C2 for acetophenones and phenylacetic acids (two carbon chains attached to the phenolic ring), and C6-C3 for hydroxycinnamic acids (HCA) (3 carbon chains attached to the phenol ring) [52–54].

HBAs includes gallic acid, p-hydroxybenzoic, vanillic, and syringic acids and protocathecins, among others, in which there are variations in the basic structure of the HBA, including the hydroxylation and methoxylation of the aromatic ring [52–54]. Although these can be detected as free acids, they occur mainly as conjugates [52–54]. For example, gallic acid and its dimer ellagic acid may be esterified with a sugar (usually glucose) to produce hydrolysable tannins [52–54].

Hydroxycinnamic acids (HCA) are trans-phenyl-3-propenoicacids, differing in their ring constitution [52]. These HCA derivatives include caffeic (3,4-dihydroxycinnamic), ferulic (3-methoxy-4-hydroxy), sinapic (3,5-dimethoxy-4-hydroxy), and p-coumaric (4-hydroxy) acids, with a wide distribution of these compounds as conjugates, mainly as esters of quinic acid (chlorogenic acids [CGA]) [52–54]. Additionally, depending on the identity, number, and position of the acyl residue, this acids can be subdivided in several groups: (1) mono-esters of caffeic, ferulic, and p-coumaric; (2) di-, tri-, and tetra-esters of caffeic acids; (3) mixed di-esters of caffeic–ferulic acid or caffeic–sinapic acids; and (4) mixed esters of caffeic acid with dibasic aliphatic acids, such as oxalic or succinic [52–54]. Furthermore, cinnamic acids can condense with molecules other than quinic acid, including rosmaric and malic, with aromatic amino acids and choline, among others [52–54].

In seaweeds, there are some studies that have proven the presence of the PAs. However, these studies are scarce and mainly in phenolic characterization without any bioactivities studied [52–55]. In green seaweeds, coumarins have been identified in species such as Dasycladus vermicularis, as well as some vanilic acid derivatives in the green macroalgae Cladophora socialis [56]. In brown seaweeds, HBAs, rosmarinic acid, and quinic acid derivatives have been characterized in *Ascophyllum nodosum*, *Bifurcaria bifurcata,* and *Fucus vesiculosus* [57]. In addition, Pas has been characterized in the genus *Gracilaria*, as well as benzoic acid, p-hydroxybenzoic acid, salicylic acid, gentisic acid, protocatechuic acid, vanillic acid, gallic acid, and syringic acid [58–60].

2.2. Phlorotannins

Of all the seaweed phenolic metabolites, the main attention has been focused on tannins due to their interesting bioactive properties [49].

Phlorotannins (Figure 1) can be found within the cell in vesicles called physodes, which are located both in the periphery of the cell and in perinuclear regions, where they are formed [48,49,61]. Phlorotannins are oligomers of phloroglucinol, which is restricted to brown seaweeds, where they exert functions as primary and secondary metabolites [48]. The monomeric unit of phlorotannins, phloroglucinol, is assumed to be formed through the acetate–malonate (polyketide) pathway, in the Golgi apparatus [48,61]. Two molecules of acetyl-CoA, in the presence of carbon dioxide, are converted into malonyl-CoA. The polyketomethylene precursor formed by 3 malonyl-CoA blocks is subjected to a "Claisen type" cyclization reaction, forming a hexacyclic ring system, which is not thermodynamically stable. Then, this molecule undergoes tautomerization, forming a more stable molecule of phloroglucinol [62,63]. The phloroglucinol residues bind through C–C and/or C–O–C residues to form polymeric molecules of phloroglucinol, which results in molecules ranging between 10 and 100 kDa whose heterogeneity is attributed to the variability of structural linkages between phloroglucinol and the hydroxyl groups present [62,63]. As such, phlorotannins can be subdivided into six groups, according to the nature of the structural linkage: (1) phloretols (aryl–ether bonds); (2) fucols (aryl–aryl bonds); (3) fucophloretols (ether or phenyl linage); (4) eckols (dibenzo-1,4-dioxin linkages); (5) fuhalols (*ortho-/para-* arranged ether bridges containing an additional hydroxyl group on one unit), and (6) carmalols (dibenzodioxin moiety) [48,62,63].

Additionally, the binding of monomers to the phloroglucinol ring can take place at different positions within each class, leading to the formation of structural isomers in addition to the conformational isomers [62,63]. As the structural complexity increases, it is necessary to use other criteria for classification. As such, compounds for each class can be classified as linear phlorotannins (in which C–C/C–O–C oxidative couplings have two terminal phloroglucinol residues) or branched phlorotannins, if they are bound to three or more monomers [62,63].

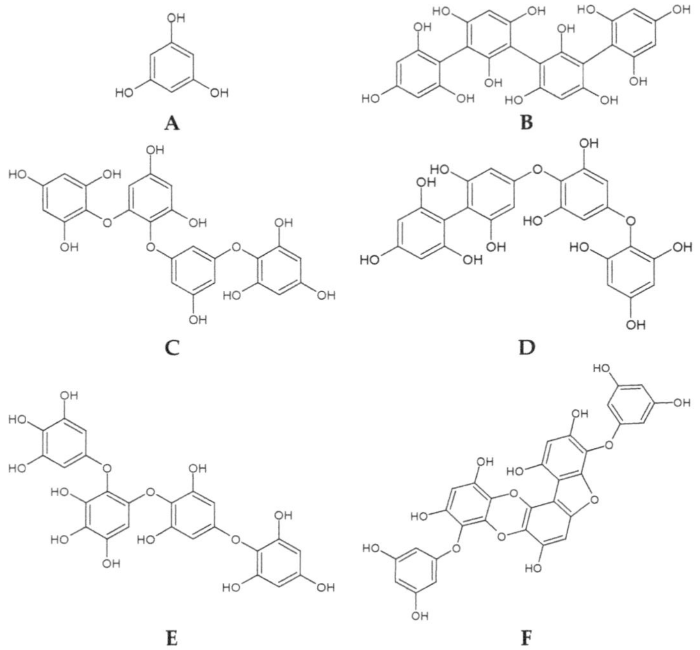

Figure 1. Chemical structures of phlorotannins: (**A**) Phloroglucinol; (**B**) Tetrafucol A; (**C**) Tetraphlorethol B; (**D**) Fucodiphlorethol A; (**E**) Tetrafuhalol A; and (**F**) Phlorofucofuroeckol.

2.3. Bromophenols

Bromophenols (BP) (Figure 2) are secondary metabolites with ecological functions, such as chemical defense and deterrence, with studies revealing a wide variety of beneficial ecological activities [33,61]. BP are common to all major algal groups [61]; they were first isolated from the red algae *Neorhodomela larix* (formerly known as *Rhodomela larix*) [64] and thereafter identified and isolated from all taxonomic groups of marine macroalgae, such as red [64–66], green [67–69], and brown algae [70–73].

Bromophenols are characterized by the presence of phenolic groups with varying degrees of bromination. Many seaweed species contain haloperoxidases, which are capable of halogenating organic substrates in the presence of halide ions and hydrogen peroxide. The bromoperoxidase activity, isolation, and characterization of haloperoxidases have been demonstrated from seaweed [33,74], but information regarding the biosynthesis of bromophenols is limited [74]. Bromoperoxidases can brominate phenol, resulting in the formation of bromophenols; however, the precursor of such bromophenols is still not established, with some authors suggesting tyrosine as a precursor for such formation [75].

Compared to phlorotannins, less is known about bromophenols due to the limited quantity of these compounds in the seaweeds, with a consequently lower isolation and bioactive characterization. More work is needed to isolate and chemically characterize this group of molecules. Yet, there are already some studies correlating the isolated compound with bioactivities and mechanisms of action [33]. Some of these studies have been conducted with synthetic bromophenols developed from the chemical characterization of the seaweed bromophenols that are mainly identified in red seaweeds [76,77]. The bromophenols are vital for the seaweed and seafood flavor, namely 2-bromophenol, 4-bromophenol, 2,4-dibromophenol, 2,6-dibromophenol, and 2,4,6-tribromophenol [72].

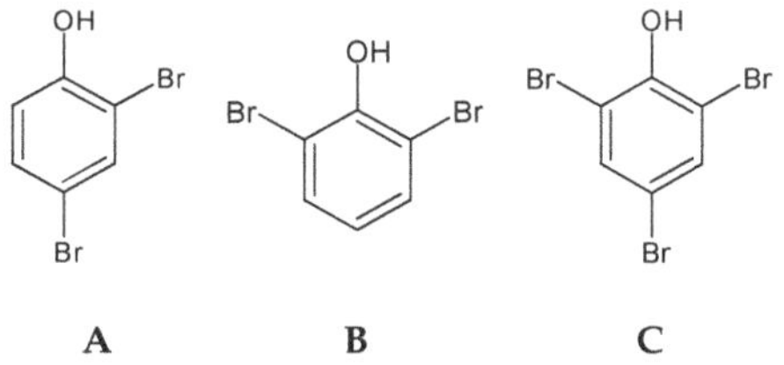

Figure 2. Chemical structures of bromophenols: (**A**) 2,4-bromophenol; (**B**) 2,6-bromophenol; (**C**) 2,4,6-tribromophenol.

2.4. Flavonoids

Flavonoids (Figure 3) are phenolic compounds structurally characterized by heterocyclic oxygen bound to two aromatic rings, which then can vary according to the degree of hydrogenation [46,51].

These compounds are widely distributed in terrestrial plants, with over 2000 compounds reported, which have been subdivided into major categories such as flavones, flavanol, flavanones, flavonols, anthocyanins, and isoflavones [51]. There are many studies on the flavonolic content in terrestrial plants, but flavonoid content studies in algae are scarce [45]. Some studies report that seaweeds are a rich source of catechins and other flavonoids. Flavonoids such as rutin, quercitin and hesperidin, among others, were detected in several Chlorophyta, Rhodophyta, and Phaeophyceae species [48] and compounds restricted to several macroalgae have been identified, such as hesperidin, kaempferol, catechin, and quercetin [78]. Isoflavones, such as daidzein or genistein, are present in red macroalgae *Chondrus crispus* and *Porphyra/Pyropia* spp. and in brown seaweeds, such as *Sargassum muticum* and *Sargassum vulgare*. A high number of flavonoid glycosides have been found in the brown macroalgae *Durvillae antarctica*, *Lessonia spicata*, and *Macrocystis pyrifera* (formerly known as *Macrocystis integrifolia*) [48]. There are already studies that support the presence of flavonoids C-glycosides in the green seaweed *Nitell Hookeri* [79].

It is a little hard to have a full understanding of the bibliography of this phenolic compound's class in the seaweed, due to the scarce bibliographic support about the seaweed flavonoids, but there are also contradictions between the studies done, where the isolation and characterization of the seaweed's flavonoids need to be more explored. For example, the work of Yonekura-Sakakibara et al. [80] states that in general, algae (micro and macro-algae) do not present flavonoid content, due to the lack of two primary enzymes for the main flavonoid biosynthesis; however, the genes encoding enzymes of the shikimate pathway were described in algae [80,81]. However, the work of Goiris et al. [82] reveals the presence of flavones, isoflavones, and flavonols in various microalgae evolutionary lineages, such in Rodophyta, Chlorophyta and Ochrophyta.

This also happens in the seaweed flavonoid biosynthesis, where there is the need to research to have a full understanding of the seaweed cell mechanism to produce these specific phenolic compounds. There are some studies that try to explain the flavonoid synthesis by the shikimate–acetate pathway [83]. Yonekura-Sakakibara et al. [80] explain that the p-coumaroyl-CoA, derived from the phenylpropanoid pathway (with tyrosine biosynthesized by the shikimate pathway), and malonyl-CoA, from the acetate–malonate (polyketide) pathway (identical to the phlorotannins pathway), are converted into naringenin chalcone by the chalcone synthase, and spontaneously, the naringenin chalcone is catalyzed by the chalcone isomerase into naringenin. The naringenin is the primary and initial precursor of the all flavonoids synthetized. Afterwards, naringenin is converted into dihydrokaempferol by the flavanone 3-hydroxylase, and there, the enzymatic mechanism transforms this molecule into the diverse flavonoid molecules, such as catechins, flavonols, anthocyanidin, flavones, and others [80].

The quantitative measurement of flavonoid content is a recurrent analysis that is used mainly as a part of biochemical characterization (together with phenolic content analysis) of the seaweeds extracts in the bibliography analyzed. However, there is a lack on the studies about the flavonoid characterization, isolation, and their specific bioactivity analysis, in the bibliography analyzed. There are only suppositions by the high content of flavonoids present in the sample analyzed and the bioactivity analyzed [35,78,84–87]. From the recent studies using various seaweeds species, the data obtained support the presence of the flavonoid compounds, with various extraction techniques, and these extracts have antioxidant and radical scavenging activity positively correlated with flavonoid concentration [12,88,89]. However, the study of Abirami and Kowsalya [90] did not detect flavonoids in *Ulva lactuca* (Chlorophyta) and *Kappaphycus alvarezii* (Rhodophyta). The detection of flavonoids and respective differences can be derived by the geographical location, time of harvest/collection, season of collection, and the methodology employed in the laboratory [12]. The study from Yoshie et al. [91] reveals the presence of catechin, epicatechin, epigallocatechin, catechin gallate, epicatechin gallate, or epigallocatechin gallate in *Acetabularia ryukyuensis, Ecklonia bicyclis, Padina arborescens, Padina minor, Neopyropia yezoensis* (as *Porphyra yezoensis*), *Gelidium elegans,* and *Portieria hornemannii* (as *Chondrococcus hornemannii*). However, this work did not detect flavonoids in *Undaria pinnatifida, Monostroma nitidum, Caulerpa serrulata, Caulerpa racemosa, Valonia macrophysa, Chondrus verrucosus,* and *Actinotrichia fragilis.*

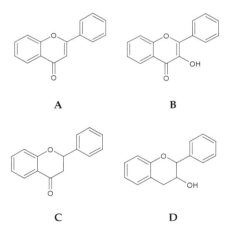

Figure 3. Main classes of flavonoids found in algae: (**A**) Flavones; (**B**) Flavonols; (**C**) Flavanones; (**D**) Flavan-3-ol.

2.5. Phenolic Terpenoids

Phenolic terpenoids (Figure 4) have been detected and characterized in brown and red seaweeds [61]. Brown seaweed phenolic terpenoids have been mainly characterized as meroditerpenoids, which are divided in plastoquinones, chromanols and chromenes, and these are found almost solely in the Sargassaceae. These meroditerpenoids consist of a polyprenyl chain bound to a hydroquinone ring moiety [92].

Diterpenes and sesquiterpenes were identified and isolated in Rhodomelaceae, and a macrolide formation under secondary cyclization was reported for the red seaweed *Callophycus serratus* (bromophycolides) [93].

Even though these compounds have been identified and isolated, the pathways for their formation have yet to be identified, as there is no evidence that these compounds follow the same biosynthesis pathway as other terpenes and terpenoids. Meroditerpenoids are partially formed from mevalonic acid pathways [94], but further studies on biosynthesis pathways should be pursued.

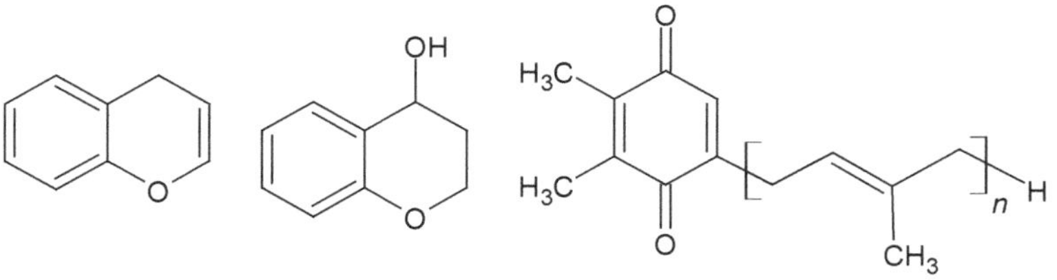

Figure 4. Main classes of phenolic terpenoids found in algae: (**A**) Chromene; (**B**) Chromanol; (**C**) Plastoquinone.

2.6. Mycosporine-Like Aminoacids (MAA)

Mycosporine-like amino acids (MAA) (Figure 5) are a group of UV-absorbing compounds that are present in a wide range of aquatic organisms, whose main function is to reduce UV-induced cellular damage [61,95–98]. These compounds were first identified in fungi, with a role in UV-induced sporulation. Thereafter, a wide range of MAAs have been found in a diverse variety of aquatic organisms, including Cyanobacteria, micro-, and macro-algae [96]. These compounds were detected in *Rhodophyta* spp., and there is still some debate on the detection of MAAs in seaweeds belonging to green and brown seaweed species [61].

Geographically, these compounds are found ubiquitously, occurring in a wide range of environments. Intracellularly, these compounds are found distributed in the cell cytoplasm [96].

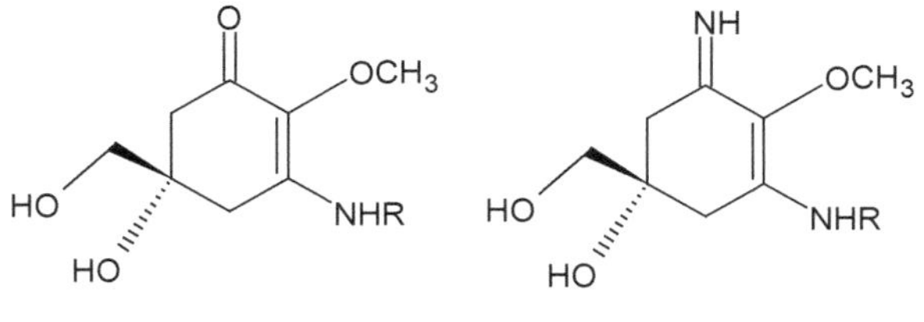

Figure 5. Mycosporine-like amino acids (MAA); (**A**) Aminocyclohexenone; (**B**) Aminocyclohexeniminone.

These compounds are water soluble, with low molecular weights (<400 Da). The chemical structure is based on a ciclehexenone or cyclohexenine ring, with amino acid substituents. The conjugated bonds within the molecule result in broadband absorptions of different wavelengths, according to the substituents in the chemical structure. Evidence suggests that most of the MAAs are synthesized via the Shikimate pathway [96,97].

2.7. Non-Typical Phenolic Compounds

Non-typical phenolic compounds (Figure 6) have been characterized in macro-algal species [61]. Colpol was found in the brown seaweed *Colpomenia sinuosa* [99]. Phenylpropanoid derivatives, such as tichocarpols, were identified in the red macroalgae *Tichocarpus crinitus* [100]. Lignin, a polymerized hydroxicinnamyl alcohol, commonly identified and thought to be restricted to vascular plants, was also identified in red seaweed *Calliarthron cheilosporioides* [101].

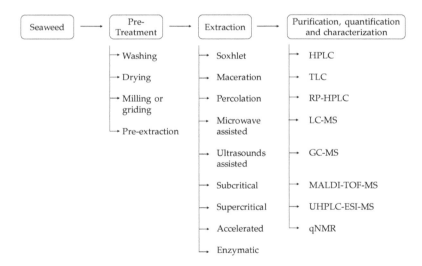

Figure 6. Other phenolic compounds in algae: (**A**) Colpol; (**B**) Tichocarpol.

3. Phenolic Compounds Extractions and Purification Methodologies

For seaweed phenolic compounds investigation, from the pre-treatment to their characterization, there are many methodologies that could be employed (Figure 7). Initially, it is necessary to select the target seaweed species and then locate and identify the compounds to be extracted—intracellular or extracellular—in order to define the strategies to perform the extraction, isolation, and assessment of the phenolic compound bioavailability and bioactivities [36,42,102–105]. To develop bio-based products with seaweed phenolic compounds, it is pivotal to develop practical and efficient analytical methods.

Seaweed	→	Pre-Treatment	→	Extraction	→	Purification, quantification and characterization

Pre-Treatment:
- → Washing
- → Drying
- → Milling or griding
- → Pre-extraction

Extraction:
- → Soxhlet
- → Maceration
- → Percolation
- → Microwave assisted
- → Ultrasounds assisted
- → Subcritical
- → Supercritical
- → Accelerated
- → Enzymatic

Purification, quantification and characterization:
- → HPLC
- → TLC
- → RP-HPLC
- → LC-MS
- → GC-MS
- → MALDI-TOF-MS
- → UHPLC-ESI-MS
- → qNMR

Figure 7. Schematic representation of possible methodologies for seaweed phenolic compounds quest.

3.1. Pre-Treatment

Seaweed pre-treatment is recommended, such as a washing step to remove stones, sand, epiphytes, or other impurities. Therefore, the algal biomass could be used fresh, dried—air drying or at 30–40 °C with aeration during 3–5 days—or freeze drying [106]. Freeze-dried is a better option because it guarantees the integrity of the biomolecules and allows better extraction yields [107].

Furthermore, a milling or grinding process is suggested to reduce the particle size, which is going to increase the exposure area between the seaweed biomass and the solvent used for extraction [108]. This will consequently increase the extraction yield.

Usually, a pre-extraction process is required to avoid the co-extraction of pigments or fatty acids [48] with low polar solvents—*n*-hexane [109], *n*-hexane:acetone [110], *n*-hexane:ethyl acetate [111], or dichloromethane [112]—which have been shown to be effective to extract phenolic compounds.

For example, an efficient pre-treatment using acetone:water (7:3) was applied to the brown seaweed *Fucus vesiculosus* before the extraction of phlorotannins [113].

3.2. Extraction

The further step is to select an extraction method, since these methodologies are widely variable.

Traditional extraction techniques include Soxhlet, solid–liquid, and liquid–liquid extractions. Commonly, in the cited methodologies, organic solvents are used (e.g., hexane, petroleum ether, cyclohexane, ethanol, methanol, acetone, benzene, dichloromethane, ethyl acetate, chloroform). Nevertheless, nowadays, the solvent applied in the extraction methods should be non-toxic and low cost [114]. From an industrial point of view, ethanol is preferred as a solvent for the extraction due to its lower cost [106].

Maceration is a classical method in which the compounds are extracted by submerging the seaweed biomass in an appropriate solvent/solvent mixture [115].

The classical Soxhlet extraction method provides several advantages because it is a continuous process, the solvent can be recycled, and it is less time and solvent consuming than maceration and percolation techniques [116]. However, the extract is constantly being heated at the boiling point of the solvent, and this can damage thermolabile compounds and affect further analysis [45]. Nevertheless, the mentioned extraction methods are not efficient and environmental friendly, due to the high quantities of organic solvent required [23].

With technological advances, these methods have evolved over time to improve the extraction efficiency and sustainability. Currently, ultrasounds and microwave-assisted extraction are low-cost technologies that are feasible at a large scale [117]. These techniques cause a physical effect that leads to cellular membrane disruption and facilitates the liberation of the target compounds [118].

Ultrasound-assisted extraction uses acoustic cavitation to disrupt the cell walls, leading to the reduction of size particle and consequently enhancing the contact between the solvent and the target compound [119].

Still, microwave-assisted extraction involves the utilization of microwave radiation to heat solvents in contact with a sample. Since the algal cell wall is highly susceptible to microwave irradiation, it is reported the rapid internal heating causes cell disruption, leading to the release of target compounds to the cold solvent [120].

Pressurized fluids can also be applied as extraction agents that lead to the development of several techniques, such as subcritical water extraction, supercritical fluid extraction, and accelerated solvent extraction [114].

The correct selection of the extraction solvent, temperature, pressure, static time, and number of cycles are variables that influence directly the total phenolic yield and rate [121]. Further research shows that the total phenol yield increases with the subcritical water extraction method (or pressurized hot water extraction) under higher temperatures; temperatures of 100 °C and 200 °C were tested [122]. This phenomenon could be explained by the increase of mass transfer due to the higher solubility of the cellular membrane, which is caused by the increment of temperature [123]. In this study, the cited trend was verified in all the seaweeds under study, namely *Sargassum vulgare, Sargassum muticum, Porphyra/Pyropia* spp., *Undaria pinnatifida,* and *Halopithys incurva* [122]. However, the increase of the temperature could also lead to the polymerization or the oxidation of some phenolic substances.

The supercritical liquid is characterized by being environmentally friendly, because it uses supercritical carbon dioxide instead of an organic solvent. This methodology is currently the most employed technique for phenolic compounds extraction, and it is defined by the supercritical state, which occurs when a fluid is above critical temperature and pressure—for example, when it is between the state transition (e.g., gaseous and liquid state) [124]. This extraction is commonly performed at low temperatures, so it is suitable for compounds susceptible to high temperatures. This methodology presents several advantages, because supercritical fluids have a lower viscosity and higher diffusion rate than liquid solvents. For this reason, the mass transfer is faster, and the extraction is more efficient [124]. In opposition, as this method involves carbon dioxide, this technique is only applicable to compounds that are lipid soluble and with low polarity.

Relative to accelerated solvent extraction, it also enhances the extraction speed, has low solvent requirements, and is possible to achieve the highest phenolic compounds extraction yield in comparison with conventional approaches. Thus, it is not suitable for bulk extraction [114].

Extraction using enzymes is not feasible at an industrial scale due to the high cost of the necessary enzymes. Nevertheless, it retains several advantages due to the high selectivity of the enzymatic extraction methods [114].

Enzymes can also be applied to promote cell wall disruption, and extraction using enzymes is an advantageous technique due to the selectivity of the degraded compounds, which is important for fragile and unstable substances. In some cases, enzymes are capable of converting insoluble compounds in water into water-soluble ones [25]. For instance, an enzymatic hydrolysate rich in polyphenols was extracted from the brown seaweed *Ecklonia cava*, in which it obtained a polyphenol yield of 20% with the enzyme Celluclast (Novo Nordisk, Bagsvaerd, Denmark) [125].

Usually, after the extraction process, seaweed extracts are concentrated using a rotary evaporator [58].

3.3. Purification, Quantification, and Characterization

Following the extraction process, it is necessary to proceed to the isolation and quantification of the target phenolic compound. Several methodologies could be applied, according to the typology of the compound to be isolated.

Generally, the analysis of phenolic compounds is affected by their source, the extraction and purification techniques employed, the sample particle size, the storage conditions, and the presence of interfering substances in extracts such as fatty acids or pigments [22].

Classically, the quantification of phenolic content is performed by colorimetric methods, namely Folin–Ciocalteu, Folin–Denis, or Prussian blue assays [14]. The most applied assay for phenolic compounds assessment is Folin–Ciocalteu—the redox reaction with the reagent Folin–Ciocalteu allows the spectrophotometric quantification assessment of phenolic compounds. However, the disadvantage of this technique is in the interference of non-phenolic reducing substances [21].

Nowadays, the isolation of phenolic compounds is made through preparative chromatography techniques, namely column chromatography, high-pressure liquid chromatography (HPLC), or thin-layer chromatography (TLC). Nevertheless, these chromatographic techniques have evolved in order to be also used for the separation, isolation, purification, identification, and quantification of distinct phenolic compounds [14].

HPLC, coupled with appropriate detectors, is a very efficient automated analytical methodology that allows the separation, purification, and characterization of a wide range of chemical samples [126]. It presents several advantages because it is a quick method, it requires a low amount of extract sample, and the equipment is easy to operate [48]. Moreover, further research reports that by the HPLC technique, it was possible to identify and quantify nine phenolic compounds (gallic acid, 4-hydroxybenzoic acid, catechin hydrate, epicatechin, catechin gallate, epicatechin gallate, epigallocatechin, epigallocatechin gallate, and pyrocatechol) in brown edible seaweeds (Phaeophyceae)—*Eisenia bicyclis* (formerly known as *Eisenia arborea* f. *bicyclis*), *Sargassum fusiforme* (formerly known as *Hizikia fusiformis*), *Saccharina japonica* (formerly known as *Laminaria japonica*), *Undaria pinnatifida*—and in red edible seaweeds, (Rhodophyta)—*Palmaria palmata* and *Pyropia tenera* (formerly known as *Porphyra tenera*) [103].

The reversed-phase liquid chromatography (RP-HPLC), in which the analysis requires a non-polar stationary phase and a polar hydro-organic mobile phase [126], increases the retention with the hydrophobicity of the solutes, the hydrophobicity of the stationary phase, and the polarity of the mobile phase [127,128]. Thus, the separation is accomplished through the partitioning process and the adsorption of the compounds [129]. Still, the identification and quantification of phlorotannins is usually performed by RP-HPLC with methanol/acetonitrile and water (buffer) solvent combinations and the detection in the UV range of the spectrum [16]. For instance, in a study conducted with the red seaweed *Rhodomela confervoides*, the RP-HPLC method provided bromophenols' identification and characterization (3-bromo-4, 5-dihydroxy benzoic acid methyl ester and 3-bromo-4,5-dihydroxy-benzaldehyde) [130].

Alternatively, thin-layer chromatography is a methodology for compound identification and isolation in which the stationary phase is an adsorbent layer of fine particles. Overall, the layer is placed on a closed chamber, and the extract sample is applied on the lower side of the layer. Inside of this chamber is the mobile phase, which is characterized by a mixture of solvents. Then, the distance covered is marked for the calculus of the retention factor (Rf), enabling the compound identification [131]. In fact, using this method in a dichloromethane/methanol/water (65:35:10, *v/v/v*) solvent system, researchers isolated phlorotannins (phlorofucofuroeckol, dieckol, and dioxinodehydroeckol) from *Ecklonia stolonifera* (Phaeophyceae) [132].

The coupling of liquid or gas chromatography with mass spectrometry also enabled the characterization of phenolic compounds [62,133]. Liquid chromatography-mass spectrometry (LC-MS) allowed the analysis of phlorotannins with different degrees of polymerization, from three brown seaweeds—*Durvillaea antarctica*, *Lessonia spicata*, and *Macrocystis integrifolia* (currently known as *Macrocystis pyrifera*) [134]. For instance, gas chromatography-mass spectrometry (GC-MS) allowed the identification of coumarin and flavones on crude extracts of *Padina tetrastromatica* (Phaeophyceae) [135]. However, this methodology is not appropriate for non-volatile compounds.

Matrix-assisted laser desorption/ionization time-of-flight mass spectrometry (MALDI-TOF-MS) is also a technique that provides the identification and structural characterization of biomolecules [136]. Previous research used this technique in order to detect the presence of phloroglucinol derived from the brown seaweed *Sargassum wightii* [137]. Merged with the previous technique, ultrahigh performance liquid chromatography–electrospray ionization tandem mass spectrometry (UHPLC-ESI-MS) provides relevant information for phenolic compounds characterization relative to their size and isomeric variations [48]. Through this technique, a team of researchers was able to identity and characterize 22 phlorotannins from *Fucus* spp. [138].

Recently, quantitative nuclear magnet resonance (qNMR) has shown the efficiency for metabolites identification and quantification [139,140]. In general, NMR spectrum is derived from the measurement of Fourier transformation signals and translated to radio-frequency impulses. Thus, in comparison with other spectroscopy methods, NMR has a lower mass sensitivity [141].

This method was applied after optimized accelerated solvent extraction, and it was possible to observe the phenolic profile of *Ulva intestinalis* (Chlorophyta) [142].

4. Seaweed Phenolic Compounds and their Bioactivities

The characterization of some phenolic extracts showed interesting results (Figure 8). The correlation between the specific compound and bioactivity potential is commonly achieved (Figure 9); however, there are some phenolic-enriched extracts that have interesting properties but have not been chemically characterized. In this topic, we describe the compounds isolated and their multi-role activity, focusing on recent studies, so there is a long road toward the development of a final product or solution.

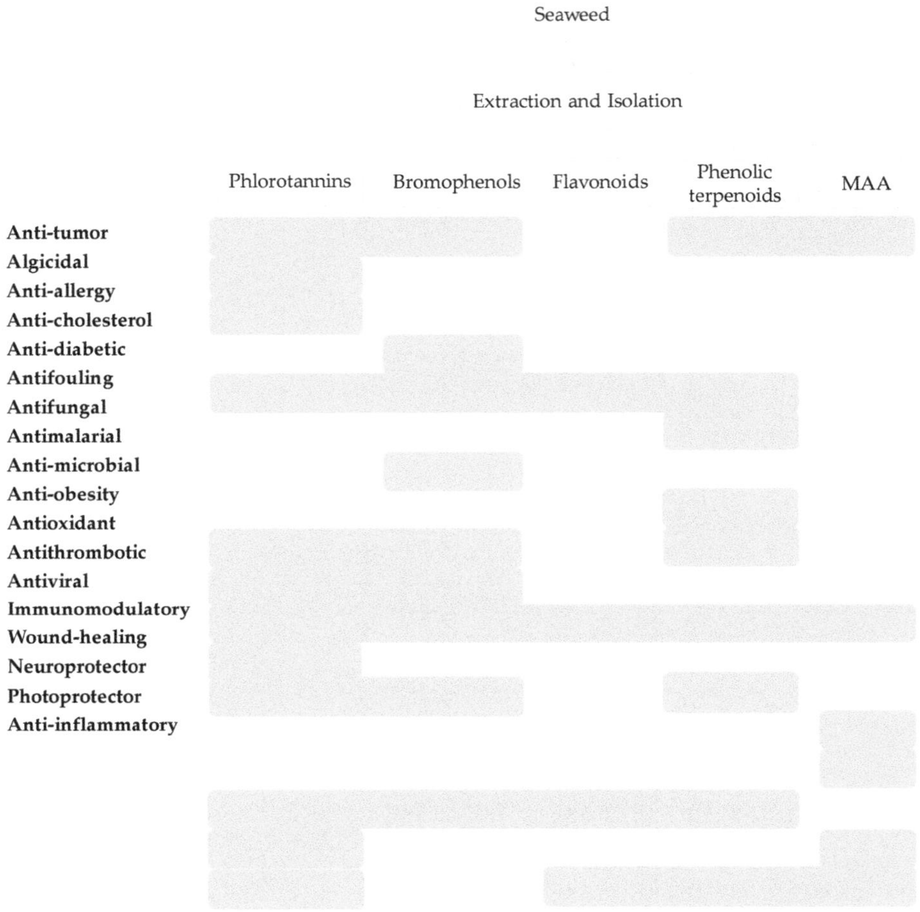

Seaweed

Extraction and Isolation

Figure 8. Biological activities from the different seaweed phenolic compounds reported in the literature.

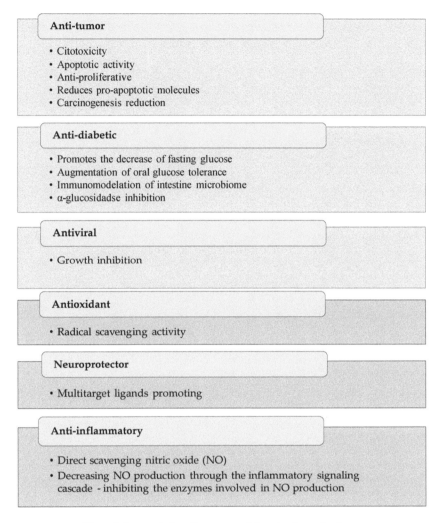

Figure 9. Seaweed phenolic compounds' mechanisms of action.

4.1. Green Seaweeds

Bromophenols and flavonoids of green seaweeds (Figure 10) have antioxidant activities. In fact, species such as *Ulva clathrata*, *Ulva compressa* (formerly known as *Enteromorpha compressa*), *Ulva intestinalis*, *Ulva linza*, *Ulva flexuosa*, *Ulva australis* (formerly known as *Ulva pertusa*), *Capsosiphon fulvescens*, and *Chaetomorpha moniligera* were tested and proven to have high radical scavenging activities [143,144]. These findings will allow the development of new products in drug, cosmetics, or food industries.

Furthermore, the phenolic fraction of *Ulva clathrata* has an anti-tumoral effect [145], and this was also verified in *Ulva flexuosa* species [146]. The phenolic extract of the last not only showed cytotoxicity against breast ductal carcinoma cell line, but also antibacterial activity. This proves the interest of phenol of green seaweeds for human health.

In more recent studies, the *C. socialis* phenolic compounds, such as 2,3,8,9-tetrahydroxybenzo [c]chromen-6-one, 3,4,3′,4′-tetrahydroxy-1,1′-biphenyl, and cladophorol have been identified with interesting antibacterial activity against methicillin-resistant *Staphylococcus aureus* [147].

Figure 10. Examples of green seaweeds (Chlorophyta): (**A**)—*Dasycladus vermicularis*; (**B**)—*Ulva clathrata*; (**C**)—*Ulva compressa*; (**D**)—*Ulva intestinalis*; (**E**)—*Ulva lactuca*; (**F**)—*Ulva linza* [148].

4.1.1. Bromophenols

Bromophenols found in green seaweeds revealed interesting properties, such as the 5′-hydroxyisoavrainvilleol isolated from the *Avrainvillea nigricans*, which is cytotoxic to KB cells and demonstrated promising anti-microbial activity [68]. Another promising bromophenol from the same genus, from the species *A. rawsoni*, was isolated—the rawsonol—which had an inhibitory effect in HMG-CoA reductase activity, as it was a rate-controlling enzyme of the mevalonate pathway that produces cholesterol molecules [149]. The study of Estrada et al. [150] identified other bromophenols, isolated from *Cymopolia barbata*, with antibacterial activity against *Staphylococcus aureus* and *Pseudomonas aeruginosa*; the compound was a brominated monoterpenoid quinol.

4.1.2. Flavonoids

Other phenolic compounds present in green seaweeds, the flavonoids compounds, have been investigated from a medical perspective, such as the anti-diabetic area. Actually, a research on *Ulva prolifera* revealed that flavonoids-rich extracts under 3 KDa molecular weight promoted the decrease of fasting blood glucose, augmentation of oral glucose tolerance, and protection against liver and kidney injury with reduced inflammation in diabetic mice [151,152]. The mechanism of action also showed a modulation of the intestinal microbiome by the growth of the bacteria *Lachnospiraceae* sp. and *Alisties* sp. in overall abundance, which has a clear influence on the release of intestinal hormones that have a direct positive impact on insulin release and resistance [151].

Caleurpa spp. (Chlorophyta) have various flavonoids, such as kaempferol and quercetin. These flavonoids have been correlated with antioxidant activity [153]. *Caulerpa corynephora* has the same concentration of the rutin hydrate as the brown seaweed *Undaria pinnatifida* [35].

4.2. Red Seaweeds

The ecological function of the phenolic compounds in red seaweeds (Figure 11) has been barely investigated, but they probably have multipurpose actions in cell life, as antioxidants, chelating and defense against herbivory agents, as well as cofactors or hormones [154]. However, there is not always information on the bioactivity of the phenolics isolated, because some assays do not occur with purified groups of phenolic compounds but with an extract enriched in polyphenolics [154].

For instance, Farideh Namvar and colleagues [155] studied the effect of *Kappaphycus alvarezii* polyphenol-rich extract (ECMES) on cancer cell lines. The concentrations applied in this study did not demonstrate a cytotoxic effect on the normal cells, though it was cytotoxic to the MCF-7 cancer cell line. These results suggest that the ECME's active substance might target cancer-associated receptors, cancer cell signaling molecules, or gene expression of the MCF-7 breast cancer cells that triggers mechanisms causing cancer cell death [155].

Phenols that are applied in food and drugs have been obtained from red seaweeds; the advantages of their bioactivities will be discussed further in Section 5 of this review.

The red seaweed has been characterized with different phenolic compounds, such bromophenols, flavonoids, phenolic terpenoids, and mycosporine-like amino acid.

Figure 11. Examples of red seaweeds (Rhodophyta): (**A**)—*Asparagopsis armata*; (**B**)—*Chondrus cispus*; (**C**)–*Mastocarpus stellatus*; (**D**)—*Palmaria palmata*; (**E**)—*Solieria chordalis*; (**F**)—*Pterocladiella capillacea*; (**G**)—*Porphyra umbilicalis*; (**H**)—*Hypnea musciformis* [148].

4.2.1. Bromophenols

From red seaweeds, phenolic compounds identified as bromophenols and benzoic acids have been the most researched, from isolation to conducting an extensive characterization [154]. Bromophenols are phenolic compounds that are prominently found in red seaweeds, with bromine substituent indistinct degrees [33].

Bromophenols isolated from *Symphyocladia latiuscula* have antioxidant activity against the DPPH assay. These phenolic compounds have various and diverse high brominated groups mainly based in the 3,4-dihydroxy-2,5,6-tribromobenzyloxy [33,102,156]. The same occurred with the *Polysiphonia stricta* (formerly known as *Polysiphonia urceolata*), revealing that the antioxidant power of red seaweeds depends on brominated units and degrees of brominating of the molecules [33,102,156]. Isolated red seaweeds bromophenols are mainly studied in the oncology, diabetic, and microbial fields, to observe their properties.

One of the main research studies of the red seaweeds bromophenolic compounds is the study of the compounds of the oncology area, with a lot of studies with isolated compounds having demonstrated interesting characteristics, as we demonstrate in this topic.

The 3-bromo-4,5-dihydroxy benzoic acid methyl ester and 3-bromo-4,5-dihydroxy-benzaldehyde bromophenols isolated from *Rhodomela confervoides* demonstrated high potential against KB, Bel-7402 (Human papillomavirus-related endocervical adenocarcinoma), and A549 cancer cell lines [157].

In addition, the bis(2,3-dibromo-4,5-dihydroxybenzyl) ether has demonstrated apoptotic activity against K562 human myelogenous leukemia cells [76], as has the lanosol butenone (Colensolide A) isolated from *Vidalia colensoi* (formely *Osmundaria colensoi*), revealing selectivity against leukemia cells [77,157]. The bis(2,3-dibromo-4,5-dihydroxybenzyl) ether showed anti-angiogenesis effects in vitro and in vivo (zebrafish embryos) by reducing the HUVEC (human umbilical vein endothelial cells) cells' proliferation, migration, and tube formation; however, it did not decrease the preformed vascular tube. Overall, it indicates a potential for further studies for cancer prevention and novel therapies, due to the unique structure that is different from the current anti-angiogenesis therapeutic agents [158].

The bis(2,3-dibromo-4,5-dihydroxybenzyl) methane isolated from *R. confervoides* demonstrated activity against the BEL-7402 (Human papillomavirus-related endocervical adenocarcinoma) cancer cell line, inhibiting the cell adhesion to fibronectin and collagen as well as cell migration and invasion, demonstrating an interesting anti-metastatic activity that can be developed to understand how it can be applied in oncologic therapeutic and compound selectivity [159]. From this seaweed, 8 bromophenols have been isolated that have been assayed against various cancer cell lines, such as HCT-8 (human colon cancer), Bel7402, BGC-823 (stomach cancer), A549 (adenocarcinomic human alveolar basal epithelial cell), and A2780 (human ovarian cancer), with only four compounds demonstrating activity against the cancer cell lines tested: 2,3-dibromo-4,5-dihydroxyphenylethanol, 2,3-dibromo-4,5-dihydroxyphenylethanol sulfate, 3-bromo-4,5-dihydroxyphenylethanol sulfate, and 3-bromo2-(2,3-dibromo-4,5-dihydroxybenzyl)-4,5-dihydroxyphenylethanol sulfate [160].

The bromophenols also show interesting results regarding their anti-diabetic and anti-obesity medical aspects. The bromophenols extracted from *S. latiuscula*, such as the 2,3,6-tribromo-4,5-dihydroxybenzyl methyl ether and its derivates, inhibit α-glucosidase, which improves insulin sensitivity and glucose uptake [161]. In this species, there were also various bromophenolic compounds isolated (2,2′,3,6,6′-pentabromo-3′,4,4′,5-tetrahydroxydibenzyl ether, bis(2,3,6-tribromo-4,5-dihydroxyphenyl)methane, and 2,2′,3,5′,6-pentabromo-3′,4,4′,5-tetrahydroxy diphenylmethane) that demonstrated aldose reductase inhibitory activity, via an enzyme in the polyol pathway, which is responsible for fructose formation from glucose and has an important role in the development of diabetes [162].

Bromophenol derivatives based on 2,3-dibromo-4,5-dihydroxybenzyl units and highly brominated isolated from *Rhodomela confervoides* demonstrate activity against PTP1B (Protein tyrosine phosphatase 1B), which is a negative regulator of the insulin signalling pathway, and the compounds demonstrated a substantial decrease of the blood glucose in diabetic rats [163].

Ko et al. [66] isolated the 5-bromo-3,4-dihydroxybenzaldehyde from *Polysiphonia morrowii* and studied the effect and mechanism of action of this compound in adipogenesis and the differentiation of 3T3-L1 preadipocytes, demonstrating that this bromophenol targets the peroxisome proliferator-activated receptor-γ expression levels and also the CCAAT/enhancer-binding proteins α and sterol regulatory element-binding protein 1. However, the main mechanism is under the AMP-activated protein kinase signal pathway, inhibiting the adipogenesis, in vitro assay. So, there is a need to develop further work to understand if the effect can be replicated in vivo.

The study of Mikami et al. [164] revealed that the n-butyl 2,3-dibromo-4,5-dihydroxybenzyl ether, isolated from the red seaweed *Odonthalia corymbifera*, demonstrated inhibition against the glucose 6-phosphate dehydrogenase, which is a key enzyme in the formation of reduced nicotinamide adenine dinucleotide phosphate (NADPH). This compound is an important factor in the biosynthesis of fatty acid and cholesterol and a factor intrinsic to cancer cells lines' growth. These studies used bacterial and yeast glucose 6-phosphate dehydrogenase enzymes, so there is a long road to understand its effect in humans, but the assay has proven the bioactive potential of the bromophenolic compound.

Researchers have proven the antimicrobial action of these molecules. For instance, bromophenols isolated from *R. confervoides*, by Xu et al. [165], were tested against various bacteria species, such as *Staphylococcus aureus*, *Staphylococcus epidermidis*, *Escherichia coli*, and *Pseudomonas aeruginosa*. The results were the most promising against the *Staphylococcus* species, and the best bromophenol in the assay was bis (2,3-dibromo-4,5-dihydroxybenzyl) ether, without effect against only one *Escherichia coli* strain.

In addition, the lanosol methyl ether, lanosol butenone, and rhodomelol isolated from the *Vidalia colensoi* (formerly known as *Osmundaria colensoi*) demonstrated anti-bacterial and anti-fungal activities against various pathogens, such as various *Halomonas* species, *Pseudomonas* sp., *Vibrio alginolyticus*, *Vibrio harveyi*, *Klebsiella pneumoniae*, *Propionibacterium acnes*, *Staphylococcus aureus*, *Alternaria alternata*, and *Candida albicans*. These effects proved to be bactericidal and bacteriostatic or fungicidal and fungistatic, demonstrating the dose-dependent curve for effect against the pathogen. These results prove the multi-performance of the red seaweeds bromophenols against terrestrial and aquatic microbes [166].

The bis(2,3-dibromo-4,5-dihydroxybenzyl) ether demonstrated anti-fungal activity against various phytopathogenic fungi, such as *Botritys cinerea*, *Valsa mali*, *Fusarium graminearum*, *Coniothyrium diplodiella*, and *Colletotrichum gloeosporioides*, and it did not demonstrate activity against *Alternaria mali* and *Alternaria porri*, indicating that the research of bromophenols applications can pass also to control pathogens in the food and contribute to high food security [167].

The *S. latiuscula* bromophenol 2,3,6-tribromo-4,5-dihydroxybenzyl methyl ether had demonstrated anti-viral effect against various types of herpes simplex type 1. This assay was performed in mice with anti-viral activity at a dosage of 20 mg/kg for 6–10 days, without the presence of major secondary effects observed. The behavior of the anti-viral activity was comparable with acyclovir, in the skin lesions and the concentration of the virus particles in the brain [168].

The bromophenols isolated from *Polysiphonia morrowii*, the 3-bromo-4,5-dihydroxybenzyl methyl ether (BDME) and 3-bromo-4,5-dihydroxybenzaldehyde (BD), demonstrated anti-viral activity against the infectious hematopoietic necrosis virus (IHNV) and infectious pancreatic necrosis virus (IPNV), which are two aggressive fish pathogenic virus. The study of Kim et al. [169] demonstrated that the BDME has an effect against the two viruses; however, the BD only has an effect against IHNV. It ought to be noted that the anti-viral effect is lower than the control used (ribavirin).

These molecules, isolated from the red seaweeds, have other interesting activities. For example, 3,5-dibromo-4-hydroxyphenylethylamine, 2,2′,3,3′-tetrabromo-4,4′,5,5′-tetrahydroxydiphenylmethane, 2,3-dibromo-4,5-dihydroxybenzyl alcohol, 2,3-dibromo-4,5-dihydroxybenzyl methyl ether, 2,2′,3-tribromo-3′,4,4′,5-tetrahydroxy-6′-hydroxymethyldiphenylmethane, and 3-bromo-4-(2,3-dibromo-4,5-dihydroxybenzyl)-5-methoxymethylpyrocatechol, isolated from *Odonthalia corymbifera*, had been assayed against a fungal pathogen that affects the rice plants (*Magnaporthe grisea*). The bromophenols demonstrated efficiency in reducing the disease impact in the wild rice plants, with comparable results of rice genetic modified to be resistant to this pathogen [170].

In addition, 2,3,6-tribromo-4,5-dihydroxybenzyl alcohol and bis-(2,3,6-tribromo-4,5-dihydroxybenzyl) ether, extracted from *S. latiuscula*, act as multitarget ligands promoting neuroprotection [161].

4.2.2. Flavonoids

The flavonoid isolation from *Acanthophora specifera* demonstrates a mixture of chlorogenic acid (69.64%), caffeic acid (12.86%), vitexin-rahmnose (12.35%), quercetin (1.41%), and catechol (0.59%) [171]. The flavonoid-enriched extract has demonstrated antioxidant activity [172].

The study of Saad et al. [173] demonstrated a flavonoid-enriched extract from *Alsidium corallinum* containing flavonoids such as luteolin 5,7,3′,4′ tetramethyl ether, quercetin 3,7 dimethylether 4′ sulfate, and catechin trimethyl ether, which can be useful tools against the kidney dysfunction provoked by potassium bromate; this assay was done in mice.

4.2.3. Phenolic Terpenoids

Davyt et al. [174] isolated 11 sesquiterpenes from *Laurencia dendroidea* (formerly known as *Laurencia scoparia*) and studied the anthelmintic activity with moderate results against the parasite stage of *Nippostrongylus brasiliensis*. Meroditerpenes of the *Callophycus serratus* and *Amphiroa crassa* had been demonstrated in the initial stages of studying antimalarial activity [175].

A chromene-based molecule isolated from *Gracilaria opuntia* has been assayed and proved to have antioxidant and also anti-inflammatory activity in in vitro assays [176].

4.2.4. Mycosporine-Like Amino Acid

This class of phenolic compounds is exclusive from red seaweeds and can be found in various species, such as *Asparagopsis armata*, *Chondrus crispus*, *Mastocarpus stellatus*, *Palmaria palmata*, *Gelidium* spp., *Pyropia* spp. (formerly known as *Porphyra* spp.), *Crassiphycus corneus* (formerly known as *Gracilaria cornea*), *Solieria chordalis*, *Grateloupia lanceola*, and *Curdiea racovitzae* (Rhodophyta). Various MAAs have already been isolated—likewise, the palythine, shinorine, asterina-330, porphyra-334, palythinol, and usujirene. These types of compounds have a high antioxidant and photoprotection activity, and also anti-proliferative activities in the HeLa cancer cell line (human cervical adenocarcinoma cell line) and HaCat (human immortalized keratinocyte). This was already passed from the studies to the final product production [177–184]. Recent studies indicate that MAAs can have other important bioactivities, such as anti-inflammatory, immunomodulatory, and wound-healing properties [185–187]. So, these compounds are a natural alternative to the synthetic UV-R filters in the sunscreens. Thus, MAAs seem to be driven and focused to a specific area that permitted studying and rapidly obtaining a product that can be applied in humans.

4.3. Brown Seaweeds

Brown seaweeds (Figure 12) also have a polyphenols group of high interest: the phlorotannins, which are found in neither green nor red seaweeds. As other phenols, phlorotannins have strong antioxidant effects, especially phloroglucinol, eckol, and dieckol, which can be extracted from *Ecklonia cava*. Moreover, these molecules revealed effectiveness in protecting DNA against hydrogen peroxide, which induces damage [188]. Other phenolic compounds isolated from brown seaweeds are bromophenols, flavonoids, and phenolic terpenoids, which have been the least studied due to the high quantity of phlorotannins.

Polyphenols obtained from *F. vesiculosus*, at concentrations that were not cytotoxic, inhibited both HIV-1-induced syncytium formation and HIV-1 reverse transcriptase enzyme activity [189]. HIV-1 is not the only virus that is susceptible to Phaeophyceae seaweeds' phenolics; murine norovirus (MNV) and feline calicivirus (FCV) are other examples [190,191].

Figure 12. Examples of brown seaweed (Phaeophyceae): (**A**)—*Ascophyllum nodosum*; (**B**)—*Bifurcaria bifurcata*; (**C**)—*Colpomenia sinuosa*; (**D**)—*Treptacantha baccata*; (**E**)—*Fucus vesiculosus*; (**F**)—*Leathesia marina*; (**G**)—*Padina pavonica*; (**H**)—*Sargassum muticum*; (**I**)—*Sargassum vulgare*; (**J**)—*Undaria pinnatifida* [148].

4.3.1. Phlorotannins

Contrasting the recent developments in other phenolic compounds classes from seaweeds, early research of isolation and characterization was performed in phlorotannins accumulated by brown seaweeds. Phlorotannins are the most studied group of phenolic compounds from algae [45].

Their antioxidant power is 2 to 10 times higher when compared to ascorbic acid or tocopherol [192,193], demonstrating a hypothesis to treatments of inflammatory diseases [194]. Additionally, it was proven that phlorotannins can be applied as a protective agent against the toxicity of drug/antibiotics in humans, without the drugs losing their power, diminishing the damage from drug-based toxicity, such as gentamicin [195].

These research studies occur mainly in cell lines and model organisms, so in the future, the studies must be advanced to the next phase of trials in humans; with safeguards, they could be applied in commercial products.

One of the principal areas of study using phlorotannins is the oncology area. It was demonstrated that dioxinodehydroeckol, dieckol, and phlorofucofuroeckol, isolated from *Ecklonia cava*, have anti-proliferative, anti-tumor, anti-inflammatory, anti-adipogenic, and anti-tumorigenic activities. These properties are mainly against breast cancer cell lines MCF-7 and MDA-MB-231, SKOV-3 (ovarian cancer line), HeLa (cervical cancer cell line), HT1080 (fibrosarcoma cancer cell line),

A549 (adenocarcinomic human alveolar basal epithelial cell), and HT-29 (human colon cancer cell line) [196–200]. The dieckol demonstrated a high anticancer activity against the non-small–cell lung cancer line A549 [201]. Eckol inhibited the proliferation of SW1990 pancreatic cells induced by Reg3A (a pancreatic inflammation upregulated protein with pro-growth function) [202]. Dieckol isolated from *E. stolonifera* induced apoptosis in human hepatocellular carcinoma Hep3B cells, demonstrating potential to be a new therapeutic agent to treat liver tumor, but more research is needed in order to develop a secure form to be effective and safe, consequently, full understanding of the mechanism of action that has reported by Yoon et al. is needed [203].

Other assays in radiotherapy observed the radio-protective effects of dieckol, triphlorethol-A, and eckol from *Ecklonia* genus to γ-irradiation; thus, the phlorotannins can have multiple roles in protection from radiation aggression or use in radiotherapy. Mainly, the mechanism of action is the radical scavenging and reducing pro-apoptotic molecules, but dieckol has an impact on the increase of the enzyme manganese superoxide dismutase that prevents DNA damage and lipid peroxidation, accelerating the hematopoietic recovery [204–208]. This radiation-induced protection has also been proven with eckol (*Ecklonia* genus) in the intestinal stem cells damaged by gamma irradiation, although the mechanism of this effect was not fully understood [204].

Several studies also reported the anti-diabetic properties of the phenolics of brown seaweeds [209,210]. *Ascophyllum nodosum* [36], *Fucus distichus* [211], and *Padina pavonica* [212] are species that produce phlorotannins, which were proven to have this effect. Other in vivo investigations, testing eckol and dieckol from *Ecklonia cava* [213,214], *Ecklonia stolonifera*, and *Eisenia bicyclis* (formerly known as *Ecklonia bicyclis*) [215], demonstrated the effects of oral administrations of these phenols in diabetic models in alleviating the postprandial hyperglycemia, suggesting a reduction in insulin resistance in those animals [214]. Octaphlorethol A isolated from *Ishige foliacea* has also shown anti-diabetic activity in type 2 diabetes. The mechanism of action studied by Lee et al. [216] elucidates the clinical applications of this phlorotannin as a new drug candidate for the treatment of type 2 diabetes.

The compound isolated from *Sargassum patens*, 2-(4-(3,5-dihydroxyphenoxy)-3,5-dihydroxyphenoxy) benzene-1,3,5-triol (DDBT), suppresses in high quantity the hydrolysis of the amylopectin by human salivary and pancreatic α; with these results, it can be beneficial as a natural nutraceutical to prevent diabetes [217].

From the study of Oh et al. [218], the dieckol isolated from *E. cava* attenuates the leptin resistance in the brain, and furthermore, it can cross the blood–brain barrier, which demonstrates dieckol as a potential treatment for obesity.

The 6,6'-bieckol compound isolated from *E. cava* demonstrated an inhibition for the high-glucose-induced cytotoxicity in human umbilical vein endothelial cells (HUVECs) and insulinoma cells, showing it can be a potential therapeutic agent against the hyperglycemia-induced oxidative stress. This problem results in diabetic endothelial dysfunction [219,220].

These molecules have also shown anti-microbial effects. Dieckols (8.4'''-dieckol, 6.6''-dieckol and 8.8'-dieckol) extracted from the species *E. cava* are inhibitors of the human immunodeficiency virus (HIV-1) reverse transcriptase, with 8.8'-dieckol being the most effective against reverse transcriptase [221–223].

The dieckol (isolated from *E. cava*) can interfere in the viral replication mechanism of SARS-CoV, more concretely in SARS-CoV 3CL protease *trans/cis*-cleavage with a high association rate with a dose-dependent effect on 3CL protease hydrolysis [224]. Additionally, this compound demonstrates a potential against the influenza A virus neuraminidase, the most promising target, because it is a critical role for viral life cycle [225]; however, the phlorofucofuroeckol was also the best phlorotannin in the assay, equal to dieckol. Even further, these compounds have synergy with oseltamivir to enhance the inhibitory effects.

In recent docking studies by Gentile et al. [226] of the chemical structure of heptafuhalol A, 8,8'-bieckol, 6,6'-bieckol, and dieckol, compounds already isolated and characterized from *E. cava*, it was found that they are the most active inhibitors from the marine origin of the SARS-CoV-2 protease, revealing great potential to be further investigated against SARS-CoV-2.

Eckol, dieckol, 8,8′-bieckol, 6,6′-bieckol, and phlorofucofuroeckol-A from *E. bicyclis* demonstrated anti-viral activity against human papilloma virus [227].

Dieckol and phlorofucofuroeckol-A, both extracted from *E. bicyclis*, were demonstrated to have potent anti-viral action [190]. Moreover, phlorotannins extracted from the *Eisenia/Ecklonia* genera, such as eckol, dieckol, fucofuroeckol-A, 8,8′-Bieckol, and phlorofucofuroeckol-A, also exhibited anti-fungal activity [228] and anti-bacterial activity against Methicillin-resistant *Staphylococcus aureus* and other bacteria that are pathogenic not only in humans but also in plants [229–231].

According to several studies, these compounds also have potential to be applied in the treatment of bone diseases. Arthritis is one of the most prevalent chronic diseases; it is commonly characterized by the degradation of the cartilage matrix in bones joints. The dieckol and 1-(3′,5′-dihydroxyphenoxy)-7-(2″,4″,6″-trihydroxyphenoxy) 2,4,9-trihydroxydibenzo-1,4-dioxin (isolated from *E. cava*) demonstrated a reduction of the inflammatory response and cell differentiation in an in vitro assay, demonstrating potential to minimize the impact of arthritis symptoms [232].

Phlorofucofuroeckol A isolated from *E. cava* promotes osteoblastogenesis, which can be used in bone remodelling and reduce osteoporosis-related complication, mainly age-related bone disorders [233].

Other than these, dieckol extracted from *E. cava* demonstrated an interesting bioactivity related to chronic diseases, because it demonstrated anti-neuroinflammatory properties that can be essential to auto-inflammatory diseases [234]. This was also proven with dieckol extracted from *E. stolonifera* [235], demonstrating potential for therapeutic application in hepatotoxicity. Still, more studies are needed to develop this area. However, in the study of Lee and Jun [236], the 8,8′-bieckol from *E. cava* demonstrated a high inhibitory effect against β-Secretase and acetylcholinesterase, which are the principal factors for the development of the Alzheimer's disease, with interesting docking assays to advance this compound even further as a drug candidate for the therapeutics of Alzheimer's disease. Furthermore, Wang et al. [237] demonstrated in vitro the potential of eckmaxol, isolated from *E. maxima*, as a therapeutic multi-action mechanism to treat the neurotoxicity that develops in the Alzheimer's disease.

The study of Seong et al. [238] focused on the action of phlorotannins in the mechanisms of anti-depressant and anti-Parkinson's disease. Dieckol and phlorofucofuroeckol-A (isolated from *Ecklonia stolonifera*) demonstrated high inhibition against the monoamine oxidases and demonstrated full agonists with high potency at the dopamine 3 and 4 receptors, proving that these compounds can have a multi-role intervention in the potential treatment of psychological disorders and Parkinson's disease.

These molecules have other potential biomedical properties. Dioxinodehydroeckol and phlorofucofuroeckol A extracted from *E. stolonifera* present an anti-allergenic effect [239]. Phlorofucofuroeckol-B isolated from *Eisenia arborea* also demonstrates an anti-allergy effect with the inhibition of histamine release [111]. More importantly, dieckol from *E. cava* demonstrates suppressing the immunoglobulin E-mediated mast cell and the passive anaphylactic reaction, diminishing the type I allergic responses; however, it was needed a relatively high dosage to suppress the hypersensitivity, so more studies are needed to fully understand it [240].

Lee et al. [241] demonstrated the suppression mechanism of action of dieckol (isolated from *E. cava*) in liver fibrosis, proving the potential of dieckol to be further evaluated to treat chronic liver inflammation provoked by alcohol abuse, metabolic diseases, viral hepatitis, cholestatic liver diseases, and autoimmune diseases.

Hypertension is one of the most common cardiovascular problems that origins/develops into other dangerous diseases in the world, where the angiotensin-converting enzyme (ACE) plays a major role in augmenting health risks. The work of Jung et al. [242] and Wijesinghe [243] demonstrated that phlorofucofuroeckol A (isolated from *E. stolonifera*) and dieckol (isolated from *E. cava*) have a high inhibition against the ACE in in vitro assays, ameliorating the hypertension symptoms and the risk of development of more life-risking treats. One explanation that was proposed is that the closed-ring dibenzo-1,4-dioxin moiety can be essential for the bioactivity as well as the molecule polymerization level [242].

Fucophlorethol C isolated from *Dactylosiphon bullosus* (formerly known as *Colpomenia bullosa*) has a lipoxygenase inhibition activity, in soybean, with comparable power to the nordihydroguaiaretic acid (obtained from Zygophyllaceae plants), which is the most well-known inhibitor. This inhibitor is expected to suppress various lipoxygenase-related diseases, such as psoriasis, asthma, rhinitis, and arthritis [244].

Phlorotannins can act as an anti-UVB protective agent; dioxinodehydroeckol (isolated from *E. cava*) demonstrated a protection on the HaCat cells, reducing the apoptosis provoked by UVB [245]. Additionally, phlorotannins are being researched as whitening and/or anti-wrinkling agents for cosmeceuticals, such as dieckol, dioxinodehydroeckol, eckol, eckstolonol, phlorofucofuroeckol A, and 7-phloroeckol (isolated from various brown seaweeds). These have shown promising inhibition of tyrosinase and hyaluronidase [5,246–251]. In addition, the 7-phloroeckol (isolated from *E. cava*) has proven that it can be a hair growth promoter agent by the study of Bak et al. [252].

Although more research needs to be done, these phlorotannins can be a key for natural biomarkers for various applications [253]. Hydroxytrifuhalol A, 7-hydroxyeckol, diphloroethol, fucophloroethol, and dioxinodehydroeckol can be applied as food biomarkers, because after intake, they can be detected by plasma or urine samples. Having a short half-life, they are considered a good short-term biomarker.

The molecules could also be applied in the control of the algal blooms, as the phlorofucofuroeckol isolated from *Ecklonia kurome* demonstrated a high algicidal activity comparable to epigallocatechin [254]. In the swine's diseases, the dieckol, 7-phloroeckol, phlorofucofuroeckol, and eckol isolated from *E. cava* demonstrated high activity against porcine epidemic diarrhoea virus (PEDV), but it is not known whether the compound interacts in the viral entry or in viral replication [255].

4.3.2. Bromophenols

In terms of brown seaweeds' bromophenols compounds isolated and their respective bioactivity analysis, the information is generally weak, having less research published, and being mainly about the brown seaweed *Leathesia marina* (formerly known as *Leathesia nana*) [33]. Various bromophenols identified in red seaweeds also appear in brown seaweeds, such as bis-(2,3-dibromo-4,5-dihydroxy-phenyl)-methane [256].

The studies of Xu et al. [257] and Shi et al. [258] isolated various bromophenol compounds from *L. marina* and conducted assays against various cancer lines *in vitro* and in mice. Some of the cancer lines were A549 (lung adenocarcinoma), BGC-823 (stomach cancer), MCF-7 (breast cancer), Bel7402 (hepatoma), and HCT-8 (human colon cancer), with five of the compounds showing cytotoxic effects against various cancer cell lines. For example, 6-(2,3-Dibromo-4,5-dihydroxybenzyl)-2,3-dibromo-4,5-dihydroxybenzyl methyl ether, bis(2,3-dibromo-4,5-dihydroxybenzyl) ether, and 3-bromo-4-(2,3-dibromo-4,5-dihydroxybenzyl)-5-methoxy methylpyrocatechol presented the best results in anti-tumor activity. Additionally, the bromophenols from *L. marina* demonstrated a growth inhibition of Sarcoma 180 tumors in vivo (mice), demonstrating a potential for further investigation.

On the other hand, (+)-3-(2,3-dibromo-4,5-dihydroxy-phenyl)-4-bromo-5,6-dihydroxy-1,3-dihydroiso -benzofuran, isolated from *L. marina,* demonstrated an inhibitory activity against thrombin in in vitro and in vivo assays [70,259]. The compound reveals a potential to treat cardiovascular diseases, where thrombin can play a key role in the development of the diseases, such as thrombosis and thromboembolism, due to thrombin being a proteinase that plays a key role in the procoagulant factors.

4.3.3. Flavonoids

The work of Agregán et al. [57] identified flavonoids in *Ascophyllum nodosum*, *Bifurcaria bifurcata*, and *Fucus vesiculosus*, which are mainly acacetin derivatives, gallocatechin derivatives, and hispidulin. This work demonstrates that the isolation work of the seaweeds' flavonoids is only beginning, and some of the flavonoid's structures are comparable with terrestrial plants, such as oregano (*Folium origani cretici*) or basil (*Ocimum basilicum*).

The myricetin isolated from *Turbinaria ornata* demonstrated attenuating the effect of rotenone-induced neuronal degeneration in *Drosophila melanogaster*. This result demonstrates that this flavonoid can have a positive impact on Parkinson's disease in the muscular coordination and memory of the fly model used [260].

4.3.4. Phenolic Terpenoids

Meroditerpenoids have been isolated and characterized from the *Treptacantha baccata* (formerly known as *Cystoseira baccata*), all of the compounds isolated have a bicycle (4.3.0) nonane ring; mainly, they have antifouling activity against algal settlements and mussel phenol oxidase, but they appear to be non-toxic and non-bactericidal for the larvae of sea urchins and oysters [261]. Nine tetraprenyltoluquinol-based meroterpenoids isolated from *Halidrys siliquosa* present antifouling properties and some have also demonstrated anti-bacterial activity [262].

Thus, meroditerpenoids isolated from *Stypopodium flabelliforme* have been assayed against the NCI-H460 (human lung cancer cell line) presenting a moderate response [263]. Meroditerpenoids, such as epitaondiol and stypodiol, isolated from *Stypopodium flabelliforme* demonstrated anti-bacterial activity against *Enterococcus faecalis*. These compounds demonstrated anticancer activity against Caco-2 (human epithelial colorectal adenocarcinoma), SH-SY5Y (neuroblastoma), and RBL-2H3 (Rat Basophilic Leukemia cells), and they especially affected the cancer line RAW.267 (Abelson murine leukemia virus-induced tumor), but the potential to be used as therapeutic is interesting, because they appear to be non-toxic against non-cancer cell line V79 [264]. It can be used as a selective anticancer drug, so there is more research to be done to fully understand the mechanism and how this unique characteristic can be explored. The epitaondiol from *Stypopodium zonale* has been assayed against human metapneumovirus with a prominent viricidal activity [265].

Zonarol, a sesquiterpene from *Dictyopteris undulata*, provides neuroprotection by protecting the neuronal cells from the oxidative stress, which is one of the main mechanisms of the phenolic compounds, so this is a candidate to be further studied to see if it is positively applicable as neurodegenerative diseases therapeutic [266].

Ali et al. [267] isolated plastoquinones, sargahydroquinoic acid, sargachromenol acid, and sargaquinoic acid from *Sargassum serratifolium* and demonstrated that these compounds have activity against protein tyrosine phosphatase 1B, so these compounds can be further explored for the prevention and treatment of type 2 diabetes.

5. Seaweed Phenolics: Commercial and Potential New Applications

The phenolic compounds that have been isolated from seaweeds are scarce, and further research will enlarge the biochemical library and improve the chance to discover new potential compounds to different industries or areas, so this area is still evolving along the road from isolation to application. This subtopic will describe the work done on the bioactivities described above and potential novel applications.

The major problem of these compounds to be inserted in real commercial applications is mainly the compound concentration in seaweed, due to the low extraction efficiency and seaweed biomass availability [15,45,105]. The solution goes through seaweed aquaculture that already exists but at a low level [268].

Although the seaweeds' polyphenols are a target of ongoing research, the in vivo effects are scarce and unclear to their correspondence in vitro, which can be explained by the diversity of the methodology used [45].

Until now, the bioavailability of seaweeds has not been completely reinvestigated. More studies and research are needed in this field. Most of the seaweed phenolic pharmaceutical and biomedical bioavailability studies have been supported in mouse-model systems. Proof of the protecting effects of seaweed phenols against other diseases has been derived from animal experiments and in vitro studies. Consequently, new research studies are needed to examine and fully understand their bioavailability in humans (percentage of the substance that enters in the human circulation system and has an active effect) [45].

Phenolic compounds are the most researched seaweed compounds and are already applied in commercial solutions (e.g., cosmetic products). Normally, the phenolic compounds are not isolated, because the commercial products of seaweed extracts have a considerable quantity of phenols.

5.1. Food Applications

The antioxidant potential of these compounds could be seen as a natural and non-harmful food stabilizer and preservative, attracting interest in food industries. However, there is a need to take into account that oxidized phenolic compounds can react with amino acids to form insoluble complexes, which may inhibit proteolytic enzymes and thus decrease the nutritive values of a food product [269]. An investigation demonstrated a negative correlation between the phenolic content of the seaweeds *Ulva lactuca*, *Hypnea charoides,* and *Hypnea japonica* (phenolic content 38.8% ± 0.5%, 16.9% ± 1.0%, and 16.3% ± 0.03%, respectively) and the digestibility and amino acid bioavailability of 85.7% ± 1.9%, 88.7% ± 0.7%, and 88.9% ± 1.4%, respectively [269]. This is likely to be a greater issue in brown seaweed species, as they are typically higher in phenolic content, including catechins, flavonoids, and phlorotannins [19,270].

Nevertheless, the restrictions of the synthetic ingredients in the food industry can be the turning point for the exploitation of seaweed compounds as safe alternatives [45], as they also have anti-microbial activities against major food spoilage and food pathogenic microorganisms [271]. Extracts that have seaweed phenolic antioxidants have been applied as enhancers of the oxidative stability and to conserve or increase the intrinsic quality and nutritional value of foods [272,273].

Their antioxidant potential is useful in the food industry, not only as nutraceutical compounds for functional food products, in which they are of indubitable valuable for improving health (as food supplements), but also to extend the shelf-life period when applied in processed food (functional foods) [11,274]. Additionally, the antimicrobial potential from the seaweed phenolics demonstrates that they can be useful in the food industry [275].

Moreover, bromophenols from *Ulva lactuca* and *Pterocladiella capillacea* were studied as "marine flavor" agents in farmed fish and other aquatic organisms, because the farming final products can have different flavor from the wild catch, and this can be inserted as a feed ingredient or a seaweed bromophenol-enriched sauce [276].

Seapolynol™ (Botamedi Inc, Seoul, Korea) is a food supplement that has been approved by the European Food Safety Authority [277]. This supplement is based on dieckol and other polyphenols from *E. cava*; it has been tested and showed promising results as an anti-hyperlipidemic [278] and cardioprotective agent against doxorubicin-induced cardiotoxicity [279]. Furthermore, Seapolynol™ affected the sensitivity of insulin in type 2 diabetes and could play a key role in the prevention of metabolic disorders [280–282]. However, these assays were performed in mice.

5.2. Cosmetic Applications

The lack of toxicity of phlorotannins when compared with other natural antioxidants, and their effects in prevention the loss of skin elasticity of aged skin, is attracting the attention of researchers to develop novel natural cosmeceutical and pharmaceutical formulations [34,231,283].

Currently, some seaweed extracts containing the respective phenolic compounds, namely phlorotannins, are present in cosmetics, such as skin care and anti-aging products. Seaweeds are already used for this purpose; for instance, in *Saccharina japonica* (formerly known as *Laminaria japonica*), its extract (dasima extract produced Natural Solutions, in South Korea) is the ingredient utilized in facial masks, and it has anti-inflammatory, antioxidant, and anti-microbial effects. *Ecklonia cava* is an anti-inflammatory, antioxidant, anti-bacterial, and UV-protecting agent; furthermore, it is also anti-bacterial specifically against *Propionibacterium acnes*, which is a bacteria that causes acne. Thus, these phlorotannins are included in various cosmetic formulations as well [284–287].

The cosmetic industry is one of the main areas that drives the insertion of seaweed phenolic compounds in commercial applications, such as natural UV screening (Helioguard® 365, produced by Mibelle Biochemistry in Buchs, Switzerland; Aethic Sôvée® produced with Photamin, phenolic extracted from seaweeds, produced by AETHIC® in London, UK from *Porphyra umbilicalis*) [95,288–290], or an anti-aging agent (ECKLEXT® BG, produced by NOF Group, product obtained by phlorotannins enriched-extraction of *Ecklonia kurome*, harvested in Japan) [291]. An *Asparagopsis armata* extract containing MAAs is integrated in lotions with anti-aging properties [292].

Another part of cosmetics is the formulation of bioactive extracts for their incorporation in commercial formulas. The Natural Solution produces two registered extracts based on seaweeds that have phenolic compounds as active ingredients [293]:

- *Ulva compressa* (formerly known as *Enteromorpha compressa)* Extract (Green Confertii Extract-NS by Natural Solution in Flemington, NJ, USA) contains flavonoids, tannins, polysaccharides, and acrylic acid as active compounds. It displays an antioxidant effect and anti-allergic effect and acts as an anti-microbial, antioxidant, and anti-allergic agent [294].
- *Fucus vesiculosus* Extract (Bladderwrack Extract-NS by Natural Solution) contains fucoidan and phlorotannins as active compounds and acts as an anti-aging, antioxidant, anti-fungal, and anti-bacterial agent [295].

At last, there are new studies and developments in this area focusing on new phenolic compounds and cosmetic utilizations, with a view to patent the new compounds and formulations [296–298].

For this wide range of bioactivities and properties, the usage of natural anti-aging products derived from seaweeds is gaining reputation and attracting researchers' consideration [34].

5.3. Pharmaceutical and Biomedical Applications

Seaweeds have been used for centuries as a normal medicine for diverse health diseases in folk medicine [299]. This attribute was considered as early as 300 BC in Asian cultures, and the Celtic, British, and Roman populations located near the sea used them for 1000 years for healing wounds, as vermifuge, or as anthelmintic, so the modern search in pharmaceutical and biomedical areas is evolving and in continuous progression [300,301].

Seaweed's phenols have a wide range of possible applications in the pharmaceutical and biomedical areas due to their various bioactivities. Phenols, mostly the phlorotannins, have an outstanding antioxidant power due to their ability as chelating agents with reactive oxygen species and consequently preventing oxidative stress and cell damage [4,5]. Therefore, the scavenging of oxidants is important to control various diseases; thus, phenolic compounds of seaweeds are extremely valuable as a natural source of antioxidant agents.

Phenolic compounds are being researched for their application in human health to ameliorate or cure some of the main disease problems nowadays, such as cardiovascular, diabetes, neurodegenerative, and mental disorders [4,5].

Free radicals' occurrence against macromolecules (e.g., membrane lipids, proteins, enzymes, DNA, and RNA) plays a pivotal role in several health disorders such as cancer, diabetes, neurodegenerative, and inflammatory diseases. Therefore, antioxidants may have a beneficial effect on human health by preventing free radical damage [33].

5.3.1. Cardiovascular Disease

It was demonstrated that a flavonoid-enriched diet improves endothelial function and lowers the blood pressure [302]. Phlorotannins have a positive effect on the amelioration of hypertension [27]. A compound isolated from *Saragassum siliquastrum*, the sargachromenol D, demonstrated high potential to be a new drug for blood pressure control in severe hypertension in instances where it cannot be controlled by conventional combinatorial therapy [303].

Phlorotannins have been explored in the last decade by companies to obtain new products. The main targets of phlorotannin supplements in cardiovascular diseases are the arteriosclerosis prevention and the increase of protective high-density lipoprotein cholesterol (HDL-C). The products containing phlorotannins are HealSea™ (produced by Diana Naturals in Rennes, France), IdAlg™ (produced by Bio Serae in Bram, France), and Seanol™ (produced by LiveChem in Jeju-do, South Korea and distributed by Simple Health, Maitland, USA) [45,304].

5.3.2. Neurodegenerative and Mental Disorders

Some of the identified seaweed bromophenols can be included in schizophrenia and Parkinson's disease therapies [161], because they can be used as a multitarget ligand to neurosensors. Furthermore, phlorotannins have neuroprotection action and may be the key to the treatment of neurodegenerative diseases [31].

Extracts of *Eisenia bicyclis* inhibited β-amyloid cleavage enzyme activity [305]. This potentiality was also demonstrated using phlorotannins of other brown seaweeds, such as *Ishige foliacea, Ecklonia maxima,* and *E. cava*, proving that they could be useful for Alzheimer's disease treatment [31,234,237,306].

5.3.3. Anticancer Properties

The phenolic compounds and their by-products can play significant roles as anticancer metabolites, performing in different parts of the evolution of cancer such as proliferative signalling, metastasis, cell cycle, resistance to cell death, evasion, angiogenesis, and the evasion of growth suppressors [8,307–309].

Chemotherapy is one of the main therapeutic approaches for cancer treatment, and there are already various natural anticancer molecules isolated—for example, the clinically used camptothecin and taxol [310].

The phenolic compound has a promising cytotoxicity against various cancer cells lines, and the selectivity of the compounds against cancer cell lines needs to be considered, as demonstrated in Section 4. This issue is one of the major problems for the current anticancer drugs [33]. Therefore, the phenolic compound can be cytotoxic to normal cell lines, such as the human embryo lung fibroblast (HELF), because the compounds can have low selectivity and can be counter-productive in their application, so a full batch of assays is needed to have a secure anticancer compound available. For this case, chemical modification can be needed to enhance the molecule selectivity [33,130,311], and more research will be necessary until a commercial product is released.

Ganesan and colleagues have proven the antioxidant effectiveness of *Euchema* sp., *Kappaphycus* sp., *Hydropuntia edulis* (formerly known as *Gracilaria edulis*), and *Acanthophora spicifera* extracts [312]. In this way, *Hypnea musciformis, H. valentiae,* and *Jania rubens* extracts were found to have this potential and others for carcinogenesis reduction and inflammatory diseases prevention [11].

5.3.4. Diabetes

Bromophenols present anti-diabetic effects [33,161] that could be used in the research and development of a novel class of anti-diabetic drugs or in supplements and functional food products.

Red seaweeds' bromophenols anti-diabetic effects were demonstrated by Matanjun and colleagues [313] using the species *Odonthalia corymbifera* and *Symphyocladia latiuscula*. More recently, this property was also confirmed in other Rhodophyta species, such as *Laurencia similis* [314], *Rhodomela confervoides* [163], and *Grateloupia elliptica* [315].

Polyphenols have these effects by inhibiting hepatic gluconeogenesis and reducing the activity of digestive enzymes such as α-amylase, α-glucosidase, lipase, and aldose reductase [125,210]. A commercial formulation of *A. nodosum* and *F. vesiculosus* phlorotannins, InSea2™ (Rimouski, QC, Canada), promotes a 90% reduction of the postprandial blood glucose, reducing the peak insulin secretion by 40% [316]. This leads these molecules to be considered as a potential natural alternative to the existing anti-diabetic drugs that have undesirable side effects [210]. Beyond their possible incorporation in pharmacological formulations, these molecules may also be seen as novel compositions for functional food or nutraceutical products.

5.3.5. Anti-Microbial Function

The research for novel anti-microbial agents has been a "long-distance run" for many years, because the discovery of new effective drugs did not keep the pace of increasing anti-microbial resistance, especially bacteria. One problem is the lack and limitation of compounds in screening libraries [317,318]. This new research phase based in blue biotech can be fundamental to the improvement of these chemical libraries, and their extension with novel biomolecules. Initial studies had proven to be useful due to a real increment of reports of phenolic compounds with promising anti-microbial activities [33].

Purified phenolic extracts were found to have powerful antimicrobial effect against bacteria, fungi [37], and virus [221], revealing their potential for their use in pharmacotherapy.

In this field, there was not a clear passage from the assay to applications, but there is some patent work with phlorotannins-based anti-bacterial agents [319].

There is one patent comprising 6,6′-bieckol from *E. cava* for HIV-1 pharmaceutical composition, demonstrating interest to further explore the commercial use of phlorotannins from this species [320].

5.3.6. Tissue and Bone Regeneration

Tissue and bone engineering is a new biomedical area where the phlorotannins are recently being tested [321]. This area focuses on the regeneration of damaged tissues, bones, or organs where biomedical scaffolds, cells, and growth factors are combined to obtain success [322]. The scaffolds used in regenerative applications need to provide biocompatibility, biodegradability, and appropriate mechanical properties for successful application [323–325].

A study by Im et al. [321] demonstrated that the insertion of phlorotannins on the fish collagen/alginate biocomposite obtained a better result in the cell proliferation. In the in vitro assay, this new mixture showed better results in calcium deposition and osteogenesis, demonstrating that this product with phlorotannins is a potential biomaterial for the bone tissue growth.

At the commercial level, Seanol® (*E. cava* phenolic extract produced by Botamedi Inc., Seoul, Korea) was evaluated as an ingredient in hydrogels for bone regeneration with the ability to promote the anti-bacterial activity and to enhance the bone mineralization [326].

5.3.7. Anti-Inflammatory

The symptoms of inflammation embrace the occurrence of inflammatory cells or mediators in the specific or non-specific location of tissue affected by adverse stimuli, as well as wounds, allergies, irritantions, or infections [327]. Various phenolics extracts and compounds already demonstrated the power to reduce inflammation—for example, *Carpodesmia tamariscifolia* (as *Cystoseira tamariscifolia*), *Treptacantha nodicaulis* (as *Cystoseira nodicaulis*), *Treptacantha usneoides* (as *Cystoseira usneoides*) and *Fucus spiralis* phlorotannin-purified extracts, and *Neoporphyra dentata* (as *Porphyra dentata*) flavonoid-enriched extract (containing catechol, rutin and hesperidin) [37,328]. Two different studies revealed that phlorotannins purified extracts were able to significantly reduce the levels of nitric oxide (NO) in mouse (RAW 264.7) macrophage cells previously exposed to a lipopolysaccharide from *Salmonella enterica* [37,328]. The investigation conducted by Katarzyna Kazłowska and colleagues demonstrated the anti-inflammatory effect of methanolic extract of the phenolic fraction of *N. dentata*, containing catechol, rutin, and hesperidin, which suppressed NO production via NF-kappaB-dependent iNOS (inducible nitric oxide synthase) gene transcription [328].

Another anti-inflammatory effect was observed in an *E. cava* phlorotannins extract through the arachidonic-dependent pathway by the downregulation of prostaglandin E2 [329].

In short, the anti-inflammatory mechanisms of action are correlated with the modulation of NO levels, which could be by direct scavenging and/or by decreasing NO production through the inflammatory signaling cascade or by inhibiting the enzymes involved in NO production [37,328,329].

5.3.8. Other Medical Applications

Phlorotannin complements show sleep-promoting properties in mice as demonstrated in Section 4 of this review. In human trials, it prevents waking up after falling sleep in adults with sleep disturbances [330].

In a new assay, for caffeine sleep disruption problems, the effects of phlorotannin complements were analyzed against a sedative-hypnotic drug zolpidem (ZPD) in mice. The phlorotannin complement attenuated the effect of the caffeine sleep disruption effects, with identical results as the ZPD assay. Furthermore, phlorotannins did not change the delta activity during the non-rapid eye movement sleep, unlike ZPD, which decreased the delta activity, which is a negative symptom of taking ZPD. Thus, this study suggests that phlorotannins can be applied to relieve transitory insomnia symptoms [330].

In the treatment of wounds, there are not many studies, although the work of Park et al. [331] demonstrated the potential of phlorotannins extracted from *E. cava* to be incorporated in polyvinyl alcohol hydrogel for wound healing after application.

Additionally, investigations showed that dietary phenols are recognized as xenobiotics in humans, and although their absorption is very low in the small intestine (about 5–10%), they revealed prebiotic action [332]. Once they reach the gut, phenols may modify and produce variations in the microflora community by exhibiting prebiotic effects and antimicrobial action against pathogenic intestinal microflora [152,333,334].

5.4. Feed and Animal Health

Seaweed's polyphenolic compounds have been demonstrated to be bioavailable for animals from the colon [335]. These phenolic compounds can be absorbed either directly in the upper digestive tract in untouched form or in the lower intestine after alteration by bacteria in the digestive tracts [336,337]. A study by Nagayama et al. [231] identified a possible application of phlorotannins extracted from *Ecklonia kurome* as anti-bacterial drugs that can be a natural substitute for the recently banned feed antibiotics.

Phlorotannins perform as prebiotics for both ruminant and monogastric animals while the dosage is given at low doses, under 5% in the animal diet. Phlorotannins might be involved in the feed of poultry and pigs only at low levels, commonly up to 5–6% in growing animals and never above 10% to prevent the over-dosage that causes negative effects in animal health. Using the recommended dosage, there is an escalation in productivity (increase in animal growth and milk production), quality (enhanced meat quality), and safety (decrease in shedding of pathogenic microorganisms; phlorotannins are more effective than the condensed tannins from terrestrial plant sources) of animal products due to phlorotannins. An amelioration of the immune system and rise in the anti-oxidative status of the animals were also observed [231,338–345].

Ascophyllum nodosum holds high amounts of phenolic compounds (mainly phlorotannins) that are insoluble in the digestive tract of animals [344,346]. Wang et al. [344] demonstrated that the *A. nodosum* extract contained a considerable concentration of phlorotannins (up to 500 g/mL), which reduced fermentation in mixed feed and barley grain feed in an in vitro assay. The effects were linear with the phlorotannins concentration. The *A. nodosum* based meal improved the slowly degraded protein fraction and protein degradability, and it was further beneficial to the feed digestibility when complementing low-quality feed diets [347].

Tasco® (Dartmouth, Nova Scotia, Canada) from Acadian Seaplants is one of the *A. nodosum*-based feeds in the feed market [348], where various studies have been conducted to prove that phlorotannins enhance animal health [339,343,345,349].

5.5. Agriculture

The seaweed phenolic compounds that are identified in some commercial seaweed extracts can be a protection factor against plant diseases [350] because of their anti-microbial activity. Another potential role of polyphenols is the protection of plant organisms from damages caused by free radicals and other oxidants [336,351].

This field is in evolution, and normally, it is mainly explored using brown seaweeds because of the knowledge regarding phlorotannins. The commercialization of a solution with phenolic compounds as a stimulant for mycorrhizal and rhizobial symbiosis is already a patented work, which can happen as fertilization or treatment based mainly in the *Fucus* and *Ascophyllum* genus [352]. Another patent is based on liquid fertilizer with an active plant disease protection effect [353].

Some extracts produced by Maxicrop, Acadian Seaplants, and algae have a high content of humic-like polyphenols or polyphenols, which have been derived mainly from brown seaweeds in this type of industry [354]. One example is the *A. nodosum* analyzed from Norway and Nova Scotia that has 15–25% of extractable polyphenols with high molecular weight [355].

5.6. Other Applications

Phlorotannins can substitute bisphenol A (BPA), a very toxic substance, in the vinyl esters. The material designed and tested by Jaillet et al. [356] demonstrated a high thermal stability and thermomechanical properties, so this phlorotannins-based material can be applied as a substitute of BPA-based material in thermoset networks for composites. There is a long way to go before these compounds can be applied, but this demonstrates the potential of phlorotannins and the phenolic compounds of seaweeds.

6. Conclusions and Future Perspectives

The phenolic compounds discovered in seaweeds are an extensive and diverse group that are divided in four major classes and specific groups of terpenoids, where the predominant bioactivity of all is the anti-oxidative activity.

The phlorotannins represent the phenolic class in a more advanced stage of research with commercials products already available in diverse areas. However, the major compounds isolated are from the *Ecklonia* genus, indicating that research is approaching seaweeds with more biomass available, and so, the hypothesis of exploring these products commercially is more feasible. Therefore, the range of studies of phenolic compounds is wide, with the exception of only MAAs, which have focused on UV protection with large success. This bioactive compound can be applied in a wide range of industries such as food and feed, biomedical and pharmaceutical, agriculture, and electronics.

From the bibliography analyzed, this review demonstrates a good index of new developments in the pharmaceutical area and in other areas by using seaweed phenolics, but there is long road to understanding the major parts of the compounds already isolated, and there are other seaweeds that can be targeted for studies due to the easy seaweed cultivation.

We also demonstrate that the extractions and isolation methods are still being developed in order to be more ecological and intuitive to perform, with better quality, purity, and quantity of the phenolic compound extracted. So, the seaweed phenolics can be key players in the future in different areas that can help the practices of humankind be greener through being supported by natural compounds that were very difficult to obtain until the developments of the last decade.

Author Contributions: Conception and design of the idea: J.C., A.L., P.M., D.P., A.M.M.G.; Design of chemical structures: A.F.; Organization of the team: J.C.; Writing and bibliographic research: J.C., A.L., P.M., D.P.; Supervision and Manuscript Revision, A.M.M.G., A.F., G.J.d.S. and L.P. All authors have read and agreed to the published version of the manuscript.

Funding: This work is financed by national funds through FCT—Foundation for Science and Technology, I.P., within the scope of the projects UIDB/04292/2020—MARE—Marine and Environmental Sciences Centre and UIDP/50017/2020+UIDB/50017/2020 (by FCT/MTCES) granted to CESAM—Centre for Environmental and Marine Studies. João Cotas thanks to the European Regional Development Fund through the Interreg Atlantic Area Program, under the project NASPA (EAPA_451/2016). Adriana Leandro thanks FCT for the financial support provided through the doctoral grant SFRH/BD/143649/2019 funded by National Funds and Community Funds through FSE. Diana Pacheco thanks to PTDC/BIA-CBI/31144/2017—POCI-01 project -0145-FEDER-031144—MARINE INVADERS, co-financed by the ERDF through POCI (Operational Program Competitiveness and Internationalization) and by the Foundation for Science and Technology (FCT, IP). Artur Figueirinha thank FCT/MCTES, Fundação para a Ciência e Tecnologia and Ministério da Ciência, Tecnologia e Ensino Superior) through grant UIDB/50006/2020 and by Programa de Cooperación Interreg V-A España—Portugal (POCTEP) 2014–2020 (project 0377_IBERPHENOL_6_E). Ana M. M. Gonçalves acknowledges University of Coimbra for the contract IT057-18-7253.

Conflicts of Interest: The authors declare no conflict of interest.

References

1. Fallarero, A.; Peltoketo, A.; Loikkanen, J.; Tammela, P.; Vidal, A.; Vuorela, P. Effects of the aqueous extract of *Bryothamnion triquetrum* on chemical hypoxia and aglycemia-induced damage in GT1-7 mouse hypothalamic immortalized cells. *Phytomedicine* **2006**, *13*, 240–245. [CrossRef] [PubMed]

2. Lopes, G.; Pinto, E.; Andrade, P.B.; Valentão, P. Antifungal activity of phlorotannins against dermatophytes and yeasts: Approaches to the mechanism of action and influence on *Candida albicans* virulence factor. *PLoS ONE* **2013**, *8*. [CrossRef] [PubMed]

3. Wang, T.; Jónsdóttir, R.; Liu, H.; Gu, L.; Kristinsson, H.G.; Raghavan, S.; Ólafsdóttir, G. Antioxidant capacities of phlorotannins extracted from the brown algae *Fucus vesiculosus*. *J. Agric. Food Chem.* **2012**, *60*, 5874–5883. [CrossRef]

4. Audibert, L.; Fauchon, M.; Blanc, N.; Hauchard, D.; Ar Gall, E. Phenolic compounds in the brown seaweed *Ascophyllum nodosum*: Distribution and radical-scavenging activities. *Phytochem. Anal.* **2010**, *21*, 399–405. [CrossRef]

5. Ferreres, F.; Lopes, G.; Gil-Izquierdo, A.; Andrade, P.B.; Sousa, C.; Mouga, T.; Valentão, P. Phlorotannin extracts from fucales characterized by HPLC-DAD-ESI-MS n: Approaches to hyaluronidase inhibitory capacity and antioxidant properties. *Mar. Drugs* **2012**, *10*, 2766–2781. [CrossRef] [PubMed]

6. Cuesta, R.G.; González García, K.L.; Del, O.; Iglesias, R.V.; Rivera, Y.H.; Suárez, Y.A. Seaweeds as sources of bioactive compounds in the benefit of human health: A review. *Rev. Cienc. Biológicas Y La Salud* **2016**, *18*, 20–27. [CrossRef]

7. Pacheco, D.; García-Poza, S.; Cotas, J.; Gonçalves, A.M.; Pereira, L. Fucoidan—A valuable source from the ocean to pharmaceutical. *Front. Drug Chem. Clin. Res.* **2020**, *3*, 1–4. [CrossRef]

8. Wijesekara, I.; Kim, S.K.; Li, Y.X.; Li, Y.X. Phlorotannins as bioactive agents from brown algae. *Process Biochem.* **2011**, *46*, 2219–2224. [CrossRef]

9. Yuan, Y.V.; Bone, D.E.; Carrington, M.F. Antioxidant activity of dulse (*Palmaria palmata*) extract evaluated in vitro. *Food Chem.* **2005**, *91*, 485–494. [CrossRef]

10. Cox, S.; Gupta, S.; Abu-Ghannam, N. Effect of different rehydration temperatures on the moisture, content of phenolic compounds, antioxidant capacity and textural properties of edible Irish brown seaweed. *LWT Food Sci. Technol.* **2012**, *47*, 300–307. [CrossRef]

11. Chakraborty, K.; Joseph, D.; Praveen, N.K. Antioxidant activities and phenolic contents of three red seaweeds (Division: Rhodophyta) harvested from the Gulf of Mannar of Peninsular India. *J. Food Sci. Technol.* **2015**, *52*, 1924–1935. [CrossRef] [PubMed]

12. Dixit, D.C.; Reddy, C.R.K.; Balar, N.; Suthar, P.; Gajaria, T.; Gadhavi, D.K. Assessment of the nutritive, biochemical, antioxidant and antibacterial potential of eight tropical macro algae along kachchh coast, india as human food supplements. *J. Aquat. Food Prod. Technol.* **2018**, *27*, 61–79. [CrossRef]

13. Liu, L.; Heinrich, M.; Myers, S.; Dworjanyn, S.A. Towards a better understanding of medicinal uses of the brown seaweed *Sargassum* in traditional chinese medicine: A phytochemical and pharmacological review. *J. Ethnopharmacol.* **2012**, *142*, 591–619. [CrossRef] [PubMed]

14. Mekinić, I.G.; Skroza, D.; Šimat, V.; Hamed, I.; Čagalj, M.; Perković, Z.P. Phenolic content of brown algae (Pheophyceae) species: Extraction, identification, and quantification. *Biomolecules* **2019**, *9*, 244. [CrossRef]

15. Cotas, J.; Leandro, A.; Pacheco, D.; Gonçalves, A.M.M.; Pereira, L. A comprehensive review of the nutraceutical and therapeutic applications of red seaweeds (Rhodophyta). *Life* **2020**, *10*, 19. [CrossRef]

16. Domínguez, H. Algae as a source of biologically active ingredients for the formulation of functional foods and nutraceuticals. In *Functional Ingredients from Algae for Foods and Nutraceuticals*; Domínguez, H., Ed.; Woodhead Publishing: Cambridge, UK, 2013; pp. 1–19. ISBN 9780857095121.

17. Singh, I.P.; Sidana, J. Phlorotannins. In *Functional Ingredients from Algae for Foods and Nutraceuticals*; Domínguez, H., Ed.; Woodhead Publishing: Cambridge, UK, 2013; pp. 181–204. ISBN 9780857095121.

18. Dimitrios, B. Sources of natural phenolic antioxidants. *Trends Food Sci. Technol.* **2006**, *17*, 505–512. [CrossRef]

19. Wang, T.; Jónsdóttir, R.; Ólafsdóttir, G. Total phenolic compounds, radical scavenging and metal chelation of extracts from icelandic seaweeds. *Food Chem.* **2009**, *116*, 240–248. [CrossRef]

20. Swanson, B.G. Tannins and polyphenols. In *Encyclopedia of Food Sciences and Nutrition*; Caballero, B., Ed.; Academic Press: Cambridge, MA, USA, 2003; pp. 5729–5733.

21. Bravo, L. Polyphenols: Chemistry, dietary sources, metabolism, and nutritional significance. *Nutr. Rev.* **2009**, *56*, 317–333. [CrossRef]

22. Shahidi, F.; Naczk, M. *Phenolics in Food and Nutraceuticals*; CRC Press: Boca Raton, FL, USA, 2003; ISBN 9780367395094.

23. Dai, J.; Mumper, R.J. Plant phenolics: Extraction, analysis and their antioxidant and anticancer properties. *Molecules* **2010**, *15*, 7313–7352. [CrossRef]

24. Wells, M.L.; Potin, P.; Craigie, J.S.; Raven, J.A.; Merchant, S.S.; Helliwell, K.E.; Smith, A.G.; Camire, M.E.; Brawley, S.H. Algae as nutritional and functional food sources: Revisiting our understanding. *J. Appl. Phycol.* **2017**, *29*, 949–982. [CrossRef]

25. Heo, S.J.; Park, E.J.; Lee, K.W.; Jeon, Y.J. Antioxidant activities of enzymatic extracts from brown seaweeds. *Bioresour. Technol.* **2005**, *96*, 1613–1623. [CrossRef] [PubMed]

26. Corona, G.; Coman, M.M.; Guo, Y.; Hotchkiss, S.; Gill, C.; Yaqoob, P.; Spencer, J.P.E.; Rowland, I. Effect of simulated gastrointestinal digestion and fermentation on polyphenolic content and bioactivity of brown seaweed phlorotannin-rich extracts. *Mol. Nutr. Food Res.* **2017**, *61*, 1–10. [CrossRef] [PubMed]

27. Gómez-Guzmán, M.; Rodríguez-Nogales, A.; Algieri, F.; Gálvez, J. Potential role of seaweed polyphenols in cardiovascular-associated disorders. *Mar. Drugs* **2018**, *16*, 250. [CrossRef] [PubMed]

28. Fairhead, V.A.; Amsler, C.D.; McClintock, J.B.; Baker, B.J. Variation in phlorotannin content within two species of brown macroalgae (*Desmarestia anceps* and *D. menziesii*) from the Western Antarctic Peninsula. *Polar Biol.* **2005**, *28*, 680–686. [CrossRef]

29. Chen, Y.; Lin, H.; Li, Z.; Mou, Q. The anti-allergic activity of polyphenol extracted from five marine algae. *J. Ocean Univ. China* **2015**, *14*, 681–684. [CrossRef]

30. Murray, M.; Dordevic, A.; Ryan, L.; Bonham, M. The Impact of a single dose of a polyphenol-rich seaweed extract on postprandial glycaemic control in healthy adults: A randomised cross-over trial. *Nutrients* **2018**, *10*, 270. [CrossRef]

31. Pangestuti, R.; Kim, S.-K. Neuroprotective effects of marine algae. *Mar. Drugs* **2011**, *9*, 803–818. [CrossRef]

32. Holdt, S.L.; Kraan, S. Bioactive compounds in seaweed: Functional food applications and legislation. *J. Appl. Phycol.* **2011**, *23*, 543–597. [CrossRef]

33. Liu, M.; Hansen, P.E.; Lin, X. Bromophenols in marine algae and their bioactivities. *Mar. Drugs* **2011**, *9*, 1273–1292. [CrossRef]

34. Thomas, N.V.; Kim, S.K. Potential pharmacological applications of polyphenolic derivatives from marine brown algae. *Env. Toxicol. Pharm.* **2011**, *32*, 325–335. [CrossRef]

35. Tanna, B.; Brahmbhatt, H.R.; Mishra, A. Phenolic, flavonoid, and amino acid compositions reveal that selected tropical seaweeds have the potential to be functional food ingredients. *J. Food Process. Preserv.* **2019**, *43*, 1–10. [CrossRef]

36. Nwosu, F.; Morris, J.; Lund, V.A.; Stewart, D.; Ross, H.A.; McDougall, G.J. Anti-proliferative and potential anti-diabetic effects of phenolic-rich extracts from edible marine algae. *Food Chem.* **2011**, *126*, 1006–1012. [CrossRef]

37. Lopes, G.; Sousa, C.; Silva, L.R.; Pinto, E.; Andrade, P.B.; Bernardo, J.; Mouga, T.; Valentão, P. Can phlorotannins purified extracts constitute a novel pharmacological alternative for microbial infections with associated inflammatory conditions? *PLoS ONE* **2012**, *7*, e31145. [CrossRef] [PubMed]

38. Urquiaga, I.; Leighton, F. Plant polyphenol antioxidants and oxidative stress. *Biol. Res.* **2000**, *33*, 55–64. [CrossRef] [PubMed]

39. Pereira, L. *Therapeutic and Nutritional Uses of Algae*; CRC Press: Boca Raton, FL, USA, 2018; ISBN 9781498755382.

40. Ravikumar, S.; Jacob Inbaneson, S.; Suganthi, P. Seaweeds as a source of lead compounds for the development of new antiplasmodial drugs from South East coast of India. *Parasitol. Res.* **2011**, *109*, 47–52. [CrossRef] [PubMed]

41. Schultz, J.C.; Hunter, M.D.; Appel, H.M. Antimicrobial activity of polyphenols mediates plant-herbivore interactions. In *Plant Polyphenols*; Hemingway, R.W., Laks, P.E., Eds.; Springer US: Boston, MA, USA, 1992; pp. 621–637.

42. Maqsood, S.; Benjakul, S.; Shahidi, F. Emerging Role of phenolic compounds as natural food additives in fish and fish products. *Crit. Rev. Food Sci. Nutr.* **2013**, *53*, 162–179. [CrossRef]

43. Panzella, L.; Napolitano, A. Natural phenol polymers: Recent advances in food and health applications. *Antioxidants* **2017**, *6*, 30. [CrossRef]

44. Leandro, A.; Pereira, L.; Gonçalves, A.M.M. Diverse applications of marine macroalgae. *Mar. Drugs* **2019**, *18*, 17. [CrossRef]

45. Freile-Pelegrín, Y.; Robledo, D. Bioactive phenolic compounds from algae. *Bioact. Compd. Mar. Foods Plant Anim. Sources* **2013**, 113–129. [CrossRef]

46. Bilal Hussain, M.; Hassan, S.; Waheed, M.; Javed, A.; Adil Farooq, M.; Tahir, A. Bioavailability and metabolic pathway of phenolic compounds. *Plant Physiol. Asp. Phenolic Compd.* **2019**, 1–18. [CrossRef]

47. Maqsood, S.; Benjakul, S.; Abushelaibi, A.; Alam, A. Phenolic compounds and plant phenolic extracts as natural antioxidants in prevention of lipid oxidation in seafood: A detailed review. *Compr. Rev. Food Sci. Food Saf.* **2014**, *13*, 1125–1140. [CrossRef]

48. Santos, S.A.O.; Félix, R.; Pais, A.C.S.; Rocha, S.M.; Silvestre, A.J.D. The quest for phenolic compounds from macroalgae: A review of extraction and identification methodologies. *Biomolecules* **2019**, *9*, 847. [CrossRef] [PubMed]

49. Arnold, T.M.; Targett, N.M. Marine tannins: The importance of a mechanistic framework for predicting ecological roles. *J. Chem. Ecol.* **2002**, *28*, 1919–1934. [CrossRef] [PubMed]

50. Rocha-Santos, T.; Duarte, A.C. Introduction to the analysis of bioactive compounds in marine samples. In *Comprehensive Analytical Chemistry*; Rocha-Santos, T., Duarte, A.C., Eds.; Elsevier B.V.: Amsterdam, Netherlands, 2014; Volume 65, pp. 1–13. ISBN 9780444633590.

51. Mukherjee, P.K. Bioactive phytocomponents and their analysis. In *Quality Control and Evaluation of Herbal Drugs*; Mukherjee, P.K., Ed.; Elsevier Inc.: Amsterdam, Netherlands, 2019; pp. 237–328. ISBN 9780128133743.

52. Pietta, P.; Minoggio, M.; Bramati, L. Plant polyphenols: Structure, occurrence and bioactivity. *Stud. Nat. Prod. Chem.* **2003**, *28*, 257–312. [CrossRef]

53. Liwa, A.C.; Barton, E.N.; Cole, W.C.; Nwokocha, C.R. *Bioactive Plant Molecules, Sources and Mechanism of Action in the Treatment of Cardiovascular Disease*; Elsevier Inc.: Amsterdam, Netherlands, 2017; ISBN 9780128020999.

54. Luna-Guevara, M.L.; Luna-Guevara, J.J.; Hernández-Carranza, P.; Ruíz-Espinosa, H.; Ochoa-Velasco, C.E. Phenolic compounds: A good choice against chronic degenerative diseases. *Stud. Nat. Prod. Chem.* **2018**, *59*, 79–108. [CrossRef]

55. Mancini-Filho, J.; Novoa, A.V.; González, A.E.B.; de Andrade-Wartha, E.R.S.; Mancini, D.A.P. Free phenolic acids from the seaweed *Halimeda monile* with antioxidant effect protecting against liver injury. *Z. Für Nat. C* **2009**, *64*, 657–663. [CrossRef]

56. Feng, Y.; Carroll, A.R.; Addepalli, R.; Fechner, G.A.; Avery, V.M.; Quinn, R.J. Vanillic acid derivatives from the green algae *Cladophora socialis* as potent protein tyrosine phosphatase 1B inhibitors. *J. Nat. Prod.* **2007**, *70*, 1790–1792. [CrossRef]

57. Agregán, R.; Munekata, P.E.S.; Franco, D.; Dominguez, R.; Carballo, J.; Lorenzo, J.M. Phenolic compounds from three brown seaweed species using LC-DAD–ESI-MS/MS. *Food Res. Int.* **2017**, *99*, 979–985. [CrossRef]

58. Farvin, K.H.S.; Jacobsen, C.; Sabeena Farvin, K.H.; Jacobsen, C. Phenolic compounds and antioxidant activities of selected species of seaweeds from Danish coast. *Food Chem.* **2013**, *138*, 1670–1681. [CrossRef]

59. Xu, T.; Sutour, S.; Casabianca, H.; Tomi, F.; Paoli, M.; Garrido, M.; Pasqualini, V.; Aiello, A.; Castola, V.; Bighelli, A. Rapid screening of chemical compositions of *Gracilaria dura* and *Hypnea mucisformis* (Rhodophyta) from Corsican Lagoon. *Int. J. Phytocosmetics Nat. Ingred.* **2015**, *2*, 8. [CrossRef]

60. Souza, B.W.S.; Cerqueira, M.A.; Martins, J.T.; Quintas, M.A.C.; Ferreira, A.C.S.; Teixeira, J.A.; Vicente, A.A. Antioxidant potential of two red seaweeds from the brazilian coasts. *J. Agric. Food Chem.* **2011**, *59*, 5589–5594. [CrossRef] [PubMed]

61. Stengel, D.B.; Connan, S.; Popper, Z.A. Algal chemodiversity and bioactivity: Sources of natural variability and implications for commercial application. *Biotechnol. Adv.* **2011**, *29*, 483–501. [CrossRef] [PubMed]

62. Imbs, T.I.; Zvyagintseva, T.N. Phlorotannins are polyphenolic metabolites of brown algae. *Russ. J. Mar. Biol.* **2018**, *44*, 263–273. [CrossRef]

63. Achkar, J.; Xian, M.; Zhao, H.; Frost, J.W. Biosynthesis of phloroglucinol. *J. Am. Chem. Soc.* **2005**, *127*, 5332–5333. [CrossRef] [PubMed]

64. Katsui, N.; Suzuki, Y.; Kitamura, S.; Irie, T. 5,6-dibromoprotocatechualdehyde and 2,3-dibromo -4,5-dihydroxybenzyl methyl ether: New dibromophenols from *Rhodomela larix*. *Tetrahedron* **1967**, *23*, 1185–1188. [CrossRef]

65. Fan, X.; Xu, N.J.; Shi, J.G. Bromophenols from the red alga *Rhodomela confervoides*. *J. Nat. Prod.* **2003**, *66*, 455–458. [CrossRef]

66. Ko, S.C.; Ding, Y.; Kim, J.; Ye, B.R.; Kim, E.A.; Jung, W.K.; Heo, S.J.; Lee, S.H. Bromophenol (5-bromo-3,4-dihydroxybenzaldehyde) isolated from red alga *Polysiphonia morrowii* inhibits adipogenesis by regulating expression of adipogenic transcription factors and AMP-activated protein kinase activation in 3T3-L1 adipocytes. *Phyther. Res.* **2019**, *33*, 737–744. [CrossRef]

67. Flodin, C.; Helidoniotis, F.; Whitfield, F.B. Seasonal variation in bromophenol content and bromoperoxidase activity in *Ulva lactuca*. *Phytochemistry* **1999**, *51*, 135–138. [CrossRef]

68. Colon, M.; Guevara, P.; Gerwick, W.H.; Ballantine, D. 5'-hydroxyisoavrainvilleol, a new diphenylmethane derivative from the tropical green alga *Avrainvillea Nigricans*. *J. Nat. Prod.* **1987**, *50*, 368–374. [CrossRef]

69. Flodin, C.; Whitfield, F.B. 4-hydroxybenzoic acid: A likely precursor of 2,4,6-tribromophenol in *Ulva lactuca*. *Phytochemistry* **1999**, *51*, 249–255. [CrossRef]

70. Shi, D.; Li, X.; Li, J.; Guo, S.; Su, H.; Fan, X. Antithrombotic effects of bromophenol, an alga-derived thrombin inhibitor. *Chin. J. Oceanol. Limnol.* **2010**, *28*, 96–98. [CrossRef]

71. Xu, X.-L.; Fan, X.; Song, F.-H.; Zhao, J.-L.; Han, L.-J.; Yang, Y.-C.; Shi, J.-G. Bromophenols from the brown alga *Leathesia nana*. *J. Asian Nat. Prod. Res.* **2004**, *6*, 217–221. [CrossRef] [PubMed]

72. Chung, H.Y.; Ma, W.C.J.; Ang, P.O.; Kim, J.S.; Chen, F. Seasonal variations of bromophenols in brown algae (*Padina arborescens*, *Sargassum siliquastrum*, and *Lobophora variegata*) collected in Hong Kong. *J. Agric. Food Chem.* **2003**, *51*, 2619–2624. [CrossRef] [PubMed]

73. Wall, M.E.; Wani, M.C.; Manikumar, G.; Taylor, H.; Hughes, T.J.; Gaetano, K.; Gerwick, W.H.; McPhail, A.T.; McPhail, D.R. Plant antimutagenic agents, 7. Structure and antimutagenic properties of cymobarbatol and 4-isocymobarbatol, new cymopols from green alga (*Cymopolia barbata*). *J. Nat. Prod.* **1989**, *52*, 1092–1099. [CrossRef]

74. Flodin, C.; Whitfield, F.B. Biosynthesis of bromophenols in marine algae. *Water Sci. Technol.* **1999**, *40*, 53–58. [CrossRef]

75. Peng, J.; Li, J.; Hamann, M.T. The marine bromotyrosine derivatives. *Alkaloids Chem. Biol.* **2005**, *61*, 59–262. [CrossRef]

76. Liu, M.; Zhang, W.; Wei, J.; Qiu, L.; Lin, X. Marine bromophenol bis(2,3-dibromo-4,5-dihydroxybenzyl) ether, induces mitochondrial apoptosis in K562 cells and inhibits topoisomerase I in vitro. *Toxicol. Lett.* **2012**, *211*, 126–134. [CrossRef]

77. Jesus, A.; Correia-da-Silva, M.; Afonso, C.; Pinto, M.; Cidade, H. Isolation and potential biological applications of haloaryl secondary metabolites from macroalgae. *Mar. Drugs* **2019**, *17*, 73. [CrossRef]

78. Yoshie-Stark, Y.; Hsieh, Y. Distribution of flavonoids and related compounds from seaweeds in Japan. *Tokyo Univ. Fish.* **2003**, *89*, 1–6.

79. Markham, K.R.; Porter, L.J. Flavonoids in the green algae (chlorophyta). *Phytochemistry* **1969**, *8*, 1777–1781. [CrossRef]

80. Yonekura-Sakakibara, K.; Higashi, Y.; Nakabayashi, R. The origin and evolution of plant flavonoid metabolism. *Front. Plant Sci.* **2019**, *10*, 1–16. [CrossRef]

81. Bowman, J.L.; Kohchi, T.; Yamato, K.T.; Jenkins, J.; Shu, S.; Ishizaki, K.; Yamaoka, S.; Nishihama, R.; Nakamura, Y.; Berger, F.; et al. Insights into land plant evolution garnered from the *Marchantia polymorpha* genome. *Cell* **2017**, *171*, 287–304.e15. [CrossRef] [PubMed]

82. Goiris, K.; Muylaert, K.; Voorspoels, S.; Noten, B.; De Paepe, D.; E Baart, G.J.; De Cooman, L. Detection of flavonoids in microalgae from different evolutionary lineages. *J. Phycol.* **2014**, *50*, 483–492. [CrossRef] [PubMed]

83. Wen, W.; Alseekh, S.; Fernie, A.R. Conservation and diversification of flavonoid metabolism in the plant kingdom. *Curr. Opin. Plant Biol.* **2020**, *55*, 100–108. [CrossRef] [PubMed]

84. Osuna-Ruíz, I.; Salazar-Leyva, J.A.; López-Saiz, C.M.; Burgos-Hernández, A.; Hernández-Garibay, E.; Lizardi-Mendoza, J.; Hurtado-Oliva, M.A. Enhancing antioxidant and antimutagenic activity of the green seaweed *Rhizoclonium riparium* by bioassay-guided solvent partitioning. *J. Appl. Phycol.* **2019**, *31*, 3871–3881. [CrossRef]

85. Kumar, J.G.S.; Umamaheswari, S.; Kavimani, S.; Ilavarasan, R. Pharmacological potential of green algae *Caulerpa*: A review. *Int. J. Pharm. Sci. Res.* **2019**, *10*, 1014. [CrossRef]

86. Haq, S.H.; Al-Ruwaished, G.; Al-Mutlaq, M.A.; Naji, S.A.; Al-Mogren, M.; Al-Rashed, S.; Ain, Q.T.; Al-Amro, A.A.; Al-Mussallam, A. Antioxidant, anticancer activity and phytochemical analysis of green algae, *Chaetomorpha* collected from the Arabian Gulf. *Sci. Rep.* **2019**, *9*, 1–7. [CrossRef]

87. Pangestuti, R.; Getachew, A.T.; Siahaan, E.A.; Chun, B.S. Characterization of functional materials derived from tropical red seaweed *Hypnea musciformis* produced by subcritical water extraction systems. *J. Appl. Phycol.* **2019**, *31*, 2517–2528. [CrossRef]

88. Suganya, S.; Ishwarya, R.; Jayakumar, R.; Govindarajan, M.; Alharbi, N.S.; Kadaikunnan, S.; Khaled, J.M.; Al-anbr, M.N.; Vaseeharan, B. New insecticides and antimicrobials derived from *Sargassum wightii* and *Halimeda gracillis* seaweeds: Toxicity against mosquito vectors and antibiofilm activity against microbial pathogens. *S. Afr. J. Bot.* **2019**, *125*, 466–480. [CrossRef]

89. Ismail, M.M.; Gheda, S.F.; Pereira, L. Variation in bioactive compounds in some seaweeds from Abo Qir bay, Alexandria, Egypt. *Rend. Lincei* **2016**, *27*, 269–279. [CrossRef]

90. Abirami, R.G.; Kowsalya, S. Nutrient and nutraceutical potentials of seaweed biomass *Ulva lactuca* and *Kappaphycus alvarezii*. *Agric. Sci. Technol.* **2011**, *5*, 109.

91. Yoshie, Y.; Wang, W.; Petillo, D.; Suzuki, T. Distribution of catechins in Japanese seaweeds. *Fish. Sci.* **2000**, *66*, 998–1000. [CrossRef]

92. Reddy, P.; Urban, S. Meroditerpenoids from the southern Australian marine brown alga *Sargassum fallax*. *Phytochemistry* **2009**, *70*, 250–255. [CrossRef] [PubMed]

93. Lane, A.L.; Stout, E.P.; Lin, A.S.; Prudhomme, J.; Le Roch, K.; Fairchild, C.R.; Franzblau, S.G.; Hay, M.E.; Aalbersberg, W.; Kubanek, J. Antimalarial bromophyceolides J-Q from the Fijian red alga *Callophycus serratus*. *J. Org. Chem.* **2009**, *74*, 2736–2742. [CrossRef] [PubMed]

94. Perveen, S. Introductory chapter: Terpenes and terpenoids. In *Terpenes and Terpenoids*; Perveen, S., Ed.; IntechOpen: London, UK, 2018.

95. Cardozo, K.H.M.; Guaratini, T.; Barros, M.P.; Falcão, V.R.; Tonon, A.P.; Lopes, N.P.; Campos, S.; Torres, M.A.; Souza, A.O.; Colepicolo, P.; et al. Metabolites from algae with economical impact. *Comp. Biochem. Physiol. C Toxicol. Pharm.* **2007**, *146*, 60–78. [CrossRef] [PubMed]

96. Llewellyn, C.A.; Airs, R.L. Distribution and abundance of MAAs in 33 species of microalgae across 13 classes. *Mar. Drugs* **2010**, *8*, 1273–1291. [CrossRef] [PubMed]

97. Carreto, J.I.; Carignan, M.O. Mycosporine-like amino acids: Relevant secondary metabolites. Chemical and ecological aspects. *Mar. Drugs* **2011**, *9*, 387–446. [CrossRef]

98. Wada, N.; Sakamoto, T.; Matsugo, S. Mycosporine-like amino acids and their derivatives as natural antioxidants. *Antioxidants* **2015**, *4*, 603–646. [CrossRef]

99. Green, D.; Kashman, Y.; Miroz, A. Colpol, a new cytotoxic C6-C4-C6 metabolite from the alga *Colpomenia sinuosa*. *J. Nat. Prod.* **1993**, *56*, 1201–1202. [CrossRef]

100. Ishii, T.; Okino, T.; Suzuki, M.; Machiguchi, Y. Tichocarpols A and B, two novel phenylpropanoids with feeding-deterrent activity from the reel alga *Tichocarpus crinitus*. *J. Nat. Prod.* **2004**, *67*, 1764–1766. [CrossRef]

101. Martone, P.T.; Estevez, J.M.; Lu, F.; Ruel, K.; Denny, M.W.; Somerville, C.; Ralph, J. Discovery of lignin in seaweed reveals convergent evolution of cell-wall architecture. *Curr. Biol.* **2009**, *19*, 169–175. [CrossRef] [PubMed]

102. Duan, X.J.; Li, X.M.; Wang, B.G. Highly brominated mono- and bis-phenols from the marine red alga *Symphyocladia latiuscula* with radical-scavenging activity. *J. Nat. Prod.* **2007**, *70*, 1210–1213. [CrossRef] [PubMed]

103. Machu, L.; Misurcova, L.; Ambrozova, J.V.; Orsavova, J.; Mlcek, J.; Sochor, J.; Jurikova, T. Phenolic content and antioxidant capacity in algal food products. *Molecules* **2015**, *20*, 1118–1133. [CrossRef] [PubMed]

104. Liu, N.; Fu, X.; Duan, D.; Xu, J.; Gao, X.; Zhao, L. Evaluation of bioactivity of phenolic compounds from the brown seaweed of *Sargassum fusiforme* and development of their stable emulsion. *J. Appl. Phycol.* **2018**, *30*, 1955–1970. [CrossRef]

105. D'Archivio, M.; Filesi, C.; Varì, R.; Scazzocchio, B.; Masella, R. Bioavailability of the polyphenols: Status and controversies. *Int. J. Mol. Sci.* **2010**, *11*, 1321–1342. [CrossRef]

106. Stengel, D.B.; Connan, S. Natural products from marine algae: Methods and protocols. *Nat. Prod. Mar. Algae Methods Protoc.* **2015**, *1308*, 1–439. [CrossRef]

107. Pádua, D.; Rocha, E.; Gargiulo, D.; Ramos, A.A. Bioactive compounds from brown seaweeds: Phloroglucinol, fucoxanthin and fucoidan as promising therapeutic agents against breast cancer. *Phytochem. Lett.* **2015**, *14*, 91–98. [CrossRef]

108. Michalak, I. Experimental processing of seaweeds for biofuels. *Wiley Interdiscip. Rev. Energy Env.* **2018**, *7*, 1–25. [CrossRef]

109. Koivikko, R.; Loponen, J.; Pihlaja, K.; Jormalainen, V. High-performance liquid chromatographic analysis of phlorotannins from the brown alga *Fucus vesiculosus*. *Phytochem. Anal.* **2007**, *18*, 326–332. [CrossRef]

110. Onofrejová, L.; Vašíčková, J.; Klejdus, B.; Stratil, P.; Mišurcová, L.; Kráčmar, S.; Kopecký, J.; Vacek, J. Bioactive phenols in algae: The application of pressurized-liquid and solid-phase extraction techniques. *J. Pharm. Biomed. Anal.* **2010**, *51*, 464–470. [CrossRef]

111. Sugiura, Y.; Matsuda, K.; Yamada, Y.; Nishikawa, M.; Shioya, K.; Katsuzaki, H.; Imai, K.; Amano, H. Isolation of a new anti-allergic phlorotannin, phlorofucofuroeckol-B, from an edible brown alga, *Eisenia arborea*. *Biosci. Biotechnol. Biochem.* **2006**, *70*, 2807–2811. [CrossRef] [PubMed]

112. Hartmann, A.; Ganzera, M.; Karsten, U.; Skhirtladze, A.; Stuppner, H. Phytochemical and analytical characterization of novel sulfated coumarins in the marine green macroalga *Dasycladus vermicularis* (Scopoli) krasser. *Molecules* **2018**, *23*, 2753. [CrossRef] [PubMed]

113. Koivikko, R.; Loponen, J.; Honkanen, T.; Jormalainen, V. Contents of soluble, cell-wall-bound and exuded phlorotannins in the brown alga *Fucus vesiculosus*, with implications on their ecological functions. *J. Chem. Ecol.* **2005**, *31*, 195–212. [CrossRef]

114. Michalak, I.; Chojnacka, K. Algal extracts: Technology and advances. *Eng. Life Sci.* **2014**, *14*, 581–591. [CrossRef]

115. Kim, S.M.; Kang, S.W.; Jeon, J.S.; Jung, Y.J.; Kim, W.R.; Kim, C.Y.; Um, B.H. Determination of major phlorotannins in *Eisenia bicyclis* using hydrophilic interaction chromatography: Seasonal variation and extraction characteristics. *Food Chem.* **2013**, *138*, 2399–2406. [CrossRef] [PubMed]

116. Ospina, M.; Castro-Vargas, H.I.; Parada-Alfonso, F. Antioxidant capacity of Colombian seaweeds: 1. extracts obtained from *Gracilaria mammillaris* by means of supercritical fluid extraction. *J. Supercrit. Fluids* **2017**. [CrossRef]

117. Ibañez, E.; Herrero, M.; Mendiola, J.A.; Castro-Puyana, M. Extraction and characterization of bioactive compounds with health benefits from marine resources: Macro and micro algae, cyanobacteria, and invertebrates. In *Marine Bioactive Compounds*; Hayes, M., Ed.; Springer: Boston, MA, USA, 2012; Volume 9781461412, pp. 55–98.

118. Vinatoru, M. An overview of the ultrasonically assisted extraction of bioactive principles from herbs. *Ultrason. Sonochem.* **2001**, *8*, 303–313. [CrossRef]

119. Rajbhar, K.; Dawda, H.; Mukundan, U. Polyphenols: Methods of extraction. *Sci. Revs. Chem. Commun.* **2015**, *5*, 1–6.

120. Yuan, Y.; Zhang, J.; Fan, J.; Clark, J.; Shen, P.; Li, Y.; Zhang, C. Microwave assisted extraction of phenolic compounds from four economic brown macroalgae species and evaluation of their antioxidant activities and inhibitory effects on α-amylase, α-glucosidase, pancreatic lipase and tyrosinase. *Food Res. Int.* **2018**, *113*, 288–297. [CrossRef]

121. Kadam, S.U.; Tiwari, B.K.; O'Donnell, C.P. Application of novel extraction technologies for bioactives from marine algae. *J. Agric. Food Chem.* **2013**, *61*, 4667–4675. [CrossRef]

122. Plaza, M.; Amigo-Benavent, M.; del Castillo, M.D.; Ibáñez, E.; Herrero, M. Facts about the formation of new antioxidants in natural samples after subcritical water extraction. *Food Res. Int.* **2010**, *43*, 2341–2348. [CrossRef]

123. Huie, C.W. A review of modern sample-preparation techniques for the extraction and analysis of medicinal plants. *Anal. Bioanal. Chem.* **2002**, *373*, 23–30. [CrossRef] [PubMed]

124. Díaz-Reinoso, B.; Moure, A.; Domínguez, H.; Parajó, J.C. Supercritical CO2 extraction and purification of compounds with antioxidant activity. *J. Agric. Food Chem.* **2006**, *54*, 2441–2469. [CrossRef] [PubMed]

125. Lee, S.H.; Park, M.H.; Han, J.S.; Jeong, Y.; Kim, M.; Jeon, Y.J. Bioactive compounds extracted from gamtae (*Ecklonia cava*) by using enzymatic hydrolysis, a potent α-glucosidase and α-amylase inhibitor, alleviates postprandial hyperglycemia in diabetic mice. *Food Sci. Biotechnol.* **2012**, *21*, 1149–1155. [CrossRef]

126. Žuvela, P.; Skoczylas, M.; Jay Liu, J.; Bączek, T.; Kaliszan, R.; Wong, M.W.; Buszewski, B. Column characterization and selection systems in reversed-phase high-performance liquid chromatography. *Chem. Rev.* **2019**, *119*, 3674–3729. [CrossRef]

127. Moll, F. Principles of Adsorption Chromatography. The Separation of Nonionic Organic Compounds. Vol. 3, Chromatographic Science Series. VonL. R. Snyder. 424 S. Marcel Dekker, Inc., New York 1968. Preis: $ 17.50. *Arch. Pharm. (Weinh.)* **1969**, *302*, 475. [CrossRef]

128. Snyder, L.R. Role of the solvent in liquid-solid chromatography—A review. *Anal. Chem.* **1974**, *46*, 1384–1393. [CrossRef]

129. Rafferty, J.L.; Zhang, L.; Siepmann, J.I.; Schure, M.R. Retention mechanism in reversed-phase liquid chromatography: A molecular perspective. *Anal. Chem.* **2007**, *79*, 6551–6558. [CrossRef]

130. Lijun, H.; Nianjun, X.; Jiangong, S.; Xiaojun, Y.; Chengkui, Z. Isolation and pharmacological activities of bromophenols from *Rhodomela confervoides*. *Chin. J. Oceanol. Limnol.* **2005**, *23*, 226–229. [CrossRef]

131. Sherma, J.; Fried, B. Thin layer chromatographic analysis of biological samples. A review. *J. Liq. Chromatogr. Relat. Technol.* **2005**, *28*, 2297–2314. [CrossRef]

132. Kim, A.R.; Shin, T.S.; Lee, M.S.; Park, J.Y.; Park, K.E.; Yoon, N.Y.; Kim, J.S.; Choi, J.S.; Jang, B.C.; Byun, D.S.; et al. Isolation and identification of phlorotannins from *Ecklonia stolonifera* with antioxidant and anti-inflammatory properties. *J. Agric. Food Chem.* **2009**, *57*, 3483–3489. [CrossRef] [PubMed]

133. Ge, L.; Li, S.P.; Lisak, G. Advanced sensing technologies of phenolic compounds for pharmaceutical and biomedical analysis. *J. Pharm. Biomed. Anal.* **2020**, *179*, 112913. [CrossRef] [PubMed]

134. Olate-Gallegos, C.; Barriga, A.; Vergara, C.; Fredes, C.; García, P.; Giménez, B.; Robert, P. Identification of polyphenols from chilean brown seaweeds extracts by LC-DAD-ESI-MS/MS. *J. Aquat. Food Prod. Technol.* **2019**, *28*, 375–391. [CrossRef]

135. Maheswari, M.U.; Reena, A.; Sivaraj, C. Analysis, antioxidant and antibacterial activity of the brown algae, *Padina tetrastromatica*. *Int. J. Pharm. Sci. Res* **2018**, *9*, 298–304.

136. Principles of MALDI-TOF Mass Spectrometry: SHIMADZU (Shimadzu Corporation). Available online: https://www.shimadzu.com/an/lifescience/maldi/princpl1.html (accessed on 14 April 2020).

137. Karthik, R.; Manigandan, V.; Sheeba, R.; Saravanan, R.; Rajesh, P.R. Structural characterization and comparative biomedical properties of phloroglucinol from Indian brown seaweeds. *J. Appl. Phycol.* **2016**, *28*, 3561–3573. [CrossRef]

138. Lopes, G.; Barbosa, M.; Vallejo, F.; Gil-Izquierdo, Á.; Andrade, P.B.; Valentão, P.; Pereira, D.M.; Ferreres, F. Profiling phlorotannins from Fucus spp. of the Northern Portuguese coastline: Chemical approach by HPLC-DAD-ESI/MS and UPLC-ESI-QTOF/MS. *Algal Res.* **2018**, *29*, 113–120. [CrossRef]

139. Nerantzaki, A.A.; Tsiafoulis, C.G.; Charisiadis, P.; Kontogianni, V.G.; Gerothanassis, I.P. Novel determination of the total phenolic content in crude plant extracts by the use of 1H NMR of the -OH spectral region. *Anal. Chim. Acta* **2011**, *688*, 54–60. [CrossRef]

140. Pauli, G.F.; Jaki, B.U.; Lankin, D.C. Quantitative 1H NMR: Development and potential of a method for natural products analysis. *J. Nat. Prod.* **2005**, *68*, 133–149. [CrossRef]

141. Blümich, B.; Singh, K. Desktop NMR and its applications from materials science to organic chemistry. *Angew. Chem. Int. Ed.* **2018**, *57*, 6996–7010. [CrossRef]

142. Wekre, M.E.; Kåsin, K.; Underhaug, J.; Jordheim, M.; Holmelid, B. Quantification of polyphenols in seaweeds: A case study of *Ulva intestinalis*. *Antioxidants* **2019**, *8*, 612. [CrossRef]

143. Farasat, M.; Khavari-Nejad, R.A.; Nabavi, S.M.B.; Namjooyan, F. Antioxidant activity, total phenolics and flavonoid contents of some edible green seaweeds from northern coasts of the Persian Gulf. *Iran. J. Pharm. Res.* **2014**, *13*, 163–170. [CrossRef] [PubMed]

144. Cho, M.; Kang, I.-J.; Won, M.-H.; Lee, H.-S.; You, S. The antioxidant properties of ethanol extracts and their solvent-partitioned fractions from various green seaweeds. *J. Med. Food* **2010**, *13*, 1232–1239. [CrossRef] [PubMed]

145. Tang, H.; Inoue, M.; Uzawa, Y.; Kawamura, Y. Anti-tumorigenic components of a sea weed, *Enteromorpha clathrata*. *BioFactors* **2004**, *22*, 107–110. [CrossRef] [PubMed]

146. Khanavi, M.; Gheidarloo, R.; Sadati, N.; Shams Ardekani, M.R.; Bagher Nabavi, S.M.; Tavajohi, S.; Ostad, S.N. Cytotoxicity of fucosterol containing fraction of marine algae against breast and colon carcinoma cell line. *Pharm. Mag.* **2012**, *8*, 60–64. [CrossRef]

147. Lavoie, S.; Sweeney-Jones, A.M.; Mojib, N.; Dale, B.; Gagaring, K.; McNamara, C.W.; Quave, C.L.; Soapi, K.; Kubanek, J. Antibacterial oligomeric polyphenols from the green alga *Cladophora socialis*. *J. Org. Chem.* **2019**, *84*, 5035–5045. [CrossRef]

148. Pereira, L. Portuguese Seaweeds Website (MACOI). Available online: http://www.flordeutopia.pt/macoi/default.php (accessed on 7 June 2020).

149. Carte, B.K.; Troupe, N.; Chan, J.A.; Westley, J.W.; Faulkner, D.J. Rawsonol, an inhibitor of HMG-CoA reductase from the tropical green alga *Avrainvillea rawsoni*. *Phytochemistry* **1989**, *28*, 2917–2919. [CrossRef]

150. Estrada, D.M.; Martin, J.D.; Perez, C. A new brominated monoterpenoid quinol from *Cymopolm barbata*. *J. Nat. Prod.* **1987**, *50*, 735–737. [CrossRef]

151. Yan, X.; Yang, C.; Lin, G.; Chen, Y.; Miao, S.; Liu, B.; Zhao, C. Antidiabetic potential of green seaweed *Enteromorpha prolifera* flavonoids regulating insulin signaling pathway and gut microbiota in type 2 diabetic mice. *J. Food Sci.* **2019**, *84*, 165–173. [CrossRef]

152. Lin, G.; Liu, X.; Yan, X.; Liu, D.; Yang, C.; Liu, B.; Huang, Y.; Zhao, C. Role of green macroalgae *Enteromorpha prolifera* polyphenols in the modulation of gene expression and intestinal microflora profiles in type 2 diabetic mice. *Int. J. Mol. Sci.* **2018**, *20*, 25. [CrossRef]

153. Tanna, B.; Choudhary, B.; Mishra, A. Metabolite profiling, antioxidant, scavenging and anti-proliferative activities of selected tropical green seaweeds reveal the nutraceutical potential of *Caulerpa* spp. *Algal Res.* **2018**, *36*, 96–105. [CrossRef]

154. Torres, P.; Santos, J.P.; Chow, F.; dos Santos, D.Y.A.C. A comprehensive review of traditional uses, bioactivity potential, and chemical diversity of the genus *Gracilaria* (Gracilariales, Rhodophyta). *Algal Res.* **2019**, *37*, 288–306. [CrossRef]

155. Namvar, F.; Mohamed, S.; Fard, S.G.; Behravan, J.; Mustapha, N.M.; Alitheen, N.B.M.; Othman, F. Polyphenol-rich seaweed (*Eucheuma cottonii*) extract suppresses breast tumour via hormone modulation and apoptosis induction. *Food Chem.* **2012**, *130*, 376–382. [CrossRef]

156. Choi, J.S.; Park, H.J.; Jung, H.A.; Chung, H.Y.; Jung, J.H.; Choi, W.C. A cyclohexanonyl bromophenol from the red alga *Symphyocladia latiuscula*. *J. Nat. Prod.* **2000**, *63*, 1705–1706. [CrossRef]

157. Popplewell, W.L.; Northcote, P.T. Colensolide A: A new nitrogenous bromophenol from the New Zealand marine red alga *Osmundaria colensoi*. *Tetrahedron Lett.* **2009**, *50*, 6814–6817. [CrossRef]

158. Qi, X.; Liu, G.; Qiu, L.; Lin, X.; Liu, M. Marine bromophenol bis (2,3-dibromo-4,5-dihydroxybenzyl) ether, represses angiogenesis in HUVEC cells and in zebrafish embryos via inhibiting the VEGF signal systems. *Biomed. Pharm.* **2015**, *75*, 58–66. [CrossRef] [PubMed]

159. Wu, N.; Luo, J.; Jiang, B.; Wang, L.; Wang, S.; Wang, C.; Fu, C.; Li, J.; Shi, D. Marine bromophenol bis (2,3-dibromo-4,5-dihydroxy-phenyl)-methane inhibits the proliferation, migration, and invasion of hepatocellular carcinoma cells via modulating β1-integrin/FAK signaling. *Mar. Drugs* **2015**, *13*, 1010–1025. [CrossRef]

160. Ma, M.; Zhao, J.; Wang, S.; Li, S.; Yang, Y.; Shi, J.; Fan, X.; He, L. Bromophenols coupled with methyl γ-ureidobutyrate and bromophenol sulfates from the red alga *Rhodomela confervoides*. *J. Nat. Prod.* **2006**, *69*, 206–210. [CrossRef]

161. Paudel, P.; Seong, S.; Park, H.; Jung, H.; Choi, J. Anti-Diabetic activity of 2,3,6-tribromo-4,5-dihydroxybenzyl derivatives from *Symphyocladia latiuscula* through PTP1B downregulation and α-glucosidase inhibition. *Mar. Drugs* **2019**, *17*, 166. [CrossRef]

162. Wang, W.; Okada, Y.; Shi, H.; Wang, Y.; Okuyama, T. Structures and aldose reductase inhibitory effects of bromophenols from the red alga *Symphyocladia latiuscula*. *J. Nat. Prod.* **2005**, *68*, 620–622. [CrossRef]

163. Shi, D.; Xu, F.; He, J.; Li, J.; Fan, X.; Han, L. Inhibition of bromophenols against PTP1B and anti-hyperglycemic effect of *Rhodomela confervoides* extract in diabetic rats. *Chin. Sci. Bull.* **2008**, *53*, 2476–2479. [CrossRef]

164. Mikami, D.; Kurihara, H.; Ono, M.; Kim, S.M.; Takahashi, K. Inhibition of algal bromophenols and their related phenols against glucose 6-phosphate dehydrogenase. *Fitoterapia* **2016**, *108*, 20–25. [CrossRef] [PubMed]

165. Xu, N.; Fan, X.; Yan, X.; Li, X.; Niu, R.; Tseng, C.K. Antibacterial bromophenols from the marine red alga *Rhodomela confervoides*. *Phytochemistry* **2003**, *62*, 1221–1224. [CrossRef]
166. Barreto, M.; Meyer, J.J.M. Isolation and antimicrobial activity of a lanosol derivative from *Osmundaria serrata* (Rhodophyta) and a visual exploration of its biofilm covering. *S. Afr. J. Bot.* **2006**, *72*, 521–528. [CrossRef]
167. Liu, M.; Wang, G.; Xiao, L.; Xu, X.; Liu, X.; Xu, P.; Lin, X. Bis (2,3-dibromo-4,5-dihydroxybenzyl) ether, a marine algae derived bromophenol, inhibits the growth of *botrytis cinerea* and interacts with DNA molecules. *Mar. Drugs* **2014**, *12*, 3838–3851. [CrossRef]
168. Park, H.-J.; Kurokawa, M.; Shiraki, K.; Nakamura, N.; Choi, J.-S.; Hattori, M. Antiviral activity of the marine alga *Symphyocladia latiuscula* against *Herpes simplex* virus (HSV-1) in Vitro and its therapeutic efficacy against HSV-1 infection in mice. *Biol. Pharm. Bull.* **2005**, *28*, 2258–2262. [CrossRef]
169. Kim, S.Y.; Kim, S.R.; Oh, M.J.; Jung, S.J.; Kang, S.Y. In Vitro antiviral activity of red alga, *Polysiphonia morrowii* extract and its bromophenols against fish pathogenic infectious hematopoietic necrosis virus and infectious pancreatic necrosis virus. *J. Microbiol.* **2011**, *49*, 102–106. [CrossRef] [PubMed]
170. Lee, H.S.; Lee, T.H.; Ji, H.L.; Chae, C.S.; Chung, S.C.; Shin, D.S.; Shin, J.; Oh, K.B. Inhibition of the pathogenicity of *Magnaporthe grisea* by bromophenols, isocitrate lyase inhibitors, from the red alga *Odonthalia corymbifera*. *J. Agric. Food Chem.* **2007**, *55*, 6923–6928. [CrossRef]
171. Jeyaprakash, R.R.K. HPLC Analysis of flavonoids in acanthophora specifera (red seaweed) collected from Gulf of Mannar, Tamilnadu, India. *Int. J. Sci. Res.* **2017**, *6*, 69–72.
172. Bilanglod, A.; Petthongkhao, S.; Churngchow, N. Phenolic and flavonoid contents isolated from the red seaweed, *Acanthophora spicifera*. In Proceedings of the 31st Annual Meeting of the Thai Society for Biotechnology and International Conference (TSB2019), Phuket, Thailand, 10–12 November 2019; pp. 189–196.
173. Ben Saad, H.; Gargouri, M.; Kallel, F.; Chaabene, R.; Boudawara, T.; Jamoussi, K.; Magné, C.; Mounir Zeghal, K.; Hakim, A.; Ben Amara, I. Flavonoid compounds from the red marine alga *Alsidium corallinum* protect against potassium bromate-induced nephrotoxicity in adult mice. *Env. Toxicol.* **2017**, *32*, 1475–1486. [CrossRef]
174. Davyt, D.; Fernandez, R.; Suescun, L.; Mombrú, A.W.; Saldaña, J.; Domínguez, L.; Coll, J.; Fujii, M.T.; Manta, E. New sesquiterpene derivatives from the red alga *Laurencia scoparia*. Isolation, structure determination, and anthelmintic activity. *J. Nat. Prod.* **2001**, *64*, 1552–1555. [CrossRef]
175. Stout, E.P.; Prudhomme, J.; Le Roch, K.; Fairchild, C.R.; Franzblau, S.G.; Aalbersberg, W.; Hay, M.E.; Kubanek, J. Unusual antimalarial meroditerpenes from tropical red macroalgae. *Bioorg. Med. Chem. Lett.* **2010**, *20*, 5662–5665. [CrossRef] [PubMed]
176. Makkar, F.; Chakraborty, K. Highly oxygenated antioxidative 2H-chromen derivative from the red seaweed *Gracilaria opuntia* with pro-inflammatory cyclooxygenase and lipoxygenase inhibitory properties. *Nat. Prod. Res.* **2018**, *32*, 2756–2765. [CrossRef] [PubMed]
177. Guihéneuf, F.; Gietl, A.; Stengel, D.B. Temporal and spatial variability of mycosporine-like amino acids and pigments in three edible red seaweeds from western Ireland. *J. Appl. Phycol.* **2018**, *30*, 2573–2586. [CrossRef]
178. Torres, P.; Santos, J.P.; Chow, F.; Pena Ferreira, M.J.; dos Santos, D.Y.A.C. Comparative analysis of in vitro antioxidant capacities of mycosporine-like amino acids (MAAs). *Algal Res.* **2018**, *34*, 57–67. [CrossRef]
179. Athukorala, Y.; Trang, S.; Kwok, C.; Yuan, Y. Antiproliferative and antioxidant activities and mycosporine-like amino acid profiles of wild-harvested and cultivated edible canadian marine red macroalgae. *Molecules* **2016**, *21*, 119. [CrossRef]
180. Yuan, Y.V.; Westcott, N.D.; Hu, C.; Kitts, D.D. Mycosporine-like amino acid composition of the edible red alga, *Palmaria palmata* (dulse) harvested from the west and east coasts of Grand Manan Island, New Brunswick. *Food Chem.* **2009**, *112*, 321–328. [CrossRef]
181. Reef, R.; Kaniewska, P.; Hoegh-Guldberg, O. Coral skeletons defend against ultraviolet radiation. *PLoS ONE* **2009**, *4*, e7995. [CrossRef]
182. Bedoux, G.; Hardouin, K.; Burlot, A.S.; Bourgougnon, N. Bioactive components from seaweeds. In *Advances in Botanical Research*; Bourgougnon, N., Ed.; Academic Press Inc.: Cambridge, MA, USA, 2014; Volume 71, pp. 345–378.
183. Ying, R.; Zhang, Z.; Zhu, H.; Li, B.; Hou, H. The protective effect of mycosporine-like amino acids (MAAs) from *Porphyra yezoensis* in a mouse model of UV irradiation-induced photoaging. *Mar. Drugs* **2019**, *17*, 470. [CrossRef]

184. Suh, S.S.; Oh, S.K.; Lee, S.G.; Kim, I.C.; Kim, S. Porphyra-334, a mycosporine-like amino acid, attenuates UV-induced apoptosis in HaCaT cells. *Acta Pharm.* **2017**, *67*, 257–264. [CrossRef]

185. Lawrence, K.P.; Long, P.F.; Young, A.R. Mycosporine-Like Amino Acids for Skin Photoprotection. *Curr. Med. Chem.* **2017**, *25*, 5512–5527. [CrossRef]

186. Orfanoudaki, M.; Hartmann, A.; Alilou, M.; Gelbrich, T.; Planchenault, P.; Derbré, S.; Schinkovitz, A.; Richomme, P.; Hensel, A.; Ganzera, M. Absolute configuration of mycosporine-like amino acids, their wound healing properties and in vitro anti-aging effects. *Mar. Drugs* **2020**, *18*, 35. [CrossRef] [PubMed]

187. Becker, K.; Hartmann, A.; Ganzera, M.; Fuchs, D.; Gostner, J.M. Immunomodulatory effects of the mycosporine-like amino acids shinorine and porphyra-334. *Mar. Drugs* **2016**, *14*, 119. [CrossRef] [PubMed]

188. Kang, K.A.; Lee, K.H.; Chae, S.; Koh, Y.S.; Yoo, B.S.; Kim, J.H.; Ham, Y.M.; Baik, J.S.; Lee, N.H.; Hyun, J.W. Triphlorethol-A from *Ecklonia cava* protects V79-4 lung fibroblast against hydrogen peroxide induced cell damage. *Free Radic. Res.* **2005**, *39*, 883–892. [CrossRef] [PubMed]

189. Béress, A.; Wassermann, O.; Bruhn, T.; Béress, L.; Kraiselburd, E.N.; Gonzalez, L.V.; de Motta, G.E.; Chavez, P.I. A new procedure for the isolation of anti-HIV compounds (polysaccharides and polyphenols) from the marine alga *Fucus vesiculosus*. *J. Nat. Prod.* **1993**, *56*, 478–488. [CrossRef]

190. Eom, S.-H.; Moon, S.-Y.; Lee, D.-S.; Kim, H.-J.; Park, K.; Lee, E.-W.; Kim, T.H.; Chung, Y.-H.; Lee, M.-S.; Kim, Y.-M. In vitro antiviral activity of dieckol and phlorofucofuroeckol-A isolated from edible brown alga *Eisenia bicyclis* against murine norovirus. *ALGAE* **2015**, *30*, 241–246. [CrossRef]

191. Choi, Y.; Kim, E.; Moon, S.; Choi, J.-D.; Lee, M.-S.; Kim, Y.-M. Phaeophyta Extracts Exhibit Antiviral Activity against Feline Calicivirus. *Fish. Aquat. Sci.* **2014**, *17*, 155–158. [CrossRef]

192. Shibata, T.; Ishimaru, K.; Kawaguchi, S.; Yoshikawa, H.; Hama, Y. Antioxidant activities of phlorotannins isolated from Japanese Laminariaceae. *J. Appl. Phycol.* **2008**, *20*, 705–711. [CrossRef]

193. Besednova, N.N.; Zvyagintseva, T.N.; Kuznetsova, T.A.; Makarenkova, I.D.; Smolina, T.P.; Fedyanina, L.N.; Kryzhanovsky, S.P.; Zaporozhets, T.S. Marine Algae Metabolites as Promising Therapeutics for the Prevention and Treatment of HIV/AIDS. *Metabolites* **2019**, *9*, 87. [CrossRef]

194. Kim, A.R.; Lee, M.S.; Shin, T.S.; Hua, H.; Jang, B.C.; Choi, J.S.; Byun, D.S.; Utsuki, T.; Ingram, D.; Kim, H.R. Phlorofucofuroeckol A inhibits the LPS-stimulated iNOS and COX-2 expressions in macrophages via inhibition of NF-κB, Akt, and p38 MAPK. *Toxicol. Vitr.* **2011**, *25*, 1789–1795. [CrossRef]

195. Chang, M.Y.; Byon, S.H.; Shin, H.C.; Han, S.E.; Kim, J.Y.; Byun, J.Y.; Lee, J.D.; Park, M.K. Protective effects of the seaweed phlorotannin polyphenolic compound dieckol on gentamicin-induced damage in auditory hair cells. *Int. J. Pediatr. Otorhinolaryngol.* **2016**, *83*, 31–36. [CrossRef]

196. Kong, C.S.; Kim, J.A.; Yoon, N.Y.; Kim, S.K. Induction of apoptosis by phloroglucinol derivative from *Ecklonia Cava* in MCF-7 human breast cancer cells. *Food Chem. Toxicol.* **2009**, *47*, 1653–1658. [CrossRef] [PubMed]

197. Kim, E.-K.; Tang, Y.; Kim, Y.-S.; Hwang, J.-W.; Choi, E.-J.; Lee, J.-H.; Lee, S.-H.; Jeon, Y.-J.; Park, P.-J. First Evidence that *Ecklonia cava*-Derived Dieckol Attenuates MCF-7 Human Breast Carcinoma Cell Migration. *Mar. Drugs* **2015**, *13*, 1785–1797. [CrossRef] [PubMed]

198. Ahn, J.H.; Yang, Y.I.; Lee, K.T.; Choi, J.H. Dieckol, isolated from the edible brown algae *Ecklonia cava*, induces apoptosis of ovarian cancer cells and inhibits tumor xenograft growth. *J. Cancer Res. Clin. Oncol.* **2014**, *141*, 255–268. [CrossRef] [PubMed]

199. Lee, Y.J.; Park, J.H.; Park, S.A.; Joo, N.R.; Lee, B.H.; Lee, K.B.; Oh, S.M. Dieckol or phlorofucofuroeckol extracted from *Ecklonia cava* suppresses lipopolysaccharide-mediated human breast cancer cell migration and invasion. *J. Appl. Phycol.* **2020**, *32*, 631–640. [CrossRef]

200. LI, Y.; Qian, Z.-J.; Kim, M.-M.; Kim, S.-K. Cytotoxic activities of phlorethol and fucophlorethol derivates isolated from Laminariaceae *Ecklonia cava*. *J. Food Biochem.* **2011**, *35*, 357–369. [CrossRef]

201. Wang, C.; Li, X.; Jin, L.; Zhao, Y.; Zhu, G.; Shen, W. Dieckol inhibits non-small–cell lung cancer cell proliferation and migration by regulating the PI3K/AKT signaling pathway. *J. Biochem. Mol. Toxicol.* **2019**, *33*. [CrossRef]

202. Zhang, M.; Zhou, W.; Zhao, S.; Li, S.; Yan, D.; Wang, J. Eckol inhibits Reg3A-induced proliferation of human SW1990 pancreatic cancer cells. *Exp. Med.* **2019**, *18*, 2825–2832. [CrossRef]

203. Yoon, J.S.; Kasin Yadunandam, A.; Kim, S.J.; Woo, H.C.; Kim, H.R.; Kim, G. Do Dieckol, isolated from *Ecklonia stolonifera*, induces apoptosis in human hepatocellular carcinoma Hep3B cells. *J. Nat. Med.* **2013**, *67*, 519–527. [CrossRef]

204. Moon, C.; Kim, S.-H.; Kim, J.-C.; Hyun, J.W.; Lee, N.H.; Park, J.W.; Shin, T. Protective effect of phlorotannin components phloroglucinol and eckol on radiation-induced intestinal injury in mice. *Phyther. Res.* **2008**, *22*, 238–242. [CrossRef]

205. Park, E.; Lee, N.H.; Joo, H.G.; Jee, Y. Modulation of apoptosis of eckol against ionizing radiation in mice. *Biochem. Biophys. Res. Commun.* **2008**, *372*, 792–797. [CrossRef]

206. Zhang, R.; Kang, K.A.; Piao, M.J.; Ko, D.O.; Wang, Z.H.; Lee, I.K.; Kim, B.J.; Jeong, I.Y.; Shin, T.; Park, J.W.; et al. Eckol protects V79-4 lung fibroblast cells against γ-ray radiation-induced apoptosis via the scavenging of reactive oxygen species and inhibiting of the c-Jun NH2-terminal kinase pathway. *Eur. J. Pharm.* **2008**, *591*, 114–123. [CrossRef] [PubMed]

207. Park, E.; Ahn, G.; Yun, J.S.; Kim, M.J.; Bing, S.J.; Kim, D.S.; Lee, J.; Lee, N.H.; Park, J.W.; Jee, Y. Dieckol rescues mice from lethal irradiation by accelerating hemopoiesis and curtailing immunosuppression. *Int. J. Radiat. Biol.* **2010**, *86*, 848–859. [CrossRef] [PubMed]

208. Kang, K.A.; Zhang, R.; Lee, K.H.; Chae, S.; Kim, B.J.; Kwak, Y.S.; Park, J.W.; Lee, N.H.; Hyun, J.W. Protective Effect of Triphlorethol-A from *Ecklonia cava* against Ionizing Radiation in vitro. *J. Radiat. Res.* **2006**, *47*, 61–68. [CrossRef] [PubMed]

209. Yuan, Y.; Zheng, Y.; Zhou, J.; Geng, Y.; Zou, P.; Li, Y.; Zhang, C. Polyphenol-Rich Extracts from Brown Macroalgae *Lessonia trabeculata* Attenuate Hyperglycemia and Modulate Gut Microbiota in High-Fat Diet and Streptozotocin-Induced Diabetic Rats. *J. Agric. Food Chem.* **2019**, *67*, 12472–12480. [CrossRef] [PubMed]

210. Lopes, G.; Andrade, P.; Valentão, P. Phlorotannins: Towards New Pharmacological Interventions for Diabetes Mellitus Type 2. *Molecules* **2016**, *22*, 56. [CrossRef] [PubMed]

211. Kellogg, J.; Grace, M.; Lila, M. Phlorotannins from Alaskan Seaweed Inhibit Carbolytic Enzyme Activity. *Mar. Drugs* **2014**, *12*, 5277–5294. [CrossRef]

212. Husni, A.; Wijayanti, R. Ustadi Inhibitory activity of α-Amylase and α-Glucosidase by *Padina pavonica* extracts. *J. Biol. Sci.* **2014**, *14*, 515–520. [CrossRef]

213. Lee, S.H.; Park, M.H.; Heo, S.J.; Kang, S.M.; Ko, S.C.; Han, J.S.; Jeon, Y.J. Dieckol isolated from *Ecklonia cava* inhibits α-glucosidase and α-amylase in vitro and alleviates postprandial hyperglycemia in streptozotocin-induced diabetic mice. *Food Chem. Toxicol.* **2010**, *48*, 2633–2637. [CrossRef]

214. Lee, S.-H.; Kim, S.-K. Biological Phlorotannins of Eisenia bicyclis. In *Marine Algae Extracts*; Kim, S., Chojnacka, K., Eds.; Wiley-VCH Verlag GmbH & Co. KGaA: Weinheim, Germany, 2015; Volume 2, pp. 453–464.

215. Moon, H.E.; Islam, M.N.; Ahn, B.R.; Chowdhury, S.S.; Sohn, H.S.; Jung, H.A.; Choi, J.S. Protein tyrosine phosphatase 1B and α-glucosidase inhibitory phlorotannins from edible brown algae, *Ecklonia stolonifera* and *Eisenia bicyclis*. *Biosci. Biotechnol. Biochem.* **2011**, *75*, 1472–1480. [CrossRef]

216. Lee, S.H.; Ko, S.C.; Kang, M.C.; Lee, D.H.; Jeon, Y.J. Octaphlorethol A, a marine algae product, exhibits antidiabetic effects in type 2 diabetic mice by activating AMP-activated protein kinase and upregulating the expression of glucose transporter 4. *Food Chem. Toxicol.* **2016**, *91*, 58–64. [CrossRef]

217. Kawamura-Konishi, Y.; Watanabe, N.; Saito, M.; Nakajima, N.; Sakaki, T.; Katayama, T.; Enomoto, T. Isolation of a new phlorotannin, a potent inhibitor of carbohydrate-hydrolyzing enzymes, from the brown alga *Sargassum patens*. *J. Agric. Food Chem.* **2012**, *60*, 5565–5570. [CrossRef] [PubMed]

218. Oh, S.; Son, M.; Choi, J.; Choi, C.H.; Park, K.Y.; Son, K.H.; Byun, K. Phlorotannins from *Ecklonia cava* Attenuates Palmitate-Induced Endoplasmic Reticulum Stress and Leptin Resistance in Hypothalamic Neurons. *Mar. Drugs* **2019**, *17*, 570. [CrossRef] [PubMed]

219. Park, M.H.; Heo, S.J.; Park, P.J.; Moon, S.H.; Sung, S.H.; Jeon, B.T.; Lee, S.H. 6,6'-Bieckol isolated from *Ecklonia cava* protects oxidative stress through inhibiting expression of ROS and proinflammatory enzymes in high-glucose-induced human umbilical vein endothelial cells. *Appl. Biochem. Biotechnol.* **2014**, *174*, 632–643. [CrossRef] [PubMed]

220. Park, M.H.; Heo, S.J.; Kim, K.N.; Ahn, G.; Park, P.J.; Moon, S.H.; Jeon, B.T.; Lee, S.H. 6,6'-Bieckol protects insulinoma cells against high glucose-induced glucotoxicity by reducing oxidative stress and apoptosis. *Fitoterapia* **2015**, *106*, 135–140. [CrossRef] [PubMed]

221. Ahn, M.J.; Yoon, K.D.; Min, S.Y.; Lee, J.S.; Kim, J.H.; Kim, T.G.; Kim, S.H.; Kim, N.G.; Huh, H.; Kim, J. Inhibition of HIV-1 reverse transcriptase and protease by phlorotannins from the brown alga *Ecklonia cava*. *Biol. Pharm. Bull.* **2004**. [CrossRef]

222. Artan, M.; Li, Y.; Karadeniz, F.; Lee, S.H.; Kim, M.M.; Kim, S.K. Anti-HIV-1 activity of phloroglucinol derivative, 6,6′-bieckol, from *Ecklonia cava*. *Bioorg. Med. Chem.* **2008**, *16*, 7921–7926. [CrossRef] [PubMed]

223. Karadeniz, F.; Kang, K.-H.; Park, J.W.; Park, S.-J.; Kim, S.-K. Anti-HIV-1 activity of phlorotannin derivative 8,4′′′-dieckol from Korean brown alga *Ecklonia cava*. *Biosci. Biotechnol. Biochem.* **2014**, *78*, 1151–1158. [CrossRef]

224. Park, J.Y.; Kim, J.H.; Kwon, J.M.; Kwon, H.J.; Jeong, H.J.; Kim, Y.M.; Kim, D.; Lee, W.S.; Ryu, Y.B. Dieckol, a SARS-CoV 3CLpro inhibitor, isolated from the edible brown algae *Ecklonia cava*. *Bioorg. Med. Chem.* **2013**, *21*, 3730–3737. [CrossRef]

225. Ryu, Y.B.; Jeong, H.J.; Yoon, S.Y.; Park, J.Y.; Kim, Y.M.; Park, S.J.; Rho, M.C.; Kim, S.J.; Lee, W.S. Influenza virus neuraminidase inhibitory activity of phlorotannins from the edible brown alga *Ecklonia cava*. *J. Agric. Food Chem.* **2011**, *59*, 6467–6473. [CrossRef]

226. Gentile, D.; Patamia, V.; Scala, A.; Sciortino, M.T.; Piperno, A.; Rescifina, A. Putative Inhibitors of SARS-CoV-2 Main Protease from A Library of Marine Natural Products: A Virtual Screening and Molecular Modeling Study. *Mar. Drugs* **2020**, *18*, 225. [CrossRef]

227. Kim, E.; Kwak, J. Antiviral phlorotannin from *Eisenia bicyclis* against human papilloma virus in vitro. *Planta Med.* **2015**, *81*, PW_22. [CrossRef]

228. Lee, M.H.; Lee, K.B.; Oh, S.M.; Lee, B.H.; Chee, H.Y. Antifungal activities of dieckol isolated from the marine brown alga *Ecklonia cava* against *Trichophyton rubrum*. *J. Appl. Biol. Chem.* **2010**, *53*, 504–507. [CrossRef]

229. Choi, J.-G.; Kang, O.-H.; Brice, O.-O.; Lee, Y.-S.; Chae, H.-S.; Oh, Y.-C.; Sohn, D.-H.; Park, H.; Choi, H.-G.; Kim, S.-G.; et al. Antibacterial Activity of *Ecklonia cava* Against Methicillin-Resistant *Staphylococcus aureus* and *Salmonella* spp. *Foodborne Pathog. Dis.* **2010**, *7*, 435–441. [CrossRef]

230. Lee, D.S.; Kang, M.S.; Hwang, H.J.; Eom, S.H.; Yang, J.Y.; Lee, M.S.; Lee, W.J.; Jeon, Y.J.; Choi, J.S.; Kim, Y.M. Synergistic effect between dieckol from *Ecklonia stolonifera* and β-lactams against methicillin-resistant *Staphylococcus aureus*. *Biotechnol. Bioprocess Eng.* **2008**, *13*, 758–764. [CrossRef]

231. Nagayama, K.; Iwamura, Y.; Shibata, T.; Hirayama, I.; Nakamura, T. Bactericidal activity of phlorotannins from the brown alga *Ecklonia kurome*. *J. Antimicrob. Chemother.* **2002**, *50*, 889–893. [CrossRef] [PubMed]

232. Ryu, B.M.; Li, Y.; Qian, Z.J.; Kim, M.M.; Kim, S.K. Differentiation of human osteosarcoma cells by isolated phlorotannins is subtly linked to COX-2, iNOS, MMPs, and MAPK signaling: Implication for chronic articular disease. *Chem. Biol. Interact.* **2009**, *179*, 192–201. [CrossRef] [PubMed]

233. Oh, J.H.; Ahn, B.-N.; Karadeniz, F.; Kim, J.-A.; Lee, J.I.; Seo, Y.; Kong, C.-S. Phlorofucofuroeckol A from Edible Brown Alga *Ecklonia Cava* Enhances Osteoblastogenesis in Bone Marrow-Derived Human Mesenchymal Stem Cells. *Mar. Drugs* **2019**, *17*, 543. [CrossRef] [PubMed]

234. Lee, S.; Youn, K.; Kim, D.; Ahn, M.-R.; Yoon, E.; Kim, O.-Y.; Jun, M. Anti-Neuroinflammatory Property of Phlorotannins from *Ecklonia cava* on Aβ25-35-Induced Damage in PC12 Cells. *Mar. Drugs* **2018**, *17*, 7. [CrossRef]

235. Jung, H.A.; Kim, J.-I.; Choung, S.Y.; Choi, J.S. Protective effect of the edible brown alga *Ecklonia stolonifera* on doxorubicin-induced hepatotoxicity in primary rat hepatocytes. *J. Pharm. Pharm.* **2014**, *66*, 1180–1188. [CrossRef]

236. Lee, J.; Jun, M. Dual BACE1 and Cholinesterase Inhibitory Effects of Phlorotannins from *Ecklonia cava* —An In Vitro and In Silico Study. *Mar. Drugs* **2019**, *17*, 91. [CrossRef]

237. Wang, J.; Zheng, J.; Huang, C.; Zhao, J.; Lin, J.; Zhou, X.; Naman, C.B.; Wang, N.; Gerwick, W.H.; Wang, Q.; et al. Eckmaxol, a Phlorotannin Extracted from *Ecklonia maxima*, Produces Anti-β-amyloid Oligomer Neuroprotective Effects Possibly via Directly Acting on Glycogen Synthase Kinase 3β. *Acs Chem. Neurosci.* **2018**, *9*, 1349–1356. [CrossRef] [PubMed]

238. Seong, S.H.; Paudel, P.; Choi, J.-W.; Ahn, D.H.; Nam, T.-J.; Jung, H.A.; Choi, J.S. Probing Multi-Target Action of Phlorotannins as New Monoamine Oxidase Inhibitors and Dopaminergic Receptor Modulators with the Potential for Treatment of Neuronal Disorders. *Mar. Drugs* **2019**, *17*, 377. [CrossRef] [PubMed]

239. Shim, S.Y.; Choi, J.S.; Byun, D.S. Inhibitory effects of phloroglucinol derivatives isolated from *Ecklonia stolonifera* on FcεRI expression. *Bioorg. Med. Chem.* **2009**, *17*, 4734–4739. [CrossRef] [PubMed]

240. Ahn, G.; Amagai, Y.; Matsuda, A.; Kang, S.-M.; Lee, W.; Jung, K.; Oida, K.; Jang, H.; Ishizaka, S.; Matsuda, K.; et al. Dieckol, a phlorotannin of *Ecklonia cava*, suppresses IgE-mediated mast cell activation and passive cutaneous anaphylactic reaction. *Exp. Derm.* **2015**, *24*, 968–970. [CrossRef]

241. Lee, S.Y.; Lee, J.; Lee, H.; Kim, B.; Lew, J.; Baek, N.; Kim, S.H. MicroRNA134 Mediated Upregulation of JNK and Downregulation of NFkB Signalings Are Critically Involved in Dieckol Induced Antihepatic Fibrosis. *J. Agric. Food Chem.* **2016**, *64*, 5508–5514. [CrossRef]

242. Jung, H.A.; Hyun, S.K.; Kim, H.R.; Choi, J.S. Angiotensin-converting enzyme I inhibitory activity of phlorotannins from *Ecklonia stolonifera*. *Fish. Sci.* **2006**, *72*, 1292–1299. [CrossRef]

243. Wijesinghe, W.A.J.P.; Ko, S.C.; Jeon, Y.J. Effect of phlorotannins isolated from *Ecklonia cava* on angiotensin I-converting enzyme (ACE) inhibitory activity. *Nutr. Res. Pract.* **2011**, *5*, 93–100. [CrossRef]

244. Kurihara, H.; Konno, R.; Takahashi, K. Fucophlorethol C, a phlorotannin as a lipoxygenase inhibitor. *Biosci. Biotechnol. Biochem.* **2015**, *79*, 1954–1956. [CrossRef]

245. Ryu, B.M.; Ahn, B.N.; Kang, K.H.; Kim, Y.S.; Li, Y.X.; Kong, C.S.; Kim, S.K.; Kim, D.G. Dioxinodehydroeckol protects human keratinocyte cells from UVB-induced apoptosis modulated by related genes Bax/Bcl-2 and caspase pathway. *J. Photochem. Photobiol. B Biol.* **2015**, *153*, 352–357. [CrossRef]

246. Shibata, T.; Fujimoto, K.; Nagayama, K.; Yamaguchi, K.; Nakamura, T. Inhibitory activity of brown algal phlorotannins against hyaluronidase. *Int. J. Food Sci. Technol.* **2002**, *37*, 703–709. [CrossRef]

247. Kang, H.S.; Kim, H.R.; Byun, D.S.; Son, B.W.; Nam, T.J.; Choi, J.S. Tyrosinase inhibitors isolated from the edible brown alga *Ecklonia stolonifera*. *Arch. Pharm. Res.* **2004**, *27*, 1226–1232. [CrossRef] [PubMed]

248. Yoon, N.Y.; Eom, T.K.; Kim, M.M.; Kim, S.K. Inhibitory effect of phlorotannins isolated from *Ecklonia cava* on mushroom tyrosinase activity and melanin formation in mouse B16F10 melanoma cells. *J. Agric. Food Chem.* **2009**, *57*, 4124–4129. [CrossRef] [PubMed]

249. Lee, S.H.; Kang, S.M.; Sok, C.H.; Hong, J.T.; Oh, J.Y.; Jeon, Y.J. Cellular activities and docking studies of eckol isolated from *Ecklonia cava* (Laminariales, Phaeophyceae) as potential tyrosinase inhibitor. *Algae* **2015**, *30*, 163–170. [CrossRef]

250. Heo, S.J.; Ko, S.C.; Cha, S.H.; Kang, D.H.; Park, H.S.; Choi, Y.U.; Kim, D.; Jung, W.K.; Jeon, Y.J. Effect of phlorotannins isolated from *Ecklonia cava* on melanogenesis and their protective effect against photo-oxidative stress induced by UV-B radiation. *Toxicol. In Vitro* **2009**, *23*, 1123–1130. [CrossRef] [PubMed]

251. Manandhar, B.; Wagle, A.; Seong, S.H.; Paudel, P.; Kim, H.R.; Jung, H.A.; Choi, J.S. Phlorotannins with Potential Anti-tyrosinase and Antioxidant Activity Isolated from the Marine Seaweed *Ecklonia stolonifera*. *Antioxidants* **2019**, *8*, 240. [CrossRef]

252. Bak, S.S.; Sung, Y.K.; Kim, S.K. 7-Phloroeckol promotes hair growth on human follicles in vitro. *Naunyn. Schmiedebergs. Arch. Pharm.* **2014**, *387*, 789–793. [CrossRef]

253. Xi, M.; Dragsted, L.O. Biomarkers of seaweed intake. *Genes Nutr.* **2019**, *14*, 24. [CrossRef]

254. Nagayama, K.; Shibata, T.; Fujimoto, K.; Honjo, T.; Nakamura, T. Algicidal effect of phlorotannins from the brown alga *Ecklonia kurome* on red tide microalgae. *Aquaculture* **2003**, *218*, 601–611. [CrossRef]

255. Kwon, H.J.; Ryu, Y.B.; Kim, Y.M.; Song, N.; Kim, C.Y.; Rho, M.C.; Jeong, J.H.; Cho, K.O.; Lee, W.S.; Park, S.J. In vitro antiviral activity of phlorotannins isolated from *Ecklonia cava* against porcine epidemic diarrhea coronavirus infection and hemagglutination. *Bioorg. Med. Chem.* **2013**, *21*, 4706–4713. [CrossRef]

256. Hussain, E.; Wang, L.J.; Jiang, B.; Riaz, S.; Butt, G.Y.; Shi, D.Y. A review of the components of brown seaweeds as potential candidates in cancer therapy. *RSC Adv.* **2016**, *6*, 12592–12610. [CrossRef]

257. Xu, X.; Song, F.; Wang, S.; Li, S.; Xiao, F.; Zhao, J.; Yang, Y.; Shang, S.; Yang, L.; Shi, J. Dibenzyl bromophenols with diverse dimerization patterns from the brown alga *Leathesia nana*. *J. Nat. Prod.* **2004**, *67*, 1661–1666. [CrossRef] [PubMed]

258. Shi, D.; Li, J.; Guo, S.; Su, H.; Fan, X. The antitumor effect of bromophenol derivatives in vitro and *Leathesia nana* extract in vivo. *Chin. J. Oceanol. Limnol.* **2009**, *27*, 277–282. [CrossRef]

259. Shi, D.; Li, J.; Guo, S.; Han, L. Antithrombotic effect of bromophenol, the alga-derived thrombin inhibitor. *J. Biotechnol.* **2008**, *136*, S579. [CrossRef]

260. Karuppaiah, J.; Dhanraj, V.; Karuppaiah, J.; Balakrishnan, R.; Elangovan, N. Myricetin attenuates neurodegeneration and cognitive impairment in Parkinsonism. *Front. Biosci.* **2018**, *10*, 835. [CrossRef]

261. Mokrini, R.; Ben Mesaoud, M.; Daoudi, M.; Hellio, C.; Maréchal, J.P.; El Hattab, M.; Ortalo-Magné, A.; Piovetti, L.; Culioli, G. Meroditerpenoids and derivatives from the brown alga *Cystoseira baccata* and their antifouling properties. *J. Nat. Prod.* **2008**, *71*, 1806–1811. [CrossRef] [PubMed]

262. Gerald Culioli, A.O.M.; Valls, R.; Hellio, C.; Clare, A.S.; Piovetti, L. Antifouling activity of meroditerpenoids from the marine brown alga *Halidrys siliquosa*. *J. Nat. Prod.* **2008**, *71*, 1121–1126. [CrossRef]

263. Sabry, O.M.M.; Andrews, S.; McPhail, K.L.; Goeger, D.E.; Yokochi, A.; LePage, K.T.; Murray, T.F.; Gerwick, W.H. Neurotoxic meroditerpenoids from the tropical marine brown alga *Stypopodium flabelliforme*. *J. Nat. Prod.* **2005**, *68*, 1022–1030. [CrossRef]

264. Pereira, D.M.; Cheel, J.; Areche, C.; San-Martin, A.; Rovirosa, J.; Silva, L.R.; Valentao, P.; Andrade, P.B. Anti-proliferative activity of meroditerpenoids isolated from the brown alga *Stypopodium flabelliforme* against several cancer cell lines. *Mar. Drugs* **2011**, *9*, 852–862. [CrossRef]

265. Mendes, G.; Soares, A.R.; Sigiliano, L.; Machado, F.; Kaiser, C.; Romeiro, N.; Gestinari, L.; Santos, N.; Romanos, M.T.V. In vitro anti-HMPV activity of meroditerpenoids from marine alga *Stypopodium zonale* (Dictyotales). *Molecules* **2011**, *16*, 8437–8450. [CrossRef]

266. Shimizu, H.; Koyama, T.; Yamada, S.; Lipton, S.A.; Satoh, T. Zonarol, a sesquiterpene from the brown algae *Dictyopteris undulata*, provides neuroprotection by activating the Nrf2/ARE pathway. *Biochem. Biophys. Res. Commun.* **2015**, *457*, 718–722. [CrossRef]

267. Ali, Y.; Kim, D.H.; Seong, S.H.; Kim, H.R.; Jung, H.A.; Choi, J.S. α-Glucosidase and protein tyrosine phosphatase 1b inhibitory activity of plastoquinones from marine brown alga *Sargassum serratifolium*. *Mar. Drugs* **2017**, *15*, 368. [CrossRef] [PubMed]

268. Gomez-Zavaglia, A.; Prieto Lage, M.A.; Jimenez-Lopez, C.; Mejuto, J.C.; Simal-Gandara, J. The potential of seaweeds as a source of functional ingredients of prebiotic and antioxidant value. *Antioxidants* **2019**, *8*, 406. [CrossRef] [PubMed]

269. Wong, K.H.; Cheung, P.C.K. Nutritional evaluation of some subtropical red and green seaweeds Part II. In vitro protein digestibility and amino acid profiles of protein concentrates. *Food Chem.* **2001**, *72*, 11–17. [CrossRef]

270. Tibbetts, S.M.; Milley, J.E.; Lall, S.P. Nutritional quality of some wild and cultivated seaweeds: Nutrient composition, total phenolic content and in vitro digestibility. *J. Appl. Phycol.* **2016**, *28*, 3575–3585. [CrossRef]

271. Gupta, S.; Abu-Ghannam, N. Recent developments in the application of seaweeds or seaweed extracts as a means for enhancing the safety and quality attributes of foods. *Innov. Food Sci. Emerg. Technol.* **2011**, *12*, 600–609. [CrossRef]

272. Apostolidis, E.; Kwon, Y.I.; Shetty, K. Inhibition of *Listeria monocytogenes* by oregano, cranberry and sodium lactate combination in broth and cooked ground beef systems and likely mode of action through proline metabolism. *Int. J. Food Microbiol.* **2008**, *128*, 317–324. [CrossRef]

273. Lin, Y.T.; Labbe, R.G.; Shetty, K. Inhibition of *Listeria monocytogenes* in fish and meat systems by use of oregano and cranberry phytochemical synergies. *Appl. Env. Microbiol.* **2004**, *70*, 5672–5678. [CrossRef]

274. Wang, T.; Jónsdóttir, R.; Kristinsson, H.G.; Thorkelsson, G.; Jacobsen, C.; Hamaguchi, P.Y.; Ólafsdóttir, G. Inhibition of haemoglobin-mediated lipid oxidation in washed cod muscle and cod protein isolates by *Fucus vesiculosus* extract and fractions. *Food Chem.* **2010**, *123*, 321–330. [CrossRef]

275. Eom, S.H.; Kim, Y.M.; Kim, S.K. Antimicrobial effect of phlorotannins from marine brown algae. *Food Chem. Toxicol.* **2012**, *50*, 3251–3255. [CrossRef]

276. Whitfield, F.B.; Helidoniotis, F.; Shaw, K.J.; Svoronos, D. Distribution of Bromophenols in Species of Marine Algae from Eastern Australia. *J. Agric. Food Chem.* **1999**, *47*, 2367–2373. [CrossRef]

277. Origin of SEANOL. Available online: http://seanolinstitute.org/ssc/origin.html (accessed on 30 April 2020).

278. Yeo, A.R.; Lee, J.; Tae, I.H.; Park, S.R.; Cho, Y.H.; Lee, B.H.; Cheol Shin, H.; Kim, S.H.; Yoo, Y.C. Anti-hyperlipidemic effect of polyphenol extract (Seapolynol™) and dieckol isolated from *Ecklonia cava* in in vivo and in vitro models. *Prev. Nutr. Food Sci.* **2012**, *17*, 1–7. [CrossRef] [PubMed]

279. Ahn, H.-S.; Lee, D.-H.; Kim, T.-J.; Shin, H.-C.; Jeon, H.-K. Cardioprotective Effects of a Phlorotannin Extract Against Doxorubicin-Induced Cardiotoxicity in a Rat Model. *J. Med. Food* **2017**, *20*, 944–950. [CrossRef] [PubMed]

280. Jeon, H.-J.; Yoon, K.-Y.; Koh, E.-J.; Choi, J.; Kim, K.-J.; Choi, H.-S.; Lee, B.-Y. Seapolynol and Dieckol Improve Insulin Sensitivity through the Regulation of the PI3K Pathway in C57BL/KsJ-db/db Mice. *J. Food Nutr. Res.* **2015**, *3*, 648–652. [CrossRef]

281. Jeon, H.-J.; Choi, H.-S.; Lee, Y.-J.; Hwang, J.-H.; Lee, O.-H.; Seo, M.-J.; Kim, K.-J.; Lee, B.-Y. Seapolynol Extracted from *Ecklonia cava* Inhibits Adipocyte Differentiation in Vitro and Decreases Fat Accumulation in Vivo. *Molecules* **2015**, *20*, 21715–21731. [CrossRef] [PubMed]

282. Turck, D.; Bresson, J.; Burlingame, B.; Dean, T.; Fairweather-Tait, S.; Heinonen, M.; Hirsch-Ernst, K.I.; Mangelsdorf, I.; McArdle, H.J.; Naska, A.; et al. Safety of *Ecklonia cava* phlorotannins as a novel food pursuant to Regulation (EC) No 258/97. *EFSA J.* **2017**, *15*. [CrossRef]

283. Hwang, H.J. Skin Elasticity and Sea Polyphenols. *Seanol Sci. Cent. Rev.* **2010**, *1*, 1–10.

284. Pimentel, F.; Alves, R.; Rodrigues, F.; Oliveira, M.P.P. Macroalgae-Derived Ingredients for Cosmetic Industry—An Update. *Cosmetics* **2017**, *5*, 2. [CrossRef]

285. Pereira, L. Seaweeds as Source of Bioactive Substances and Skin Care Therapy—Cosmeceuticals, Algotheraphy, and Thalassotherapy. *Cosmetics* **2018**, *5*, 68. [CrossRef]

286. Cosmetics Ingredients Database | Online Raw Materials Search. Available online: https://cosmetics.specialchem.com/ (accessed on 9 April 2020).

287. Default Web Site Page. Available online: https://sealgae.pt/ (accessed on 9 April 2020).

288. Exclusive Worldwide Licence—Aethic. Available online: https://aethic.com/aethic-granted-exclusive-worldwide-license-use-seaweed-compound/ (accessed on 13 April 2020).

289. Aethic Wins Exclusive License to Use Novel Seaweed Compound. Available online: https://www.cosmeticsbusiness.com/news/article_page/Aethic_wins_exclusive_license_to_use_novel_seaweed_compound/132704 (accessed on 13 April 2020).

290. 5 Best Eco-friendly Suncreens | The Independent. Available online: https://www.independent.co.uk/extras/indybest/fashion-beauty/sun-care-tanning/best-eco-friendly-sunscreen-natural-organic-mineral-for-face-baby-sensitive-skin-a8340881.html (accessed on 13 April 2020).

291. ECKLEXT® for Natural Cosmetic Ingredients, NOF EUROPE GmbH. Available online: https://nofeurope.com/index.php?dispatch=categories.view&category_id=270 (accessed on 10 April 2020).

292. Ingredients & Formulas | SEPPIC. Available online: https://www.seppic.com/ingredients-formulas (accessed on 13 April 2020).

293. The Garden of Naturalsolution. Available online: http://www.naturalsolution.co.kr/eng/home.php (accessed on 10 April 2020).

294. Green Confertii Extract-NS—The Garden of Naturalsolution—Datasheet. Available online: https://cosmetics.specialchem.com/product/i-natural-solution-green-confertii-extract-ns (accessed on 10 April 2020).

295. Bladderwrack Extract-NS—The Garden of Naturalsolution—Datasheet. Available online: https://cosmetics.specialchem.com/product/i-natural-solution-bladderwrack-extract-ns (accessed on 10 April 2020).

296. DE10259966A1—Cosmetic or Pharmaceutical Composition Containing Mycosporin-Like Amino Acids, Useful for Treatment and Prevention of Hypoxia and Associated Conditions, Improves Oxygen Uptake—Google Patents. Available online: https://patents.google.com/patent/DE10259966A1/en?q=mycosporine-like+amino+acid&oq=mycosporine-like+amino+acid&page=2 (accessed on 13 April 2020).

297. GB2472021A—Cosmetic Sunscreen Composition—Google Patents. Available online: https://patents.google.com/patent/GB2472021A/en?q=mycosporine-like+amino+acid&oq=mycosporine-like+amino+acid&page=2 (accessed on 13 April 2020).

298. EP1473028A1—Cosmetic Skin Care Products and Cosmetic Agents for Protecting Skin Against Premature Aging—Google Patents. Available online: https://patents.google.com/patent/EP1473028A1/en?q=mycosporine-like+amino+acid&oq=mycosporine-like+amino+acid&page=2 (accessed on 13 April 2020).

299. Jarald, E.; Joshi, S.B.; Jain, D.C. Diabetes and Herbal Medicines. *Iran. J. Pharm.* **2008**, *7*, 97–106.

300. Smit, A.J. Medicinal and pharmaceutical uses of seaweed natural products: A review. *J. Appl. Phycol.* **2004**, *16*, 245–262. [CrossRef]

301. Freile-Pelegrín, Y.; Tasdemir, D. Seaweeds to the rescue of forgotten diseases: A review. *Bot. Mar.* **2019**, *62*, 211–226. [CrossRef]

302. Hodgson, J.M.; Croft, K.D. Dietary flavonoids: Effects on endothelial function and blood pressure. *J. Sci. Food Agric.* **2006**, *86*, 2492–2498. [CrossRef]

303. Park, B.G.; Shin, W.S.; Oh, S.; Park, G.M.; Kim, N.I.; Lee, S. A novel antihypertension agent, sargachromenol D from marine brown algae, *Sargassum siliquastrum*, exerts dual action as an L-type Ca2+ channel blocker and endothelin A/B2 receptor antagonist. *Bioorg. Med. Chem.* **2017**, *25*, 4649–4655. [CrossRef] [PubMed]

304. Apostolidis, E.; Lee, C.M. Brown Seaweed-Derived Phenolic Phytochemicals and Their Biological Activities for Functional Food Ingredients with Focus on *Ascophyllum nodosum*. In *Handbook of Marine Macroalgae*; Kim, S., Ed.; John Wiley & Sons, Ltd.: Chichester, UK, 2011; pp. 356–370.

305. Jung, H.A.; Oh, S.H.; Choi, J.S. Molecular docking studies of phlorotannins from *Eisenia bicyclis* with BACE1 inhibitory activity. *Bioorg. Med. Chem. Lett.* **2010**, *20*, 3211–3215. [CrossRef]
306. Um, M.Y.; Lim, D.W.; Son, H.J.; Cho, S.; Lee, C. Phlorotannin-rich fraction from *Ishige foliacea* brown seaweed prevents the scopolamine-induced memory impairment via regulation of ERK-CREB-BDNF pathway. *J. Funct. Foods* **2018**, *40*, 110–116. [CrossRef]
307. Catarino, M.D.; Silva, A.M.S.; Mateus, N.; Cardoso, S.M. Optimization of phlorotannins extraction from *Fucus vesiculosus* and evaluation of their potential to prevent metabolic disorders. *Mar. Drugs* **2019**, *17*, 162. [CrossRef]
308. Hussain, S.P.; Hofseth, L.J.; Harris, C.C. Radical causes of cancer. *Nat. Rev. Cancer* **2003**, *3*, 276–285. [CrossRef]
309. Wijesekara, I.; Yoon, N.Y.; Kim, S.-K.K. Phlorotannins from *Ecklonia cava* (Phaeophyceae): Biological activities and potential health benefits. *BioFactors* **2010**, *36*, 408–414. [CrossRef]
310. Liu, E.-H.; Qi, L.-W.; Wu, Q.; Peng, Y.-B.; Li, P. Anticancer Agents Derived from Natural Products. *Mini-Rev. Med. Chem.* **2009**, *9*, 1547–1555. [CrossRef]
311. Xu, N.; Fan, X.; Yan, X.; Tseng, C.K. Screening marine algae from China for their antitumor activities. *J. Appl. Phycol.* **2004**, *16*, 451–456. [CrossRef]
312. Ganesan, P.; Kumar, C.S.; Bhaskar, N. Antioxidant properties of methanol extract and its solvent fractions obtained from selected Indian red seaweeds. *Bioresour. Technol.* **2008**, *99*, 2717–2723. [CrossRef] [PubMed]
313. Matanjun, P.; Mohamed, S.; Mustapha, N.M.; Muhammad, K.; Ming, C.H. Antioxidant activities and phenolics content of eight species of seaweeds from north Borneo. *J. Appl. Phycol.* **2008**, *20*, 367–373. [CrossRef]
314. Qin, J.; Su, H.; Zhang, Y.; Gao, J.; Zhu, L.; Wu, X.; Pan, H.; Li, X. Highly brominated metabolites from marine red alga *Laurencia similis* inhibit protein tyrosine phosphatase 1B. *Bioorg. Med. Chem. Lett.* **2010**, *20*, 7152–7154. [CrossRef] [PubMed]
315. Kim, K.Y.; Nam, K.A.; Kurihara, H.; Kim, S.M. Potent α-glucosidase inhibitors purified from the red alga *Grateloupia elliptica*. *Phytochemistry* **2008**, *69*, 2820–2825. [CrossRef]
316. Roy, M.C.; Anguenot, R.; Fillion, C.; Beaulieu, M.; Bérubé, J.; Richard, D. Effect of a commercially-available algal phlorotannins extract on digestive enzymes and carbohydrate absorption in vivo. *Food Res. Int.* **2011**, *44*, 3026–3029. [CrossRef]
317. Gwynn, M.N.; Portnoy, A.; Rittenhouse, S.F.; Payne, D.J. Challenges of antibacterial discovery revisited. *Ann. N. Y. Acad. Sci.* **2010**, *1213*, 5–19. [CrossRef]
318. Silver, L.L. Challenges of antibacterial discovery. *Clin. Microbiol. Rev.* **2011**, *24*, 71–109. [CrossRef]
319. JP2003277203A—Antibacterial Agent Based on Phlorotannins—Google Patents. Available online: https://patents.google.com/patent/JP2003277203A/en?q=phlorotannins&oq=phlorotannins (accessed on 10 April 2020).
320. US20130338218A1—HIV-1 Inhibiting Pharmaceutical Composition Containing an Ecklonia Cava-Derived Phloroglucinol Polymer Compound—Google Patents. Available online: https://patents.google.com/patent/US20130338218A1/en (accessed on 29 April 2020).
321. Im, J.H.; Choi, C.H.; Mun, F.; Lee, J.H.; Kim, H.; Jung, W.K.; Jang, C.H.; Kim, G.H. A polycaprolactone/fish collagen/alginate biocomposite supplemented with phlorotannin for hard tissue regeneration. *RSC Adv.* **2017**, *7*, 2009–2018. [CrossRef]
322. Khademhosseini, A.; Du, Y.; Rajalingam, B.; Vacanti, J.P.; Langer, R.S. Microscale technologies for tissue engineering. *Adv. Tissue Eng.* **2008**, *103*, 349–369. [CrossRef]
323. Yang, S.; Leong, K.F.; Du, Z.; Chua, C.K. The design of scaffolds for use in tissue engineering. Part I. Traditional factors. *Tissue Eng.* **2001**, *7*, 679–689. [CrossRef]
324. Sung, H.J.; Meredith, C.; Johnson, C.; Galis, Z.S. The effect of scaffold degradation rate on three-dimensional cell growth and angiogenesis. *Biomaterials* **2004**, *25*, 5735–5742. [CrossRef] [PubMed]
325. Hutmacher, D.W.; Schantz, T.; Zein, I.; Ng, K.W.; Teoh, S.H.; Tan, K.C. Mechanical properties and cell cultural response of polycaprolactone scaffolds designed and fabricated via fused deposition modeling. *J. Biomed. Mater. Res.* **2001**, *55*, 203–216. [CrossRef]
326. Douglas, T.E.L.; Dokupil, A.; Reczyńska, K.; Brackman, G.; Krok-Borkowicz, M.; Keppler, J.K.; Božič, M.; Van Der Voort, P.; Pietryga, K.; Samal, S.K.; et al. Enrichment of enzymatically mineralized gellan gum hydrogels with phlorotannin-rich *Ecklonia cava* extract Seanol®to endow antibacterial properties and promote mineralization. *Biomed. Mater.* **2016**, *11*, 045015. [CrossRef] [PubMed]
327. Abbas, A.K.; Lichtman, A.H.H.; Pillai, S. *Cellular and Molecular Immunology*, 8th ed.; Elsevier Saunders: Philadelphia, PA, USA, 2014; ISBN 9780323316149.

328. Kazłowska, K.; Hsu, T.; Hou, C.C.; Yang, W.C.; Tsai, G.J. Anti-inflammatory properties of phenolic compounds and crude extract from *Porphyra dentata*. *J. Ethnopharmacol.* **2010**, *128*, 123–130. [CrossRef] [PubMed]

329. Shin, H.-C.; Hwang, H.J.; Kang, K.J.; Lee, B.H. An antioxidative and antiinflammatory agent for potential treatment of osteoarthritis from *Ecklonia cava*. *Arch. Pharm. Res.* **2006**, *29*, 165–171. [CrossRef]

330. Kwon, S.; Yoon, M.; Lee, J.; Moon, K.D.; Kim, D.; Kim, S.B.; Cho, S. A standardized phlorotannin supplement attenuates caffeine-induced sleep disruption in mice. *Nutrients* **2019**, *11*, 556. [CrossRef]

331. Park, H.-H.; Ko, S.-C.; Oh, G.-W.; Heo, S.-J.; Kang, D.-H.; Bae, S.-Y.; Jung, W.-K. Fabrication and characterization of phlorotannins/poly (vinyl alcohol) hydrogel for wound healing application. *J. Biomater. Sci. Polym. Ed.* **2018**, *29*, 972–983. [CrossRef]

332. Kumar Singh, A.; Cabral, C.; Kumar, R.; Ganguly, R.; Kumar Rana, H.; Gupta, A.; Rosaria Lauro, M.; Carbone, C.; Reis, F.; Pandey, A.K. Beneficial Effects of Dietary Polyphenols on Gut Microbiota and Strategies to Improve Delivery Efficiency. *Nutrients* **2019**, *11*, 2216. [CrossRef]

333. Cardona, F.; Andrés-Lacueva, C.; Tulipani, S.; Tinahones, F.J.; Queipo-Ortuño, M.I. Benefits of polyphenols on gut microbiota and implications in human health. *J. Nutr. Biochem.* **2013**, *24*, 1415–1422. [CrossRef]

334. Tzounis, X.; Rodriguez-Mateos, A.; Vulevic, J.; Gibson, G.R.; Kwik-Uribe, C.; Spencer, J.P. Prebiotic evaluation of cocoa-derived flavanols in healthy humans by using a randomized, controlled, double-blind, crossover intervention study. *Am. J. Clin. Nutr.* **2011**, *93*, 62–72. [CrossRef]

335. Keyrouz, R.; Abasq, M.L.; Le Bourvellec, C.; Blanc, N.; Audibert, L.; Argall, E.; Hauchard, D. Total phenolic contents, radical scavenging and cyclic voltammetry of seaweeds from Brittany. *Food Chem.* **2011**, *126*, 831–836. [CrossRef]

336. Chojnacka, K. Biologically Active Compounds in Seaweed Extracts—The Prospects for the Application. *Open Conf. Proc. J.* **2012**, *3*, 20–28. [CrossRef]

337. Galleano, M.; Pechanova, O.; G. Fraga, C. Hypertension, Nitric Oxide, Oxidants, and Dietary Plant Polyphenols. *Curr. Pharm. Biotechnol.* **2010**, *11*, 837–848. [CrossRef] [PubMed]

338. Makkar, H.P.S.; Tran, G.; Heuzé, V.; Giger-Reverdin, S.; Lessire, M.; Lebas, F.; Ankers, P. Seaweeds for livestock diets: A review. *Anim. Feed Sci. Technol.* **2016**, *212*, 1–17. [CrossRef]

339. Saker, K.E.; Allen, V.G.; Fontenot, J.P.; Bagley, C.P.; Ivy, R.L.; Evans, R.R.; Wester, D.B. Tasco-Forage: II. Monocyte immune cell response and performance of beef steers grazing tall fescue treated with a seaweed extract. *J. Anim. Sci.* **2001**, *79*, 1022. [CrossRef]

340. Spiers, D.E.; Eichen, P.A.; Leonard, M.J.; Wax, L.E.; Rottinghaus, G.E.; Williams, J.E.; Colling, D.P. Benefit of dietary seaweed (*Ascophyllum nodosum*) extract in reducing heat strain and fescue toxicosis: A comparative evaluation. *J. Biol.* **2004**, *29*, 753–757. [CrossRef]

341. Bach, S.J.; Wang, Y.; McAllister, T.A. Effect of feeding sun-dried seaweed (*Ascophyllum nodosum*) on fecal shedding of *Escherichia coli* O157:H7 by feedlot cattle and on growth performance of lambs. *Anim. Feed Sci. Technol.* **2008**, *142*, 17–32. [CrossRef]

342. Braden, K.W.; Blanton, J.R.; Allen, V.G.; Pond, K.R.; Miller, M.F. *Ascophyllum nodosum* supplementation: A preharvest intervention for reducing *Escherichia coli* O157:H7 and *Salmonella* spp. in feedlot steers. *J. Food Prot.* **2004**, *67*, 1824–1828. [CrossRef]

343. Braden, K.W.; Blanton, J.R.; Montgomery, J.L.; van Santen, E.; Allen, V.G.; Miller, M.F. Tasco supplementation: Effects on carcass characteristics, sensory attributes, and retail display shelf-life. *J. Anim. Sci.* **2007**, *85*, 754–768. [CrossRef]

344. Wang, Y.; Xu, Z.; Bach, S.J.; McAllister, T.A. Effects of phlorotannins from *Ascophyllum nodosum* (brown seaweed) on in vitro ruminal digestion of mixed forage or barley grain. *Anim. Feed Sci. Technol.* **2008**, *145*, 375–395. [CrossRef]

345. Wiseman, M. Evaluation of Tasco ®as a Candidate Prebiotic in Broiler Chickens. Master's Thesis, Dalhousie University Halifax, Nova Scotia, NS, Canada, 2012.

346. McHugh, D.J. *A Guide to the Seaweed Industry*; Food and Agriculture Organization of the United Nations, Ed.; FAO Fisher.: Rome, Italy, 2003; ISBN 9251049580.

347. Leupp, J.L.; Caton, J.S.; Soto-Navarro, S.A.; Lardy, G.P. Effects of cooked molasses blocks and fermentation extract or brown seaweed meal inclusion on intake, digestion, and microbial efficiency in steers fed low-quality hay. *J. Anim. Sci.* **2005**, *83*, 2938–2945. [CrossRef] [PubMed]

348. Producers, Manufacturers, Seaweed Processors, Exporters, Specialists in Seaweed Harvesting and Cultivation, Renewable Resource Technologies and Marine Food Safety—Acadian Seaplants. Available online: https://www.acadianseaplants.com/marine-plant-seaweed-manufacturers/#environmental_sustainability (accessed on 11 April 2020).
349. Evans, F.D.; Critchley, A.T. Seaweeds for animal production use. *J. Appl. Phycol.* **2014**, *26*, 891–899. [CrossRef]
350. Wijesinghe, W.A.J.P.; Jeon, Y.-J. Exploiting biological activities of brown seaweed *Ecklonia cava* for potential industrial applications: A review. *Int. J. Food Sci. Nutr.* **2012**, *63*, 225–235. [CrossRef] [PubMed]
351. Wang, J.; Zhang, Q.; Zhang, Z.; Zhang, J.; Li, P. Synthesized phosphorylated and aminated derivatives of fucoidan and their potential antioxidant activity in vitro. *Int. J. Biol. Macromol.* **2009**, *44*, 170–174. [CrossRef]
352. WO2017032954A1—Use of Phlorotannins as a Stimulant for Mychorrhizal and Rhizobial Symbioses—Google Patents. Available online: https://patents.google.com/patent/WO2017032954A1/en (accessed on 10 April 2020).
353. JP2004189532A—Stable Liquid Organic Fertilizer Having Plant Disease Protection Effect—Google Patents. Available online: https://patents.google.com/patent/JP2004189532A/en?q=phlorotannins&oq=phlorotannins&page=2 (accessed on 10 April 2020).
354. Craigie, J.S. Seaweed extract stimuli in plant science and agriculture. *J. Appl. Phycol.* **2011**, *23*, 371–393. [CrossRef]
355. Glombitza, K.-W. Highly Hydroxylated Phenols of the Phaeophyceae. In *Marine Natural Products Chemistry*; Faulkner, D.J., Fenical, W.H., Eds.; Springer: Boston, MA, USA, 1977; pp. 191–204.
356. Jaillet, F.; Nouailhas, H.; Boutevin, B.; Caillol, S. Synthesis of novel vinylester from biobased phloroglucinol. *Green Mater.* **2016**, *4*. [CrossRef]

Article

LC-ESI-QTOF-MS/MS Characterization of Seaweed Phenolics and Their Antioxidant Potential

Biming Zhong [1], Nicholas A. Robinson [2,3], Robyn D. Warner [1], Colin J. Barrow [4], Frank R. Dunshea [1] and Hafiz A.R. Suleria [1,4,*]

[1] School of Agriculture and Food, Faculty of Veterinary and Agricultural Sciences, The University of Melbourne, Parkville, VIC 3010, Australia; bimingz@student.unimelb.edu.au (B.Z.); robyn.warner@unimelb.edu.au (R.D.W.); fdunshea@unimelb.edu.au (F.R.D.)

[2] Sustainable Aquaculture Laboratory-Temperate and Tropical (SALTT), School of BioSciences, The University of Melbourne, Parkville, VIC 3010, Australia; nicholas.robinson@nofima.no

[3] Norwegian Institute of Food, Fisheries and Aquaculture Research (Nofima), NO-1431 Ås, Norway

[4] Centre for Chemistry and Biotechnology, School of Life and Environmental Sciences, Deakin University, Waurn Ponds, VIC 3217, Australia; colin.barrow@deakin.edu.au

* Correspondence: hafiz.suleria@unimelb.edu.au; Tel.: +61-470-439-670

Received: 9 May 2020; Accepted: 21 June 2020; Published: 24 June 2020

Abstract: Seaweed is an important food widely consumed in Asian countries. Seaweed has a diverse array of bioactive compounds, including dietary fiber, carbohydrate, protein, fatty acid, minerals and polyphenols, which contribute to the health benefits and commercial value of seaweed. Nevertheless, detailed information on polyphenol content in seaweeds is still limited. Therefore, the present work aimed to investigate the phenolic compounds present in eight seaweeds [Chlorophyta (green), *Ulva* sp., *Caulerpa* sp. and *Codium* sp.; Rhodophyta (red), *Dasya* sp., *Grateloupia* sp. and *Centroceras* sp.; Ochrophyta (brown), *Ecklonia* sp., *Sargassum* sp.], using liquid chromatography electrospray ionization quadrupole time-of-flight mass spectrometry (LC-ESI-QTOF-MS/MS). The total phenolic content (TPC), total flavonoid content (TFC) and total tannin content (TTC) were determined. The antioxidant potential of seaweed was assessed using a 2,2-diphenyl-1-picrylhydrazyl (DPPH) free radical scavenging assay, a 2,2′-azino-bis-3-ethylbenzothiazoline-6-sulfonic acid (ABTS) free radical scavenging assay and a ferric reducing antioxidant power (FRAP) assay. Brown seaweed species showed the highest total polyphenol content, which correlated with the highest antioxidant potential. The LC-ESI-QTOF-MS/MS tentatively identified a total of 54 phenolic compounds present in the eight seaweeds. The largest number of phenolic compounds were present in *Centroceras* sp. followed by *Ecklonia* sp. and *Caulerpa* sp. Using high-performance liquid chromatography-photodiode array (HPLC-PDA) quantification, the most abundant phenolic compound was *p*-hydroxybenzoic acid, present in *Ulva* sp. at 846.083 ± 0.02 µg/g fresh weight. The results obtained indicate the importance of seaweed as a promising source of polyphenols with antioxidant properties, consistent with the health potential of seaweed in food, pharmaceutical and nutraceutical applications.

Keywords: seaweeds; polyphenols; antioxidant potential; LC-ESI-QTOF-MS/MS; HPLC-PDA

1. Introduction

Seaweed has been utilized as a food for humans for centuries, and the current global market is valued at more than USD 6 billion per annum with an annual volume of approximately 12 million tonnes in 2018 [1,2]. Seaweeds (macroalgae) are classified into three major groups including Chlorophyta (green algae), Rhodophyta (red algae) and Ochrophyta (brown algae) based on their color. It is estimated that 1800 different green macroalgae, 6200 red macroalgae, and 1800 brown macroalgae are found in the marine environment [3]. Like plants, they have chlorophyll for photosynthesis but also contain other pigments which may be colored red, blue, brown or gold. Seaweeds are used in many countries as a source of food especially in East Asia, seaweeds are associated with different Japanese, Koreans and Chinese cuisines [4]. Seaweed is considered an excellent source of bioactive compounds with positive health effects, including carotenoids, phenolics, chitosan, gelatin, polyunsaturated fatty acids, various vitamins and minerals [5]. Recent interest in seaweed has focused on seaweed natural bioactive compounds in the functional food, pharmaceutical and cosmeceutical industries [6]. Among these bioactives, polyphenols, which are defined as the compounds containing one or more aromatic rings bearing hydroxyl groups, have attracted considerable attention [7]. Polyphenols have been shown to exhibit antioxidant, antimicrobial, antidiabetic, anti-inflammatory and anticancer properties in in vitro and in vivo studies [8], and are categorized into subclasses of phenolic acids, flavonoids, stilbenes, and lignans, depending on the chemical structure [9].

A promising bioactive property of polyphenols relates to their antioxidant activity and redox potential, allowing them to reduce the reactive oxygen species (ROS) that are involved in a range of human disorders [10]. Strong antioxidant properties of various edible seaweeds have been reported, particularly with seaweeds with high polyphenol content, which can be as high as 20–30% of the dry weight of some brown seaweeds [11,12]. Several phenolic compounds are abundant in a range of species of seaweed, including gallic acid, protocatechuic acid, caffeic acid and epicatechin, with these species showing potential as functional foods [13]. Antioxidants in food can exhibit their activity by donating hydrogen atoms, providing electrons and chelating free metals [14]. Antioxidant compounds have been successfully extracted from seaweeds and commercialized for their health benefits or for their ability to prolong the shelf-life of food through their antioxidant potential [15,16].

Total phenolic, flavonoid and tannin contents in seaweed can be indirectly measured using assays for total phenolic content (TPC), total flavonoid content (TFC) and total tannin (TTC), respectively. The antioxidant activities of seaweed can be quantified using various assays based on different mechanisms, including 2,2-diphenyl-1-picrylhydrazyl (DPPH) and 2,2'-azino-bis-3-ethylbenzothiazoline-6-sulfonic acid (ABTS) assays based on free-radical scavenged by antioxidant compounds, and ferric reducing of antioxidant power (FRAP) assay based on the reducing capacity of antioxidants [17]. However, TPC and other colorimetric methods neither separate, nor quantify, individual compounds. High-performance liquid chromatography coupled with electrospray ionization-quadrupole-time of flight-mass spectrometry (LC-ESI-QTOF-MS/MS) has been a standard method to isolate and characterize phenolic compounds based on their molecular weight [18]. High-performance liquid chromatography photodiode array (HPLC-PDA) has been used to quantify various bioactive compounds in seaweed extracts [19].

The objectives of the current study were: (1) to extract phenolic compounds from a range of seaweeds; (2) quantify the total phenolic and antioxidant capacities of seaweed extracts using different assays and (3) apply LC-ESI-QTOF-MS/MS and HPLC-PDA to characterize and quantify individual phenolic compounds.

2. Results and Discussion

2.1. Polyphenol Estimation (TPC, TFC and TTC)

The polyphenol content was measured as TPC, TFC and TTC (Table 1). Brown seaweed *Ecklonia* sp. showed significantly higher TPC (1044 ± 2.5 µg GAE/$g_{f.w.}$) and TTC (167 ± 23.2 µg CE/$g_{f.w.}$) contents than other seaweed ($p < 0.05$). The presence of higher total phenolics in brown seaweed *Ecklonia* sp. compared to green seaweed *Ulva* sp. and red seaweed *Porphyra* sp. was previously observed by García-Casal, et al. [20]. The significant higher total phenolic and tannin content in brown seaweed *Ecklonia* sp. is proposed to be related to the presence of phlorotannins, which are restricted to brown algae, in special vesicles (physodes) within the cells [21]. Phlorotannins are highly complex compounds formed by the polymerization of phloroglucinol, which has already been characterized by LC-MS in previous studies [22,23] and supported by our current study. The highest total flavonoid content was found in red seaweed *Grateloupia* sp. (54.4 ± 0.74 µg QE/g$_{f.w.}$) ($p < 0.05$) as compared to brown and green seaweeds. However, compared to previous studies [24], the total flavonoid content of red seaweed we found was relatively low compared with that of brown and green seaweed. The inconsistency might be explained by Chan, et al. [25], who reported that the total flavonoid content of seaweeds is impacted by sunlight, climate, region and extraction solvent.

Table 1. Phenolic content estimated in the seaweeds investigated in this study.

Samples.	TPC (µg GAE/g)	TFC (µg QE/g)	TTC (µg CE/g)
Green seaweeds			
Ulva sp.	14.80 ± 0.54 [d]	9.80 ± 1.96 [de]	-
Caulerpa sp.	4.30 ± 0.45 [d]	0.73 ± 0.08 [f]	3.31 ± 7.02 [b]
Codium sp.	2.29 ± 0.26 [d]	1.11 ± 0.63 [f]	-
Red seaweeds *			
Dasya sp.	260.15 ± 2.25 [c]	29.96 ± 0.48 [c]	24.90 ± 3.46 [b]
Grateloupia sp.	524.56 ± 0.46 [b]	54.43 ± 0.74 [a]	-
Centroceras sp.	49.31 ± 2.17 [d]	42.55 ± 0.52 [b]	4.45 ± 4.37 [b]
Brown seaweeds *			
Ecklonia sp.	1044.36 ± 2.54 [a]	13.87 ± 1.18 [d]	166.87 ± 23.24 [a]
Sargassum sp.	22.27 ± 0.15 [d]	3.88 ± 0.27 [ef]	5.62 ± 0.01 [b]

The data are shown as mean ± standard error (n = 3); the superscript letters (a–f), indicate the means within a column with significant difference ($p < 0.05$) using a one-way analysis of variance (ANOVA) and Tukey's test. Data of seaweed is reported on a fresh weight basis. *: total polyphenol content of brown seaweeds was significantly higher than green and red seaweeds; total flavonoid content of red seaweeds was significantly higher than green and brown seaweeds ($p < 0.05$). The phenolic content, as measured by total phenolic content (TPC), total flavonoid content (TFC), total tannin contents (TTC). GAE stands for gallic acid equivalents, QE stands for quercetin equivalents and CE stands for catechin equivalents.

Regarding seaweed groups, brown seaweeds presented statistically higher TPC and TTC values than green and red seaweeds ($p < 0.05$). This is in agreement with previous research which reported that brown seaweed had a higher total phenolic content than red and green seaweeds [26]. In addition, a study conducted by Cox, Abu-Ghannam and Gupta [24] also indicated that the total tannin content of brown seaweeds was significantly higher than that of green and red seaweed, which is explained by the presence of the unique polyphenolic components of phlorotannin in brown seaweed [27].

2.2. Antioxidant Activities (ABTS, DPPH and FRAP)

The antioxidant activities were determined using ABTS, DPPH and FRAP assays (Table 2.). The brown seaweed *Ecklonia* sp. had a significantly higher level of antioxidant potential than other seaweeds (958 ± 0.4 µg AAE/g$_{f.w.}$ for ABTS, 510 ± 3.4 µg AAE/g$_{f.w.}$ for DPPH and 170 ± 2.0 µg AAE/g$_{f.w.}$ for FRAP, $p < 0.05$). The result was consistent with a previous study where phlorotannins were successfully isolated from *Ecklonia* sp. and exhibited strong DPPH radical scavenging activity [28]. In the present work, although *Ulva* sp., *Caulerpa* sp. and *Codium* sp. exhibited ABTS radical scavenging activities, no DPPH radical scavenging activities were detected. This might be due to limitations of the DPPH assay [29]. Firstly, unlike water-soluble ABTS$^+$, hydrophobic DPPH must be performed in organic solvent, which interferes with the hydrogen atom transfer reaction by disturbing the release of hydrogen atoms. Secondly, DPPH reacts rapidly, mainly through single electron transfer, with ascorbic acid and simple phenols with no ring adducts, but slowly with complex phenolic compounds with side chains and ring adducts. Therefore, the application of organic solvent and the complex structure of phenolic compounds in seaweed might lead to underestimation of DPPH scavenging activities.

Table 2. Antioxidant activities detected in the seaweeds investigated in this study.

Samples	ABTS (µg AAE/g)	DPPH (µg AAE/g)	FRAP (µg AAE/g)
Green seaweeds			
Ulva sp.	14.24 ± 0.93 [d]	-	4.10 ± 1.45 [bc]
Caulerpa sp.	20.93 ± 2.62 [d]	-	0.53 ± 0.05 [c]
Codium sp.	10.05 ± 6.65 [d]	-	1.07 ± 0.62 [c]
Red seaweeds			
Dasya sp.	179.63 ± 9.3 [c]	12.71 ± 0.83 [b]	27.39 ± 1.47 [bc]
Grateloupia sp.	243.06 ± 3.78 [b]	19.12 ± 0.64 [b]	35.05 ± 1.54 [b]
Centroceras sp.	27.91 ± 3.79 [d]	6.30 ± 0.73 [b]	1.86 ± 1.15 [c]
Brown seaweeds *			
Ecklonia sp.	957.85 ± 0.36 [a]	510.32 ± 3.38 [a]	170.03 ± 2.04 [a]
Sargassum sp.	42.62 ± 3.09 [d]	13.71 ± 5.67 [b]	4.76 ± 0.48 [bc]

The data are shown as mean ± standard error (n = 3); the superscript letters (a–d), indicate the means within a column with significant difference ($p < 0.05$) using a one-way analysis of variance (ANOVA) and Tukey's test. Data of seaweed is reported on a fresh weight basis. *: Antioxidant capacities of brown seaweeds are significantly higher than that of green and red seaweeds ($p < 0.05$). DPPH stands for 2,2-diphenyl-1-picrylhydrazyl, ABTS stands for 2,2′-azino-bis-3-ethylbenzothiazoline-6-sulfonic acid and FRAP stands for ferric reducing antioxidant power assay. AAE stands for ascorbic acid equivalents.

Within the seaweed groups, brown seaweed species presented significantly higher antioxidant properties for all assays than green and red seaweed species ($p < 0.05$). This result was in accordance with a previous study, which also found brown seaweed had higher ABTS radical scavenging activity than red or green seaweeds [30].

The relationship between TPC and antioxidant potential of all three type of (green, red and brown) seaweeds was confirmed by performing a regression model between the values of TPC and each antioxidant assay. Results showed a significant positive correlation between TPC and antioxidant activity ($r^2 = 0.926$ for ABTS, $r^2 = 0.714$ for DPPH and $r^2 = 0.899$ for FRAP, $p < 0.05$). A positive correlation between total phenolic content and antioxidant assay results was also supported by previous studies, suggesting that phenolics are the major contributor to the excellent antioxidant properties of seaweeds [21,30].

2.3. LC-ESI-QTOF-MS/MS Characterization of The Phenolic Compounds

LC-MS has been widely used for the characterization of the phenolic profiles of different plant and marine samples [31]. A qualitative analysis of the phenolic compounds from different seaweed extracts were achieved by LC-ESI-QTOF-MS/MS analysis in negative and positive ionization modes (Table S1, Figures S1 and S2-Supplementary Materials). Phenolic compounds present in eight different seaweeds were tentatively identified from their *m/z* value and MS spectra in both negative and positive ionization modes ([M − H]⁻/[M + H]⁺) using Agilent LC-MS Qualitative Software and Personal Compound Database and Library (PCDL). Compounds with mass error < ± 5 ppm and PCDL library score more than 80 were selected for further MS/MS identification and *m/z* characterization purposes.

In the present work, LC-MS/MS enabled the tentative identification of 54 phenolic compounds, including 22 phenolic acids, 17 flavonoids, 11 other polyphenols and 4 lignans (Table 3).

Table 3. Characterization of phenolic compounds in seaweeds by using LC-ESI-QTOF-MS/MS.

No.	Proposed Compounds	Molecular Formula	RT (min)	Ionization (ESI⁺/ESI⁻)	Molecular Weight	Theoretical (m/z)	Observed (m/z)	Mass Error (ppm)	MS/MS Product Ions	Seaweeds
Phenolic acid										
Hydroxybenzoic acids										
1	Vanillic acid 4-sulfate	$C_8H_8O_7S$	9.112	$[M-H]^-$	247.9991	246.9918	246.9925	2.83	217, 203, 167	* Sargassum sp., Centroceras sp., Ulva sp.
2	Gallic acid	$C_7H_6O_5$	9.885	** $[M-H]^-$	170.0215	169.0142	169.0138	-2.37	125	Centroceras sp.
3	4-Hydroxybenzoic acid 4-O-glucoside	$C_{13}H_{16}O_8$	11.515	$[M-H]^-$	300.0845	299.0772	299.0778	2.01	255, 137	Sargassum sp.
4	Protocatechuic acid 4-O-glucoside	$C_{13}H_{16}O_9$	13.546	** $[M-H]^-$	316.0794	315.0721	315.0719	-0.63	153	* Centroceras sp., Grateloupia sp.
5	p-Hydroxybenzoic acid	$C_7H_6O_3$	32.906	$[M-H]^-$	138.0317	137.0244	137.0240	-2.91	93	* Ulva sp., Caulerpa sp., Centroceras sp.
6	Ellagic acid glucoside	$C_{20}H_{16}O_{13}$	38.451	$[M-H]^-$	464.0591	463.0518	463.0518	0.01	301	Ecklonia sp.
Hydroxycinnamic acids										
7	3-Sinapoylquinic acid	$C_{18}H_{22}O_{10}$	7.005	** $[M-H]^-$	398.1213	397.1140	397.1144	1.01	223, 179	* Centroceras sp., Ecklonia sp.
8	Cinnamoyl glucose	$C_{15}H_{18}O_7$	8.861	** $[M-H]^-$	310.1053	309.098	309.0992	3.88	147, 131, 103	* Codium sp., Ulva sp.
9	Caffeoyl glucose	$C_{15}H_{18}O_9$	10.983	** $[M-H]^-$	342.0951	341.0878	341.0882	1.17	179, 161	* Ecklonia sp., Centroceras sp.
10	Caffeic acid 3-O-glucuronide	$C_{15}H_{16}O_{10}$	14.259	** $[M-H]^-$	356.0743	355.0670	355.0671	0.28	179	Caulerpa sp.
11	Chlorogenic acid	$C_{16}H_{18}O_9$	15.004	** $[M-H]^-$	354.0951	353.0878	353.0862	-4.53	253, 190, 144	* Centroceras sp., Caulerpa sp.
12	Caffeic acid	$C_9H_8O_4$	18.274	$[M-H]^-$	180.0423	179.0350	179.0350	0.01	151, 143, 133	Caulerpa sp.
13	Caffeic acid 4-sulfate	$C_9H_8O_7S$	18.291	$[M-H]^-$	259.9991	258.9918	258.9929	4.25	215, 179, 135	Caulerpa sp.
14	Caffeoyl tartaric acid	$C_{13}H_{12}O_9$	24.061	** $[M-H]^-$	312.0481	311.0408	311.0403	-1.61	161	* Grateloupia sp., Centroceras sp.
15	Isoferulic acid 3-sulfate	$C_{10}H_{10}O_7S$	24.520	** $[M-H]^-$	274.0147	273.0074	273.0086	4.4	193, 149	Caulerpa sp.
16	Sinapic acid	$C_{11}H_{12}O_5$	25.852	** $[M-H]^-$	224.0685	223.0612	223.0621	4.03	205, 179, 163	* Ulva sp., Caulerpa sp., Grateloupia sp.
17	Ferulic acid	$C_{10}H_{10}O_4$	32.604	$[M-H]^-$	194.0579	193.0506	193.0513	3.63	178, 149, 134	Caulerpa sp.
18	Coumaric acid	$C_9H_8O_3$	33.797	** $[M+H]^+$	164.0473	163.0400	163.0406	3.68	119	* Ulva sp., Ecklonia sp.
19	Sinapine	$C_{16}H_{24}NO_5$	88.066	$[M+H]^+$	310.1652	310.1654	310.1646	-2.58	251, 207, 175	Codium sp.
Hydroxyphenylpentanoic acids										
20	5-(3',5'-dihydroxyphenyl)-γ-valerolactone 3-O-glucuronide	$C_{17}H_{20}O_{10}$	14.855	** $[M-H]^-$	384.1056	383.0983	383.1001	4.70	221, 206, 191	* Ecklonia sp., Codium sp.
21	5-(3',4'-dihydroxyphenyl)-valeric acid	$C_{11}H_{14}O_4$	51.563	** $[M-H]^-$	210.0892	209.0819	209.0821	0.96	165, 150	Caulerpa sp.
Hydroxyphenylacetic acids										
22	2-Hydroxy-2-phenylacetic acid	$C_8H_8O_3$	6.18	** $[M+H]^+$	152.0473	153.0546	153.055	2.61	125	* Centroceras sp., Caulerpa sp., Sargassum sp.
Flavonoids										
Anthocyanins										

Table 3. Cont.

No.	Proposed Compounds	Molecular Formula	RT (min)	Ionization (ESI+/ESI−)	Molecular Weight	Theoretical (m/z)	Observed (m/z)	Mass Error (ppm)	MS/MS Product Ions	Seaweeds
23	Delphinidin 3-O-sambubioside	$C_{26}H_{29}O_{16}$	9.327	$[M + H]^+$	597.1464	597.1456	597.1473	2.85	303, 257, 229	*Grateloupia* sp.
24	Isopeonidin 3-O-arabinoside	$C_{21}H_{21}O_{10}$	41.658	$[M + H]^+$	433.1134	433.1135	433.1136	0.23	271, 253, 243	*Centroceras* sp.
25	Malvidin 3-O-glucoside	$C_{23}H_{25}O_{12}$	54.152	$[M + H]^+$	493.1343	493.1346	493.1343	−0.61	331	*Centroceras* sp.
Flavanols										
26	Gallocatechin	$C_{15}H_{14}O_7$	7.604	** $[M − H]^-$	306.0740	305.0667	305.0668	0.33	261, 219	* *Caulerpa* sp., *Ulva* sp., *Dasya* sp., *Ecklonia* sp., *Sargassum* sp.
27	3′-O-Methylcatechin	$C_{16}H_{16}O_6$	17.857	** $[M − H]^-$	304.0947	303.0874	303.0886	3.96	271, 163	*Grateloupia* sp.
28	Catechin (isomer)	$C_{15}H_{14}O_6$	45.118	$[M − H]^-$	290.0790	289.0717	289.0731	4.84	245, 205, 179	*Caulerpa* sp.
Flavonols										
29	Quercetin 3-O-(6″-malonyl-glucoside)	$C_{24}H_{22}O_{15}$	9.902	$[M − H]^-$	550.0959	549.0886	549.0887	0.18	463, 301, 161	* *Centroceras* sp., *Caulerpa* sp.
30	5,3′,4′-Trihydroxy-3-methoxy-6,7-methylenedioxyflavone 4′-O-glucuronide	$C_{23}H_{20}O_{14}$	33.878	$[M − H]^-$	520.0853	519.0780	519.0779	−0.19	343	*Ecklonia* sp.
31	3,7-Dimethylquercetin	$C_{17}H_{14}O_7$	80.642	$[M − H]^-$	330.0740	329.0667	329.0674	2.13	314, 299, 271	*Centroceras* sp.
Flavones										
32	Rhoifolin	$C_{27}H_{30}O_{14}$	44.036	** $[M − H]^-$	578.1636	577.1563	577.1588	4.33	413, 269	*Centroceras* sp.
Isoflavonoids										
33	Sativanone	$C_{17}H_{16}O_5$	4.240	$[M − H]^-$	300.0998	299.0925	299.0918	−2.34	284, 269, 225	*Ecklonia* sp.
34	Glycitein 7-O-glucuronide	$C_{22}H_{20}O_{11}$	4.454	** $[M − H]^-$	460.1006	459.0933	459.0923	−2.18	283, 268, 117	*Centroceras* sp.
35	3′,4′,5,7-Tetrahydroxyisoflavanone	$C_{15}H_{12}O_6$	4.640	** $[M − H]^-$	288.0634	287.0561	287.0556	−1.74	269, 259	* *Caulerpa* sp., *Grateloupia* sp., *Centroceras* sp.
36	3′-O-Methylequol	$C_{16}H_{16}O_4$	4.803	** $[M − H]^-$	272.1049	271.0976	271.0972	−1.48	147, 123, 121	* *Ecklonia* sp., *Grateloupia* sp.
37	Dalbergin	$C_{16}H_{12}O_4$	9.344	** $[M − H]^-$	268.0736	267.0663	267.0666	1.12	252, 224, 180	*Grateloupia* sp., *Centroceras* sp.
38	Dihydrobiochanin A	$C_{16}H_{14}O_5$	80.715	** $[M − H]^-$	286.0841	285.0768	285.0771	1.05	270	* *Codium* sp., *Centroceras* sp.
39	3′-Hydroxydaidzein	$C_{15}H_{10}O_5$	86.956	$[M − H]^-$	270.0528	269.0455	269.0457	0.74	151, 117, 107	* *Grateloupia* sp., *Centroceras* sp., *Caulerpa* sp., *Ecklonia* sp.
Other polyphenols										
Hydroxybenzaldehydes										
40	p-Hydroxybenzaldehyde	$C_7H_6O_2$	15.921	$[M − H]^-$	122.0368	121.0295	121.0295	0.01	92, 77	* *Dasya* sp., *Ecklonia* sp., *Codium* sp.
Hydroxycoumarins										
41	Urolithin A	$C_{13}H_8O_4$	4.64	$[M − H]^-$	228.0423	227.0350	227.0341	−3.96	198, 182	*Grateloupia* sp.
42	Scopoletin	$C_{10}H_8O_4$	84.705	** $[M − H]^-$	192.0423	191.0350	191.0352	1.05	176, 147	* *Codium* sp., *Grateloupia* sp., *Sargassum* sp.
Phenolic terpenes										
43	Rosmanol	$C_{20}H_{26}O_5$	24.965	$[M + H]^+$	346.1780	347.1853	347.1843	−2.88	301, 231	* *Dasya* sp., *Ulva* sp., *Grateloupia* sp., *Ecklonia* sp., *Codium* sp.

Table 3. Cont.

No.	Proposed Compounds	Molecular Formula	RT (min)	Ionization (ESI⁺/ESI⁻)	Molecular Weight	Theoretical (m/z)	Observed (m/z)	Mass Error (ppm)	MS/MS Product Ions	Seaweeds
44	Carnosol	$C_{20}H_{26}O_4$	85.931	** [M − H]⁻	330.1831	329.1758	329.1747	−3.34	287, 286, 285	* Codium sp., Caulerpa sp.
45	Carnosic acid	$C_{20}H_{28}O_4$	86.958	** [M − H]⁻	332.1988	331.1915	331.1912	−0.91	287, 269	* Ecklonia sp., Dasya sp., Codium sp., Sargassum sp.
Tyrosols										
46	Hydroxytyrosol 4-O-glucoside	$C_{14}H_{20}O_8$	36.653	** [M − H]⁻	316.1158	315.1085	315.1091	1.90	153, 123	* Centroceras sp., Dasya sp., Grateloupia sp., Sargassum sp.
47	3,4-DHPEA-EDA	$C_{17}H_{20}O_6$	87.423	[M − H]⁻	320.1260	319.1187	319.1200	4.07	301, 275, 195	Caulerpa sp.
Other polyphenols										
48	3,4-Dihydroxyphenylglycol	$C_8H_{10}O_4$	7.005	[M − H]⁻	170.0579	169.0506	169.0503	−1.77	141, 139, 123	Centroceras sp.
49	Phloroglucinol Isopropyl	$C_6H_6O_3$	14.793	[M − H]⁻	126.0317	125.0244	125.0242	−1.59	97	* Ecklonia sp., Sargassum sp.
50	3-(3,4-dihydroxyphenyl)-2-hydroxypropanoate	$C_{12}H_{16}O_5$	24.882	** [M − H]⁻	240.0998	239.0925	239.0919	−2.51	195, 155, 99	Dasya sp.
Lignans										
Lignan derivatives										
51	2'-Hydroxyenterolactone	$C_{18}H_{18}O_5$	7.781	[M − H]⁻	314.1154	313.1081	313.1082	0.32	295, 283	Grateloupia sp.
52	Arctigenin	$C_{21}H_{24}O_6$	8.131	** [M − H]⁻	372.1573	371.1500	371.1509	2.42	356, 312, 295	* Centroceras sp., Sargassum sp.
53	Dimethylmatairesinol	$C_{22}H_{26}O_6$	83.663	[M + H]⁺	386.1729	387.1802	387.1805	0.77	372, 369, 357, 329	* Caulerpa sp., Dasya sp.
54	Deoxyschisandrin	$C_{24}H_{32}O_6$	85.152	** [M + H]⁺	416.2199	417.2272	417.2286	3.36	402, 347, 316, 301	* Ecklonia sp., Codium sp., Sargassum sp.

* Compound was detected in more than one seaweed samples, data presented in this table are from asterisk sample. ** Compounds were detected in both negative [M − H]⁻ and positive [M + H]⁺ mode of ionization while only single mode data was presented. RT = stands for "retention time".

2.3.1. Phenolic Acids

Phenolic acids have been reported as the most abundant phenolic compounds in red, green and brown algae [21]. In the present work, four sub-classes of phenolic acid were detected, including hydroxybenzoic acids, hydroxycinnamic acids, hydroxyphenylpentanoic acids and hydroxyphenylacetic acids.

Hydroxybenzoic Acids Derivatives

Six hydroxybenzoic acid derivatives were detected in six out of eight seaweeds. The typical neutral losses of CO_2 (44 Da) and hexosyl moiety (162 Da) were observed in phenolic acids [32]. Compound **2** with $[M - H]^-$ *m/z* at 169.0138 was only detected from red seaweed *Centroceras* sp., and characterized as gallic acid based on the product ion at 125 *m/z*, corresponding to the loss of CO_2 (44 Da) from precursor ion [32]. Gallic acid was also previously reported as abundant in the brown seaweed *Himanthalia elongate* [33]. *p*-Hydroxybenzoic acid (Compound **5** with $[M - H]^-$ ion at *m/z* 137.0240) present in *Ulva* sp., *Caulerpa* sp. and *Centroceras* sp. was identified and confirmed by MS^2 experiments (Figure 1). In the MS^2 spectrum of *m/z* 137.0240, the product ion at *m/z* 93 was due to the loss of a CO_2 (44 Da) from the parent ion [32]. This is consistent with *p*-hydroxybenzoic acid also being found in seaweeds from the Danish coastal area [34].

4-Hydroxybenzoic acid 4-*O*-glucoside (Compound **3**, *m/z* 299.0778), protocatechuic acid 4-*O*-glucoside (Compound **4**, *m/z* 315.0719) and ellagic acid glucoside (compound **6**, *m/z* 463.0518) were identified in *Sargassum* sp., *Centroceras* sp., *Grateloupia* sp. and *Ecklonia* sp. in both modes. The molecular ions of 4-hydroxybenzoic acid 4-*O*-glucoside, protocatechuic acid 4-*O*-glucoside and ellagic acid glucoside produced the product ions at *m/z* 137, 153 and 301, respectively, indicating the loss of hexosyl moiety (162 Da) from precursor ions [32].

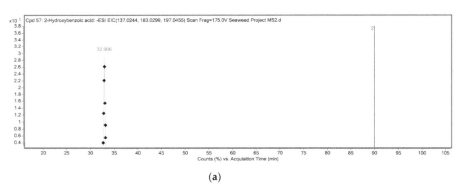

(a)

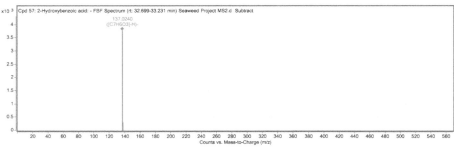

(b)

Figure 1. *Cont.*

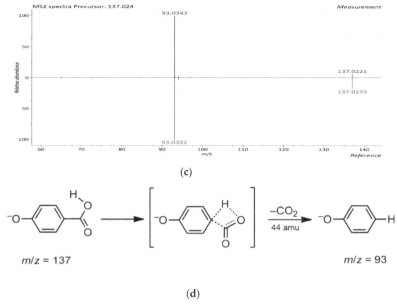

(c)

(d)

Figure 1. The LC-ESI-QTOF-MS/MS characterization of *p*-hydroxybenzoic acid; (**a**) A chromatograph of *p*-hydroxybenzoic acid (Compound 5, Table 3), Retention time (RT = 32.906 min) in the negative mode of ionization [M − H]⁻ tentatively identified in *Ulva* sp.; (**b**) Mass spectra of *p*-hydroxybenzoic acid with observed/precursor of *m/z* 137.0240 in *Ulva* sp.; (**c**) MS/MS spectrum of *p*-hydroxybenzoic acid reflecting the product ion of *m/z* 93, confirmation via online LC-MS library and database; (**d**) Fragmentation of *p*-hydroxybenzoic acid in negative mode [M − H]⁻, with observed/precursor of *m/z* 137, showing product ion of *m/z* 93 due to the loss of a CO_2 (44 Da).

Hydroxycinnamic Acids and Other Phenolic Acid Derivatives

Thirteen hydroxycinnamic acids derivatives, two hydroxyphenylpentanoic acids and one hydroxyphenylacetic acid were tentatively identified in our study.

Compound (**7**) was identified as 3-sinapoylquinic acid based on the precursor ion [M − H]⁻ at *m/z* 397.1144, with product ions at *m/z* 223 (sinapic acid ion) and *m/z* 179 (sinapic acid − COO) in *Centroceras* sp. and *Ecklonia* sp., which was previously characterized in extracts of arnica flower [35]. Cinnamoyl glucose (Compound **8**) was also found in *Codium* sp. and *Ulva* sp. The presence of cinnamoyl glucose was confirmed by a [M − H]⁻ *m/z* at 309.0992, which yielded product ions at *m/z* 147, *m/z* 131 and *m/z* 103, indicating the expected loss of hexosyl moiety (162 Da), $C_6H_{10}O_6$ (178 Da) and $C_7H_{10}O_7$ (206 Da), respectively [36].

Compound (**9**), having a precursor ion [M − H]⁻ *m/z* at 341.0882, was tentatively characterized as caffeoyl glucose and was present in *Ecklonia* sp. and *Centroceras* sp. The MS² analysis showed the product ions at *m/z* 179 [M − H − 162] and *m/z* 161 [M − H − 180], consistent with losses of hexosyl moiety and further loss of H_2O [37]. Compound **14** was tentatively characterized as caffeoyl tartaric acid found in *Grateloupia* sp. and *Centroceras* sp. based on [M − H]⁻ *m/z* at 311.0403. The identification was further supported by the MS² spectrum, which exhibited typical product ion at *m/z* 161, formed by the neutral loss of 150 mass units as a result of tartaric acid fission [38]. To the best of our knowledge, caffeoyl tartaric acid and caffeoyl glucose were previously reported primarily in fruit samples such as grape, however, it was the first time that they were reported in seaweeds [39]. For caffeic acid 3-*O*-glucuronide found in *Caulerpa* sp. (Compound **10** with [M − H]⁻ *m/z* of 355.0671), MS/MS fragmentation yielded the predominant ion at *m/z* 179 after the loss of glucuronide moiety (176 Da), indicating the presence of caffeic acid ion [37].

Compound **11** was tentatively characterized as chlorogenic acid, and only found in *Centroceras* sp. and *Caulerpa* sp. based on [M − H]⁻ *m/z* at 353.0862, and identification was further supported by the MS² spectrum. The identity of chlorogenic acid was confirmed by the product ions at *m/z* 253 [M − H − 100], 190 [M − H − 163] and 144 [M − H − 209], corresponding to the loss of three H_2O and HCOOH; three H_2O and $C_6H_5O_2$; H_2O and $C_7H_{11}O_6$, respectively [40]. Chlorogenic acid was also present in the green seaweed *Capsosiphon fulvescens* from Korea, according to previous research [41].

Four hydroxycinnamic acid derivatives (Compound **12**, **13**, **15** and **17**) were detected in *Caulerpa* sp. in both ionization modes, and were tentatively identified as caffeic acid, caffeic acid 4-sulfate, isoferulic acid 3-sulfate and ferulic acid, according to the precursor ions [M − H]⁻ at *m/z* 179.0350, 258.9929, 273.0086 and 193.0513, respectively. The identification of caffeic acid was confirmed by the product ions at *m/z* 151 [M − H − 28], *m/z* 143 [M − H − 36] and *m/z* 133 [M − H − 46], representing the loss of CO, two H_2O units and HCOOH, respectively, from the precursor ion [40]. In the MS² experiment of caffeic acid 4-sulfate, the spectra displayed the product ions at *m/z* 179, (presence of caffeic acid ion) and at *m/z* 135, corresponding to the loss of SO_3 (80 Da) and further loss of CO_2 (44 Da) from the precursor ion [42]. The similar cleavage was observed in the MS² spectra of isoferulic acid 3-sulfate, which displayed the product ions at *m/z* 193 [M − H − SO_3] and *m/z* 149 [M − H − SO_3 − CO_2], consistent with the presence of isoferulic acid ion (193 Da) and further loss of CO_2 [42], while the product ions at *m/z* 178 (M − H − 15, loss of CH_3), *m/z* 149 (M − H − 44, loss of CO_2) and *m/z* 134 (M − H − 59, loss of CH_3 and H_2O) identified ferulic acid [43]. According to a previous study, caffeic acid and ferulic acid were also found in some seaweeds [33,34].

Sinapic acid (Compound **16**) were detected in both positive (ESI⁺) and negative (ESI⁻) modes in *Ulva* sp. *Caulerpa* sp. and *Grateloupia* sp. with an observed [M − H]⁻ *m/z* at 223.0621. In the MS² spectrum of sinapic acid, the product ions at *m/z* 205, 179 and 163 were due to the loss of H_2O (18 Da), CO_2 (44 Da) and two CH_2O units (60 Da) from the parent ion, respectively, which was comparable with the fragmentation rules of sinapinic acid [42].

Coumaric acid (compound **18** with [M − H]⁻ *m/z* at 163.0406), yielding a main product ion at *m/z* 119, which corresponded to loss of CO_2 (44 Da), was found in *Caulerpa* sp. [43]. The presence of coumaric acid in marine seaweeds was also previously reported [34].

Three other phenolic acid derivatives were also detected, including two hydroxyphenylpentanoic acid derivatives and one hydroxyphenylacetic acid derivative. To our best knowledge, this is the first time these other phenolic acid derivatives have been reported in seaweeds. Phenolic acids are the predominant polyphenol compounds found in different seaweeds, which were characterized by using LC-MS in previous studies, and displayed remarkable antioxidant potential [44,45].

2.3.2. Flavonoids

Flavonoid is the main class of phenolic compounds responsible for the antioxidant and free radical scavenging properties observed in seaweed [24]. In the present study, a total of 17 flavonoids were tentatively identified, which were further divided into anthocyanins (03), flavanols (03), flavonols (03), flavone (01) and isoflavonoids (07).

Anthocyanins, Flavanols and Flavonols Derivatives

Anthocyanins are naturally occurring pigments that belong to the subclass of flavonoids, which were previously reported in brown Irish seaweeds [46]. In our study, three anthocyanin derivatives were detected only in the red seaweeds *Grateloupia* sp. and *Centroceras* sp., in positive ionization mode. This is the first time all of these anthocyanins derivatives have been reported in seaweeds.

Three flavanols (Compound **26**, **27** and **28**) were detected in all seaweeds except *Centroceras* sp. and *Codium* sp. Compound (**26**) showing precursor ion [M − H]⁻ at *m/z* 305.0668 in negative mode, was the most widely distributed flavanol and was identified as gallocatechin presenting in *Caulerpa* sp., *Ulva* sp., *Dasya* sp., *Ecklonia* sp. and *Sargassum* sp. The presence of gallocatechin derivatives in brown seaweed *ascophyllum nodosum* was reported by Agregán, Munekata, Franco, Dominguez, Carballo and Lorenzo [44] based on the production $[M - H]^-$ ion at m/z 305. In MS/MS experiment, the product ion at 261 $[M - H - 44]$ was due to the loss of CO_2 and at m/z 219 $[M - H - 86]$ was caused by the loss of C_3O_2 and H_2O [43]. 3'-*O*-methylcatechin (Compound **27** with $[M - H]^-$ m/z of 303.0886) was identified in *Grateloupia* sp. in the present study, with the product ions at m/z 271 (M − H − 32, loss of CH_3OH) and m/z 163 (M − H − 140, loss of CH_3OH and $C_6H_5O_2$) [47]. Catechin (isomer) was proposed as compound (**28**), from *Caulerpa* sp., with a precursor ion $[M - H]^-$ m/z of 289.0731. The MS^2 spectrum showed the product ions at m/z 245, m/z 205, and m/z 179, indicating the loss of CO_2 (44 Da), flavonid A ring (84 Da) and flavonid B ring (110 Da) from the precursor ion, respectively [32].

Three flavonols were detected in negative mode in *Centroceras* sp., *Caulerpa* sp. and *Ecklonia* sp. 3,7-dimethylquercetin detected in *Centroceras* sp. was assigned for compound (**31**) based on the observed $[M - H]^-$ m/z of 329.0674. The further identification of 3,7-dimethylquercetin was achieved by comparing the previous study, which characterized the same compound from *Ipomoea batatas* leaves and showed the product ions at m/z 314, m/z 299 and m/z 271, corresponding to the loss of CH_3 (15 Da), two CH_3 (30 Da) and two CH_3 plus CO unit from the precursor ion, respectively [48].

Rhoifolin (Compound **32** with $[M - H]^-$ m/z at 577.1588) was the only flavone identified in *Centroceras* sp. with the product ions at m/z 413 (M − H − 164) and m/z 269 (M − H − 308), representing the loss of rhamnose moiety and H_2O (164 Da) and hexosyl moiety plus rhamnose moiety (308 Da) from the parent ion [49]. This is the first time that all of the flavonols and flavone derivatives identified in the current study have been reported in seaweeds.

Isoflavonoids Derivatives

Isoflavonoids derivatives (a total of seven) were the most diverse flavonoids identified in seaweeds. Sativanone (Compound **33**) was only detected in *Ecklonia* sp. in negative mode with $[M - H]^-$ m/z at 299.0918. The identity was confirmed by comparing the previous study which characterized sativanone in *Dalbergia odorifera* using LC-MS/MS, and the spectrum displayed the product ions at m/z 284 (M − H − 15, loss of CH_3 from B-ring) and at m/z 269 (M − H − 30, loss of two CH_3) and at m/z 225 (M − H − 74, loss of two CH_3 and a CO_2) [50]. Compound **37** with $[M - H]^-$ m/z at 267.0666 exhibited characteristic fragment ions at m/z 252 $[M - H - CH_3]$, m/z 224 $[M - H - CH_3 - CO]$ and m/z 180 $[M - H - CH_3 - CO - CO_2]$ was identified as dalbergin [50]. To the best of our knowledge, this is the first time that isoflavonoids derivatives were identified and characterized in seaweeds. Flavonoids in different seaweeds with high antioxidant potential have already been reported, which are promising as functional food ingredients or dietary supplements for daily intake [51].

2.3.3. Other Polyphenols

Eleven other polyphenols found were classified as hydroxybenzaldehyde (01), hydroxycoumarins (02), phenolic terpenes (03), tyrosol (02) and other polyphenols (03).

Hydroxybenzaldehydes, hydroxycoumarins and hydroxyphenylpropenes Derivatives

p-Hydroxybenzaldehyde (Compound **40** with $[M - H]^-$ at m/z 121.0295, RT = 15.921 min) was the only hydroxybenzaldehyde presenting in *Dasya* sp., *Ecklonia* sp. and *Codium* sp. The MS^2 spectrum of *p*-hydroxybenzaldehyde displayed the product ions at m/z 92 and m/z 77, indicating the loss of CHO (29 Da) and CO_2 (44 Da) [52]. The presence of *p*-hydroxybenzaldehyde in Irish brown seaweed *Himanthalia elongate* was also previously reported by Rajauria, Foley and Abu-Ghannam [9]. Two hydroxycoumarins derivatives (Compound **41** and **42**) were discovered. Urolithin A with $[M - H]^-$ m/z at 227.0341 was assigned as compound **41**, from *Grateloupia* sp. MS/MS identification by product ions at m/z 198 (M − H − 29, loss of CHO) and 182 m/z (M − H − 45, loss of COOH) [53]. Scopoletin with $[M - H]^-$ m/z at 191.0352 was proposed as compound **42** found in *Codium* sp., *Grateloupia* sp. and

Sargassum sp., and was identified by the neutral loss of CH_3 (15 Da) and CO_2 (44 Da), resulting in product ions at *m/z* 176 and *m/z* 147, respectively [54].

Phenolic Terpenes Derivatives

Rosmanol (Compound **43**), showing as precursor ion at $[M + H]^+$ at *m/z* 347.1843, was detected in *Dasya* sp., *Ulva* sp., *Grateloupia* sp., *Ecklonia* sp. and *Codium* sp. The product ions at *m/z* 301 and *m/z* 231 came from the loss of a unit of H_2O and CO (46 Da), and cleavage of molecules pentene, water, and carbon monoxide [55]. Carnosic acid (Compound **45**), identified based on $[M - H]^-$ *m/z* at 331.1912, was found in *Ecklonia* sp. *Dasya* sp., *Codium* sp. and *Sargassum* sp. The molecular ion of carnosic acid (*m/z* 331.1912) produced the major fragment ion at *m/z* 287 and *m/z* 269, corresponding to the loss of CO_2 and further loss of H_2O from the parent ion [56]. Hermund, et al. [57] also confirmed the presence of carnosic and carnosol as synergistic antioxidants with radical scavenging activity in brown seaweed *Fucus vesiculosus*.

Tyrosols and Other Polyphenols Derivatives

Compounds (**46**) were present in *Centroceras* sp., *Dasya* sp., *Grateloupia* sp., and Sargassum, and was tentatively identified as hydroxytyrosol 4-*O*-glucoside based on the observed $[M - H]^-$ ions at *m/z* 315.1091. In the MS^2 spectrum of hydroxytyrosol 4-*O*-glucoside, the typical loss of hexosyl moiety (162 Da) was observed from precursor, resulting in product ions at *m/z* 153 [52]. Compound **47** with $[M - H]^-$ *m/z* at 319.1200 was only detected from *Caulerpa* sp., and characterized as 3,4-DHPEA-EDA based on the product ions at *m/z* 301, *m/z* 275 and *m/z* 195, corresponding to loss of H_2O (18 Da), CO_2 (44 Da) and $C_5H_6(CHO)_2$ (124 Da) from the precursor ion [58]. This is the first report of the presence of these tyrosol derivatives in seaweed, while 3,4-DHPEA-AC was previously reported by Gomez-Alonso, et al. [59] in Cornicabra olive oil variety.

Three other polyphenols derivatives were detected, including compound (**49**) with $[M - H]^-$ at *m/z* 125.0242, which was proposed as phloroglucinol appearing in brown seaweed *Ecklonia* sp. and *Sargassum* sp. The identity was confirmed by the MS^2 spectrum, which produced a major fragment ion at *m/z* 97, resulting from the loss of CO (28 Da) from the precursor ion [9]. The presence of phloroglucinol in Irish brown seaweed *Himanthalia elongate* was previously reported by Rajauria, Foley and Abu-Ghannam [9] according to the precursor and product ions, and further confirmed by the UV spectrum and retention time using phloroglucinol standard.

2.3.4. Lignans

Lignans were minor components present in the seaweeds. In the present study, a total of four lignans were shown to be present in seven out of eight seaweeds.

Lignans Derivatives

Compounds **52** detected in *Centroceras* sp. and *Sargassum* sp. was tentatively characterized as arctigenin according to the precursor ions at $[M - H]^-$ *m/z* 371.1509. Fragmentation of arctigenin yielded product ions at *m/z* 356, *m/z* 312 and *m/z* 295, corresponding to the loss of CH_3 (15 Da), unit of CH_3 and CO_2 (59 Da), and unit of CH_3, CO_2 and OH (76 Da), respectively [60]. Compound **54** (deoxyschisandrin) displaying the $[M + H]^+$ *m/z* at 417.2286 and was found in *Ecklonia* sp. and confirmed by the characteristic ions at *m/z* 402 $[M - H - CH_3]$, *m/z* 347 $[M - H - C_5H_{10}]$, *m/z* 316 $[M - H - C_5H_{10} - OCH_3]$ and *m/z* 301 $[M - H - C_5H_{10} - OCH_3 - CH_3]$ [61]. Lignans are abundant in seaweeds, however, the lignans in the present study have not previously been reported in seaweeds [62]. Previously, it was reported that lignans are abundant in seaweeds with various health-promoting properties, including antioxidant, anti-inflammatory and antitumor activities [62,63]. In addition, some epidemiological studies have proposed the therapeutic potential of lignans in chronic diseases, such as cardiovascular disease, type 2 diabetes and cancers [64,65].

The screening and characterization of polyphenolic compounds showed that some of the polyphenols presented in these seaweeds have strong antioxidant potential. Hydroxycinnamic acid derivatives, hydroxybenzoic acids and their derivatives, protocatechuic acid, anthocyanins, flavonoids and their derivatives, hydroxybenzaldehydes, hydroxytyrosol, phloroglucinol and quercetin derivatives are regarded as potential compounds showing considerable free radical scavenging capacity [66–71]. The presence of these antioxidant compounds indicates that seaweeds can be good sources of polyphenols and could be utilized in food, feed, and pharmaceutical industries.

2.4. HPLC Quantitative Analysis

The quantitative analysis of targeted phenolic compounds was performed based on peak area computation using the calibration of corresponding standards and the result are presented as µg/g fresh weight of seaweeds (Table 4.). In total, seven polyphenols were targeted to quantify by HPLC-PDA, including six phenolic acids (gallic acid, caftaric acid, chlorogenic acid, caffeic acid, *p*-hydroxybenzoic acid and coumaric acid) and one flavonoid (catechin).

Table 4. Quantification of targeted phenolic compounds by high-performance liquid chromatography (HPLC) in seaweeds.

No.	Compound Name	Structure Formula	RT (min)	Concentration ($\mu g/g_{f.w.}$)	Seaweed Samples
1	Gallic acid	$C_7H_6O_5$	9.685	138.887 ± 0.02	*Centroceras* sp.
2	Chlorogenic acid	$C_{16}H_{18}O_9$	15.004	122.706 ± 0.01	*Centroceras* sp.
3	Caffeic acid	$C_9H_8O_4$	18.274	612.824 ± 0.02	*Caulerpa* sp.
4	Caftaric acid	$C_{13}H_{12}O_9$	24.532	19.667 ± 0.01	*Centroceras* sp.
5	*p*-hydroxybenzoic acid	$C_7H_6O_3$	32.906	846.083 ± 0.02	*Ulva* sp.
6	Coumaric acid	$C_9H_8O_3$	33.797	505.387 ± 0.03	*Ulva* sp.
7	Catechin	$C_{15}H_{14}O_6$	64.081	29.469 ± 0.03	*Caulerpa* sp.

RT = stands for "retention time".

The most abundant targeted phenolic compound was *p*-hydroxybenzoic acid (Compound **5**), which was present in *Ulva* sp. with the concentration of 846.0 ± 0.02 $\mu g/g_{f.w.}$ The *p*-hydroxybenzoic acid content of eight green and red seaweeds in South Africa was previously reported as ranging from 0.51 ± 0.01 to 13.53 ± 0.03 µg/g dry weight (d.w.) [72], which was significantly lower than that of *Ulva* sp. in the present study. Gallic acid (Compound **1**), chlorogenic acid (Compound **2**) and caftaric acid (Compound **4**) were detected in *Centroceras* sp. with the concentration of 138.9 ± 0.02 $\mu g/g_{f.w.}$, 122.7 ± 0.01 $\mu g/g_{f.w.}$ and 19.7 ± 0.01 $\mu g/g_{f.w.}$, respectively. Coumaric acid (Compound **6**) was quantified in *Ulva* sp. with concentrations of 505.4 ± 0.03 $\mu g/g_{f.w.}$ Caffeic acid (Compound **3**) and catechin (Compound **7**) were present in *Caulerpa* sp. with a concentration of 612.9 ± 0.02 $\mu g/g_{f.w.}$ and 29.5 ± 0.03 $\mu g/g_{f.w.}$, respectively. Concentrations of gallic acid, chlorogenic acid and caffeic acid in brown seaweed *Himanthalia elongate* were also previously reported, being measured as 96.3 ± 3.12 $\mu g/g_{d.w.}$, 38.8 ± 1.94 $\mu g/g_{d.w.}$ and 44.4 ± 2.72 $\mu g/g_{d.w.}$, respectively [33]. About 10 marine-derived pharmaceutical drugs were approved by the Food and Drug Administration (FDA), and 30 candidates were in different stages of clinical trials for application in a number of disease areas [73]. The presence of these abundant polyphenols provide evidence for seaweeds as a good source of antioxidants for application in food and pharmaceutical industries, while further toxicity, pharmacological and clinical studies are needed.

3. Materials and Methods

3.1. Chemicals and Reagents

Unless otherwise stated, all chemicals used for extraction, characterization and antioxidant assays were analytical grade and purchased from Sigma-Aldrich (Castle Hill, NSW, Australia). Gallic acid, quercetin, catechin, ascorbic acid, 2,2-diphenyl-1-picrylhydrazyl (DPPH), 2,4,6-tripyridyl-s-triazine (TPTZ), aluminum chloride, iron (III) chloride, vanillin, potassium persulfate and 2,2′-azino-bis(3-ethylbenzothiazoline-6-sulphonic acid) (ABTS) were purchased from Sigma-Aldrich (Castle Hill, NSW, Australia). Sulfuric acid 98% was from RCI Labscan (Rongmuang, Thailand) and sodium carbonate anhydrous was from Chem-Supply Pty Ltd. (Adelaide, SA, Australia). Analytical-grade methanol, ethanol, hydrochloric acid, anhydrous sodium acetate and hydrated sodium acetate were from Fisher Chemical (Waltham, MA, USA). Acetic acid solution and acetonitrile, which comprised the mobile phases for HPLC and LC-MS, were from Sigma-Aldrich (St. Louis, MO, USA) and LiChrosolv (Darmstadt, Germany), respectively. The HPLC reference standards including gallic acid, caftaric acid, chlorogenic acid, caffeic acid, *p*-hydroxybenzoic acid, coumaric acid and catechin, were purchased from Sigma-Aldrich (St. Louis, MO, USA). Water was deionized to reach a resistivity of 18.2 MΩ/cm using a Millipore Milli-Q Gradient Water Purification System (Darmstadt, Germany) and was filtered through a 0.45 μm type Millipak® Express 20 Filter (Milli-Q, Darmstadt, Germany).

3.2. Sample Preparation and Extraction of Polyphenols

Eight seaweeds which were identified as Chlorophyta (green; *Ulva* sp., *Caulerpa* sp. and *Codium* sp.), Rhodophyta (Red; *Dasya* sp., *Grateloupia* sp. and *Centroceras* sp.) and Ochrophyta (Brown; *Ecklonia* sp. and *Sargassum* sp.) were freshly collected from Brighton Beach in March 2019, VIC, Australia. Seaweeds were morphologically identified to the genus level. Classifications for Rhodophyta and Chlorophyta were verified using cytochrome c oxidase subunit I (COI-5P) and Elongation factor Tu 1-*Escherichia coli* (strain K12) tufA sequence data, respectively, following the protocol of Saunders and Kucera [74].

Extracts were prepared by modifying the previous studies [75,76], 2 g of each seaweed was grounded and mixed with 10 mL of 80% ethanol followed by homogenization using an Ultra-Turrax® T25 homogenizer (Rawang, Selangor, Malaysia) at 10,000 rpm for 20 s. Then, incubation was carried out in a shaking incubator (ZWYR-240, Labwit, Ashwood, VIC, Australia) at 120 rpm at 4 °C for 16 h. Then, all the samples were centrifuged (Hettich Rotina 380R, Tuttlingen, Germany) at 10,000 rpm for 10 min. The supernatant was collected and stored at −20 °C for further analysis. For HPLC and LC-MS analysis, the extracts were filtered through a 0.45 μm syringe filter (Thermo Fisher Scientific Inc., Waltham, MA, USA).

3.3. Estimation of Polyphenols and Antioxidant Assays

For polyphenol estimation, TPC, TFC and TTC were measured, while for antioxidant potential, three different antioxidant assays, including DPPH, FRAP, and ABTS, were performed using the method of Feng, et al. [77]. The data were obtained by the Multiskan® Go microplate photometer (Thermo Fisher Scientific, Waltham, MA, USA).

3.3.1. Total Phenolic Content (TPC)

The total phenolic content of seaweed was determined using the Folin-Ciocalteu's method [13] with some modifications. Twenty-five microliters of standards and samples (supernatant), 25 μL of 25% (*v/v*) folin reagent solution and 200 μL water were added to the wells in a 96-well plate (Corning Inc., Corning, NY, USA) and incubated at 25 °C for 5 min. Then, 25 μL of 10% (*w/w*) sodium carbonate was added and further incubated for 1 h at 25°C. The absorbance was measured at 765 nm against a blank using a Multiskan® Go microplate photometer (Thermo Fisher Scientific, Waltham, MA, USA). The calibration curve was plotted using a gallic acid standard ranging from 0 to 200 μg/mL in ethanolic solution and the results were presented as microgram equivalents of gallic acid equivalents (GAE) per gram ± standard error (SE) on the basis of fresh weight (f.w.) (y = 0.0059x + 0.0593, R^2 = 0.9996).

3.3.2. Total Flavonoid Content (TFC)

The total flavonoid content was measured by aluminum chloride colorimetry according to Chan, Matanjun, Yasir and Tan [25], with some modifications. Methanolic quercetin standards and samples (80 μL) were added to the 96-well plate. Then, 80 μL of 2% (*w/v*) aluminum chloride (diluted with analytical grade ethanol) and 120 μL 50 g/L sodium acetate was added the wells in the plate followed by the incubation at 25 °C for 2.5 h in the dark. The calibration curve was plotted using quercetin standards ranging from 0 to 50 μg/mL and the results are presented as microgram equivalents of quercetin equivalents (QE)/$g_{f.w.}$ ± SE (y = 0.0195x + 0.0646, R^2 = 0.999).

3.3.3. Total Tannins Content (TTC)

Total tannin content was measured by modifying the method of Rebaya, et al. [78]. Sample/standard (25 μL of supernatant or standard), 150 μL 4% (*w/v*) methanolic vanillin solution and 25 μL 32% (*v/v*) sulfuric acid (diluted with methanol) were mixed in a 96-well plate and incubated at room temperature for 15 min. The absorbance was measured at 500 nm wavelength against a blank using the microplate reader. The calibration curve was plotted by catechin methanolic solution ranging from 0 to 1000 μg/mL and the results are presented as microgram equivalents of catechin (CE)/$g_{f.w.}$ ± SE (y = 0.0005x + 0.0578, R^2 = 0.9854).

3.3.4. 2,2-diphenyl-1-picrylhydrazyl (DPPH) Assay

DPPH radical scavenging activities of different extracts were determined based on Chan et al. [25] with some modifications. Quantities of 40 μL samples/standards and 260 μL of 0.1 mM methanolic DPPH were added to a 96-well plate. The reaction mixture was incubated for 30 min in the dark at room temperature, and the absorbance was measured under 517 nm wavelength against a blank. The standard curve was plotted by ascorbic acid aqueous solution ranging from 0 to 50 μg/mL and the results are expressed as the microgram equivalents of ascorbic acid (AAE)/$g_{f.w.}$ ± SE (y = −0.0089x + 0.5988, R^2 = 0.9708).

3.3.5. Ferric Reducing Antioxidant Power (FRAP) Assay

The ferric reducing capabilities of the samples were measured using the FRAP method described by Matanjun, et al. [79], with slight modifications. The FRAP reagent was freshly prepared by mixing 300 mM acetate buffer, 10 mM TPTZ solution and 20 mM ferric chloride in the ratio of 10:1:1 (*v/v*). 20 μL samples/standards were added into the 96-well plate and mixed with 280 μL FRAP reagent. The mixture was incubated at 37 °C in the plate reader for 10 min before absorbance was measured at 593 nm. A standard curve was generated using ascorbic acid aqueous solution ranging from 0 to 50 μg/mL and the results are expressed as the microgram AAE/$g_{f.w.}$ ± SE (y = 0.009x + 0.403, R^2 = 0.9819).

3.3.6. 2,2′-Azino-bis-3-ethylbenzothiazoline-6-sulfonic Acid (ABTS) assay

The antioxidant activities of seaweeds were also measured by an ABTS assay according to Matanjun, Mohamed, Mustapha, Muhammad and Ming [79], with some modifications. $ABTS^+$ was prepared by mixing 5 mL of 7 mM ABTS solution and 88 μL of 140 mM potassium persulfate solution, and the mixture was placed in the dark for 16 h to allow free radical generation. The stock solution was further diluted with 45 mL analytical-grade ethanol while the absorbance of the dye was fixed at approximately 0.7 at 734 nm. Quantities of 10 μL of sample/standards and 290 μL prepared dye solution were added into a 96-well plate followed by incubation at room temperature for 6 min and the absorbance was measured at 734 nm wavelength. The standard curve was plotted using ascorbic acid aqueous solution ranging from 0 to 200μ/mL and the results are expressed as the microgram AAE/$g_{f.w.}$ ± SE (y = -0.0042x + 0.6923, R^2 = 0.9962).

3.4. LC-ESI-QTOF-MS/MS Characterization of Phenolic Compounds

LC-ESI-QTOF-MS/MS analysis was performed with an Agilent 1200 series HPLC (Agilent Technologies, Santa Clara, CA, USA) equipped with an Agilent 6520 Accurate-Mass Q-TOF LC-MS (Agilent Technologies, Santa Clara, CA, USA) via an electrospray ionization source (ESI). The separation was achieved by a Synergi Hydro-RP 80 Å, LC Column (250 mm × 4.6 mm, 4 µm) (Phenomenex, Lane Cove, NSW, Australia) at room temperature and the sample temperature was set at 10 °C. LC-MS/MS analysis were performed by modifying the method of Chao et al [66]. The mobile phase consisted of water/acetic acid (98:2, *v/v*; eluent A) and acetonitrile/acetic acid/ water (50:0.5:49.5, *v/v/v*; eluent B). The gradient profile was described as follows: 10–25% B (0–25 min), 25–35% B (25–35 min), 35–40% B (35–45 min), 40–55% B (45–75 min), 55–80% B (75–79 min), 80–90% B (79–82 min), 90–100% B (82–84 min), 100–10% B (84–87 min), isocratic 10% B (87–90 min). A volume of 6 µL was injected for each standard or sample and the flow rate was set at 0.8 mL/min. Nitrogen gas nebulization was set at 45 psi with a flow rate of 5L/min at 300 °C and the sheath gas was set at 11 L/min at 250 °C. The capillary and nozzle voltage were set at 3.5 kV and 500 V, respectively. A complete mass scan ranging from *m/z* 50 to 1300 was used, MS/MS analyses were carried out in automatic mode with collision energy (10, 15 and 30 eV) for fragmentation. Peak identification was performed in both positive and negative modes while the instrument control, data acquisition and processing were performed using MassHunter workstation software (Qualitative Analysis, version B.03.01) (Agilent Technologies, Santa Clara, CA, USA).

3.5. HPLC-PDA Quantitative Analysis of Individual Phenolic Compounds

The quantitative measurement of individual phenolic compounds present in seaweed samples was performed with an Agilent 1200 HPLC equipped with a photodiode array (PDA) detector by adopting the protocol of Peng et al. [68]. The same column and conditions were used as described above in LC-ESI-QTOF-MS/MS, except for a sample injection volume of 20 µL. The compositions of extracts were detected under λ 280 nm, 320 nm, and 370 nm by PDA detector simultaneously with 1.25 scan/s (peak width = 0.2 min) spectral acquisition rate. The targeted phenolic compounds were quantified based on linear regression of external standards peak area against concentration. Data acquisition and analysis were performed by MassHunter workstation software—version B.03.01 (Agilent Technologies, Santa Clara, CA, USA).

3.6. Statistical Analysis

All analyses were performed in triplicates and the results are presented as mean ± standard error (n = 3). Data were analyzed using Tukey's one-way analysis of variance (ANOVA) by Minitab® 19 for windows (Minitab, NSW, Australia). A significant difference was considered at the level of $p \leq 0.05$ using Tukey's HSD test.

4. Conclusions

Brown seaweed species showed significantly higher polyphenolic content and potential antioxidant capacity than green and red seaweeds. The antioxidant properties varied across different species. Application of LC-ESI-QTOF-MS/MS enabled the isolation and identification of 54 phenolic compounds present in seaweeds. Quantitative analysis of targeted compounds was achieved by calibration of standards using HPLC-PDA. Seven targeted compounds were quantified in seaweeds, with *p*-hydroxybenzoic acid being the most abundant. This is the first report that applied different antioxidant assays to estimate the antioxidant potential and applied LC-MS technique to isolate and characterize the polyphenols in some abundant Australian seaweed species. The presence of the various polyphenols with antioxidant potential was identified. Further toxicity, pharmacological and clinical studies should be explored before the application of these Australian seaweeds as ingredients in food, nutraceuticals and pharmaceutical products.

Supplementary Materials: The following are available online at http://www.mdpi.com/1660-3397/18/6/331/s1, Table S1: Characterization of phenolic compounds in seaweeds by using LC-ESI-QTOF-MS/MS. Figure S1: Base peak chromatogram (BPC) for characterization of phenolic compounds of seaweeds. Figure S2. Extracted ion chromatogram and mass spectrum of "Vanillic acid 4-sulfate" detected in three different seaweeds.

Author Contributions: Conceptualization, methodology, validation and investigation, B.Z. and H.A.R.S.; resources, H.A.R.S., C.J.B.; F.R.D. and R.D.W.; writing—original draft preparation, B.Z. and H.A.R.S.; writing—review and editing, B.Z., C.J.B.; N.A.R.; H.A.R.S. and R.D.W.; supervision, F.R.D., H.A.R.S. and R.D.W.; ideas sharing, H.A.R.S.; C.J.B.; F.R.D.; R.D.W. and N.A.R.; funding acquisition, H.A.R.S., F.R.D. and R.D.W. All authors have read and agreed to the published version of the manuscript.

Funding: This research was funded by the University of Melbourne under the "McKenzie Fellowship Scheme" (Grant No. UoM-18/21) and the "Faculty Research Initiative Funds" funded by the Faculty of Veterinary and Agricultural Sciences, The University of Melbourne, Australia and "the Alfred Deakin Research Fellowship" funded by the Deakin University, Australia.

Acknowledgments: We would like to thank Nicholas Williamson, Shuai Nie and Michael Leeming from the Mass Spectrometry and Proteomics Facility, Bio21 Molecular Science and Biotechnology Institute, the University of Melbourne, VIC, Australia for providing access and support for the use of HPLC-PDA and LC-ESI-QTOF-MS/MS and data analysis. We would like to appreciate for Tristan Graham and Trevor T. Bringloe from School of BioSciences, University of Melbourne, VIC, Australia for providing and identifying the samples. We would also like to thank for Rana Dildar Khan, Chao Ma, Danying Peng, Jiafei Tang, Yuying Feng, Danwei Yang, Yasir Iqbal and Akhtar Ali from the School of Agriculture and Food, Faculty of Veterinary and Agricultural Sciences, the University of Melbourne for their incredible support.

Conflicts of Interest: The authors declare no conflict of interest.

References

1. Ferdouse, F.; Holdt, S.L.; Smith, R.; Murua, P.; Yang, Z. *The Global Status of Seaweed Production, Trade and Utilization*; Food and Agriculture Organization of the United Nations: Rome, Italy, 2018.
2. Wyrepkowski, C.C.; Costa, D.L.; Sinhorin, A.P.; Vilegas, W.; De Grandis, R.A.; Resende, F.A.; Varanda, E.A.; dos Santos, L.C. Characterization and quantification of the compounds of the ethanolic extract from caesalpinia ferrea stem bark and evaluation of their mutagenic activity. *Molecules* **2014**, *19*, 16039–16057. [CrossRef] [PubMed]
3. Mouritsen, O.G.; Mouritsen, J.D.; Johansen, M. *Seaweeds: Edible, Available & Sustainable*; The University of Chicago Press: Chicago, IL, USA; London, UK, 2013.
4. Fleurence, J. Chapter 5–Seaweeds as food. In *Seaweed in Health and Disease Prevention*; Fleurence, J., Levine, I., Eds.; Academic Press: San Diego, CA, USA, 2016; pp. 149–167.
5. Menon, V.V.; Lele, S.S. Nutraceuticals and bioactive compounds from seafood processing waste. In *Springer Handbook of Marine Biotechnology*; Springer: Berlin/Heidelberg, Germany, 2015; pp. 1405–1425.
6. Suleria, H.A.; Masci, P.; Gobe, G.; Osborne, S. Current and potential uses of bioactive molecules from marine processing waste. *J. Sci. Food Agric.* **2016**, *96*, 1064–1067. [CrossRef] [PubMed]
7. Maqsood, S.; Benjakul, S.; Shahidi, F. Emerging role of phenolic compounds as natural food additives in fish and fish products. *Crit. Rev. Food Sci. Nutr.* **2013**, *53*, 162–179. [CrossRef] [PubMed]
8. Manach, C.; Scalbert, A.; Morand, C.; Remesy, C.; Jimenez, L. Polyphenols: Food sources and bioavailability. *Am. J. Clin. Nutr.* **2004**, *79*, 727–747. [CrossRef]
9. Rajauria, G.; Foley, B.; Abu-Ghannam, N. Identification and characterization of phenolic antioxidant compounds from brown irish seaweed himanthalia elongata using lc-dad-esi-ms/ms. *Innov. Food Sci. Emerg. Technol.* **2016**, *37*, 261–268. [CrossRef]
10. Valko, M.; Leibfritz, D.; Moncol, J.; Cronin, M.T.; Mazur, M.; Telser, J. Free radicals and antioxidants in normal physiological functions and human disease. *Int. J. Biochem. Cell Biol.* **2007**, *39*, 44–84. [CrossRef]
11. Airanthi, M.K.; Hosokawa, M.; Miyashita, K. Comparative antioxidant activity of edible japanese brown seaweeds. *J. Food. Sci.* **2011**, *76*, C104–C111. [CrossRef]
12. Namvar, F.; Mohamad, R.; Baharara, J.; Zafar-Balanejad, S.; Fargahi, F.; Rahman, H.S. Antioxidant, antiproliferative, and antiangiogenesis effects of polyphenol-rich seaweed (sargassum muticum). *BioMed Res. Int.* **2013**, *2013*, 604787. [CrossRef]
13. Lopez, A.; Rico, M.; Rivero, A.; de Tangil, M.S. The effects of solvents on the phenolic contents and antioxidant activity of stypocaulon scoparium algae extracts. *Food Chem.* **2011**, *125*, 1104–1109. [CrossRef]
14. Leopoldini, M.; Russo, N.; Toscano, M. The molecular basis of working mechanism of natural polyphenolic antioxidants. *Food Chem.* **2011**, *125*, 288–306. [CrossRef]
15. Kelman, D.; Posner, E.K.; McDermid, K.J.; Tabandera, N.K.; Wright, P.R.; Wright, A.D. Antioxidant activity of hawaiian marine algae. *Mar. Drugs* **2012**, *10*, 403–416. [CrossRef] [PubMed]

16. Kulawik, P.; Ozogul, F.; Glew, R.; Ozogul, Y. Significance of antioxidants for seafood safety and human health. *J. Agric. Food Chem.* **2013**, *61*, 475–491. [CrossRef] [PubMed]

17. Kalita, P.; Tapan, B.K.; Pal, T.K.; Kalita, R. Estimation of total flavonoids content (tfc) and anti oxidant activities of methanolic whole plant extract of biophytum sensitivum linn. *JDDT* **2013**, *3*, 33–37. [CrossRef]

18. Lopes, G.; Barbosa, M.; Vallejo, F.; Gil-Izquierdo, A.; Andrade, P.B.; Valentao, P.; Pereira, D.M.; Ferreres, F. Profiling phlorotannins from fucus spp. Of the northern portuguese coastline: Chemical approach by hplc-dad-esi/msn and uplc-esi-qtof/ms. *Algal Res.* **2018**, *29*, 113–120. [CrossRef]

19. Liu, B.; Kongstad, K.T.; Wiese, S.; Jäger, A.K.; Staerk, D. Edible seaweed as future functional food: Identification of α-glucosidase inhibitors by combined use of high-resolution α-glucosidase inhibition profiling and hplc–hrms–spe–nmr. *J. Food Chem.* **2016**, *203*, 16–22. [CrossRef]

20. García-Casal, M.N.; Ramirez, J.; Leets, I.; Pereira, A.C.; Quiroga, M.F. Antioxidant capacity, polyphenol content and iron bioavailability from algae (ulva sp., sargassum sp. And porphyra sp.) in human subjects. *J. Food Chem.* **2008**, *101*, 79–85. [CrossRef]

21. Sabeena Farvin, K.H.; Jacobsen, C. Phenolic compounds and antioxidant activities of selected species of seaweeds from danish coast. *Food Chem.* **2013**, *138*, 1670–1681. [CrossRef]

22. Mekinic, I.G.; Skroza, D.; Simat, V.; Hamed, I.; Cagalj, M.; Perkovic, Z.P. Phenolic content of brown algae (pheophyceae) species: Extraction, identification, and quantification. *Biomolecules* **2019**, *9*, 244. [CrossRef]

23. Ford, L.; Theodoridou, K.; Sheldrake, G.N.; Walsh, P.J. A critical review of analytical methods used for the chemical characterisation and quantification of phlorotannin compounds in brown seaweeds. *Phytochem. Anal.* **2019**, *30*, 587–599. [CrossRef]

24. Cox, S.; Abu-Ghannam, N.; Gupta, S. An assessment of the antioxidant and antimicrobial activity of six species of edible irish seaweeds. *J. Food Chem.* **2010**, *17*, 205–220.

25. Chan, P.T.; Matanjun, P.; Yasir, S.M.; Tan, T.S. Antioxidant activities and polyphenolics of various solvent extracts of red seaweed, gracilaria changii. *J. Food Chem.* **2015**, *27*, 2377–2386. [CrossRef]

26. Wang, T.; Jonsdottir, R.; Ólafsdóttir, G. Total phenolic compounds, radical scavenging and metal chelation of extracts from icelandic seaweeds. *J. Food Chem.* **2009**, *116*, 240–248. [CrossRef]

27. Pinteus, S.; Silva, J.; Alves, C.; Horta, A.; Fino, N.; Rodrigues, A.I.; Mendes, S.; Pedrosa, R. Cytoprotective effect of seaweeds with high antioxidant activity from the peniche coast (portugal). *Food Chem.* **2017**, *218*, 591–599. [CrossRef] [PubMed]

28. Kim, A.R.; Shin, T.S.; Lee, M.S.; Park, J.Y.; Park, K.E.; Yoon, N.Y.; Kim, J.S.; Choi, J.S.; Jang, B.C.; Byun, D.S.; et al. Isolation and identification of phlorotannins from ecklonia stolonifera with antioxidant and anti-inflammatory properties. *J. Agric. Food Chem.* **2009**, *57*, 3483–3489. [CrossRef]

29. Schaich, K.M.; Tian, X.; Xie, J. Hurdles and pitfalls in measuring antioxidant efficacy: A critical evaluation of abts, dpph, and orac assays. *J. Funct. Foods* **2015**, *18*, 782–796. [CrossRef]

30. Sachindra, N.M.; Airanthi, M.K.; Hosokawa, M.; Miyashita, K. Radical scavenging and singlet oxygen quenching activity of extracts from indian seaweeds. *J. Food Sci. Technol.* **2010**, *47*, 94–99. [CrossRef]

31. Ramon-Goncalves, M.; Gomez-Mejia, E.; Rosales-Conrado, N.; Leon-Gonzalez, M.E.; Madrid, Y. Extraction, identification and quantification of polyphenols from spent coffee grounds by chromatographic methods and chemometric analyses. *Waste Manag.* **2019**, *96*, 15–24. [CrossRef]

32. Escobar-Avello, D.; Lozano-Castellon, J.; Mardones, C.; Perez, A.J.; Saez, V.; Riquelme, S.; von Baer, D.; Vallverdu-Queralt, A. Phenolic profile of grape canes: Novel compounds identified by lc-esi-ltq-orbitrap-ms. *Molecules* **2019**, *24*, 3763.

33. Rajauria, G. Optimization and validation of reverse phase hplc method for qualitative and quantitative assessment of polyphenols in seaweed. *J. Pharm. Biomed. Anal.* **2018**, *148*, 230–237. [CrossRef]

34. Dinh, T.V.; Saravana, P.S.; Woo, H.C.; Chun, B.S. Ionic liquid-assisted subcritical water enhances the extraction of phenolics from brown seaweed and its antioxidant activity. *Sep. Purif. Technol.* **2018**, *196*, 287–299. [CrossRef]

35. Lin, L.Z.; Harnly, J.M. Identification of hydroxycinnamoylquinic acids of arnica flowers and burdock roots using a standardized lc-dad-esi/ms profiling method. *J. Agric. Food Chem.* **2008**, *56*, 10105–10114. [CrossRef] [PubMed]

36. Yang, D.-z.; Sun, G.; Zhang, A.; Fu, S.; Liu, J.-h. Screening and analyzing the potential bioactive components from rhubarb, using a multivariate data processing approach and ultra-high performance liquid chromatography coupled with time-of-flight mass spectrometry. *Anal. Methods* **2015**, *7*, 650–661. [CrossRef]

37. Wang, X.; Liu, J.; Zhang, A.; Sun, H.; Zhang, Y. Chapter 23–Systematic characterization of the absorbed components of acanthopanax senticosus stem. In *Serum Pharmacochemistry of Traditional Chinese Medicine*; Wang, X., Ed.; Academic Press: Cambridge, MA, USA, 2017; pp. 313–336.

38. Al-Ayed, A.S. Integrated mass spectrometry approach to screening of phenolic molecules in hyphaene thebiaca fruits with their antiradical activity by thin-layer chromatography. *Indian J. Chem. Technol.* **2015**, *22*, 155–161.

39. De Oliveira, D.N.; de Bona Sartor, S.; Damário, N.; Gollücke, A.P.; Catharino, R.R. Antioxidant activity of grape products and characterization of components by electrospray ionization mass spectrometry. *J. Food Meas. Charact.* **2014**, *8*, 9–14. [CrossRef]

40. Lin, H.Q.; Zhu, H.L.; Tan, J.; Wang, H.; Wang, Z.Y.; Li, P.Y.; Zhao, C.F.; Liu, J.P. Comparative analysis of chemical constituents of moringa oleifera leaves from china and india by ultra-performance liquid chromatography coupled with quadrupole-time-of-flight mass spectrometry. *Molecules* **2019**, *24*, 942. [CrossRef]

41. Tong, T.; Li, J.; Ko, D.-O.; Kim, B.-S.; Zhang, C.; Ham, K.-S.; Kang, S.-G. In vitro antioxidant potential and inhibitory effect of seaweed on enzymes relevant for hyperglycemia. *Food Sci. Biotechnol.* **2014**, *23*, 2037–2044. [CrossRef]

42. Geng, C.A.; Chen, H.; Chen, X.L.; Zhang, X.M.; Lei, L.G.; Chen, J.J. Rapid characterization of chemical constituents in saniculiphyllum guangxiense by ultra fast liquid chromatography with diode array detection and electrospray ionization tandem mass spectrometry. *Int. J. Mass Spectrom.* **2014**, *361*, 9–22. [CrossRef]

43. Wang, J.; Jia, Z.; Zhang, Z.; Wang, Y.; Liu, X.; Wang, L.; Lin, R. Analysis of chemical constituents of melastoma dodecandrum lour. By uplc-esi-q-exactive focus-ms/ms. *Molecules* **2017**, *22*, 476. [CrossRef]

44. Agregán, R.; Munekata, P.E.; Franco, D.; Dominguez, R.; Carballo, J.; Lorenzo, J.M. Phenolic compounds from three brown seaweed species using lc-dad–esi-ms/ms. *Food Res. Int.* **2017**, *99*, 979–985. [CrossRef]

45. Olate-Gallegos, C.; Barriga, A.; Vergara, C.; Fredes, C.; Garcia, P.; Gimenez, B.; Robert, P. Identification of polyphenols from chilean brown seaweeds extracts by lc-dad-esi-ms/ms. *J. Aquat. Food Prod. Technol.* **2019**, *28*, 375–391. [CrossRef]

46. Rajauria, G. In-vitro antioxidant properties of lipophilic antioxidant compounds from 3 brown seaweed. *Antioxidants* **2019**, *8*, 596. [CrossRef] [PubMed]

47. Reed, K.A. *Identification of Phenolic Compounds from Peanut Skin Using hplc-ms^n*; Virginia Polytechnic Institute and State University: Blacksburg, VA, USA, 2009.

48. Zhang, L.; Tu, Z.-C.; Wang, H.; Fu, Z.-F.; Wen, Q.-H.; Chang, H.-X.; Huang, X.-Q. Comparison of different methods for extracting polyphenols from ipomoea batatas leaves, and identification of antioxidant constituents by hplc-qtoe-ms2. *Food Res. Int.* **2015**, *70*, 101–109. [CrossRef]

49. Zeng, X.; Su, W.; Zheng, Y.; Liu, H.; Li, P.; Zhang, W.; Liang, Y.; Bai, Y.; Peng, W.; Yao, H. Uflc-q-tof-ms/ms-based screening and identification of flavonoids and derived metabolites in human urine after oral administration of exocarpium citri grandis extract. *Molecules* **2018**, *23*, 895. [CrossRef] [PubMed]

50. Zhao, X.; Zhang, S.; Liu, D.; Yang, M.; Wei, J. Analysis of flavonoids in dalbergia odorifera by ultra-performance liquid chromatography with tandem mass spectrometry. *Molecules* **2020**, *25*, 389. [CrossRef] [PubMed]

51. Tanna, B.; Brahmbhatt, H.R.; Mishra, A. Phenolic, flavonoid, and amino acid compositions reveal that selected tropical seaweeds have the potential to be functional food ingredients. *J. Food Process. Preserv.* **2019**, *43*, e14266. [CrossRef]

52. Wang, Y.; Vorsa, N.; Harrington, P.d.B.; Chen, P. Nontargeted metabolomic study on variation of phenolics in different cranberry cultivars using uplc-im-hrms. *J. Agric. Food Chem.* **2018**, *66*, 12206–12216. [CrossRef]

53. Wu, S.-H.; Li, H.-B.; Li, G.-L.; Lv, N.; Qi, Y.-J. Metabolite identification of gut microflora-cassia seed interactions using uplc-qtof/ms. *Exp. Ther. Med.* **2020**, *19*, 3305–3315. [CrossRef]

54. Zeng, Y.; Lu, Y.; Chen, Z.; Tan, J.; Bai, J.; Li, P.; Wang, Z.; Du, S. Rapid characterization of components in bolbostemma paniculatum by uplc/ltq-orbitrap msn analysis and multivariate statistical analysis for herb discrimination. *Molecules* **2018**, *23*, 1155. [CrossRef]

55. Jesionek, W.; Majer-Dziedzic, B.; Horvath, G.; Moricz, A.M.; Choma, I.M. Screening of antibacterial compounds in salvia officinalis l. Tincture using thin-layer chromatography-direct bioautography and liquid chromatography-tandem mass spectrometry techniques. *JPC J. Planar Chromatogr. Mod. TLC* **2017**, *30*, 357–362. [CrossRef]

56. Pacifico, S.; Piccolella, S.; Lettieri, A.; Nocera, P.; Bollino, F.; Catauro, M. A metabolic profiling approach to an italian-sage leaf extract (soa541) defines its antioxidant and anti-acetylcholinesterase properties. *J. Funct. Foods* **2017**, *29*, 1–9. [CrossRef]

57. Hermund, D.; Jacobsen, C.; Chronakis, I.S.; Pelayo, A.; Yu, S.; Busolo, M.; Lagaron, J.M.; Jónsdóttir, R.; Kristinsson, H.G.; Akoh, C.C. Stabilization of fish oil-loaded electrosprayed capsules with seaweed and commercial natural antioxidants: Effect on the oxidative stability of capsule-enriched mayonnaise. *Eur. J. Lipd Sci. Technol.* **2019**, *121*, 1800396. [CrossRef]

58. Di Maio, I.; Esposto, S.; Taticchi, A.; Selvaggini, R.; Veneziani, G.; Urbani, S.; Servili, M. Characterization of 3,4-dhpea-eda oxidation products in virgin olive oil by high performance liquid chromatography coupled with mass spectrometry. *Food Chem.* **2013**, *138*, 1381–1391. [CrossRef]

59. Gomez-Alonso, S.; Salvador, M.D.; Fregapane, G. Phenolic compounds profile of cornicabra virgin olive oil. *J. Agric. Food Chem.* **2002**, *50*, 6812–6817. [CrossRef] [PubMed]

60. Salama, M.; El-Hawary, S.; Mousa, O.; El- Askary, N.; Esmat, A. In vivo tnf-α and il-1β inhibitory activity of phenolics isolated from trachelospermum jasminoides (lindl.) lem. *Med. Plants Res.* **2015**, *9*, 35–41.

61. Yang, S.; Shan, L.; Luo, H.; Sheng, X.; Du, J.; Li, Y. Rapid classification and identification of chemical components of schisandra chinensis by uplc-q-tof/ms combined with data post-processing. *Molecules* **2017**, *22*, 1778. [CrossRef] [PubMed]

62. Dhargalkar, V. Uses of seaweeds in the indian diet for sustenance and well-being. *Sci. Cult.* **2015**, *80*, 192–202.

63. Rodriguez-Garcia, C.; Sanchez-Quesada, C.; Toledo, E.; Delgado-Rodriguez, M.; Gaforio, J.J. Naturally lignan-rich foods: A dietary tool for health promotion? *Molecules* **2019**, *24*, 917. [CrossRef] [PubMed]

64. Peterson, J.; Dwyer, J.; Adlercreutz, H.; Scalbert, A.; Jacques, P.; McCullough, M.L. Dietary lignans: Physiology and potential for cardiovascular disease risk reduction. *Nutr. Rev.* **2010**, *68*, 571–603. [CrossRef] [PubMed]

65. Kershaw, J.; Kim, K.H. The therapeutic potential of piceatannol, a natural stilbene, in metabolic diseases: A review. *J. Med. Food* **2017**, *20*, 427–438. [CrossRef]

66. Ma, C.; Dunshea, F.R.; Suleria, H.A.R. Lc-esi-qtof/ms characterization of phenolic compounds in palm fruits (jelly and fishtail palm) and their potential antioxidant activities. *Antioxidants* **2019**, *8*, 483. [CrossRef]

67. Tang, J.; Dunshea, F.R.; Suleria, H.A.R. Lc-esi-qtof/ms characterization of phenolic compounds from medicinal plants (hops and juniper berries) and their antioxidant activity. *Foods* **2020**, *9*, 7. [CrossRef] [PubMed]

68. Peng, D.; Zahid, H.F.; Ajlouni, S.; Dunshea, F.R.; Suleria, H.A.R. Lc-esi-qtof/ms profiling of australian mango peel by-product polyphenols and their potential antioxidant activities. *Processes* **2019**, *7*, 764. [CrossRef]

69. Yang, P.; Xu, F.; Li, H.F.; Wang, Y.; Li, F.C.; Shang, M.Y.; Liu, G.X.; Wang, X.; Cai, S.Q. Detection of 191 taxifolin metabolites and their distribution in rats using hplc-esi-it-tof-msn. *Molecules* **2016**, *21*, 764. [CrossRef] [PubMed]

70. Oszmiański, J.; Kolniak-Ostek, J.; Wojdyło, A. Application of ultra performance liquid chromatography-photodiode detector-quadrupole/time of flight-mass spectrometry (uplc-pda-q/tof-ms) method for the characterization of phenolic compounds of lepidium sativum l. Sprouts. *Eur. Food Res. Technol.* **2013**, *236*, 699–706. [CrossRef]

71. Kim, I.; Lee, J. Variations in anthocyanin profiles and antioxidant activity of 12 genotypes of mulberry (morus spp.) fruits and their changes during processing. *Antioxidants* **2020**, *9*, 242. [CrossRef]

72. Rengasamy, K.R.; Amoo, S.O.; Aremu, A.O.; Stirk, W.A.; Gruz, J.; Šubrtová, M.; Doležal, K.; Van Staden, J. Phenolic profiles, antioxidant capacity, and acetylcholinesterase inhibitory activity of eight south african seaweeds. *J. Appl. Phycol.* **2015**, *27*, 1599–1605. [CrossRef]

73. Lever, J.; Brkljaca, R.; Kraft, G.; Urban, S. Natural products of marine macroalgae from south eastern australia, with emphasis on the port phillip bay and heads regions of victoria. *Mar. Drugs* **2020**, *18*, 142. [CrossRef]

74. Saunders, G.W.M.; Tanya, E. Refinements for the amplification and sequencing of red algal DNA barcode and redtol phylogenetic markers: A summary of current primers, profiles and strategies. *Algae* **2013**, *28*, 31–43. [CrossRef]

75. Heffernan, N.; Smyth, T.J.; Soler-Villa, A.; Fitzgerald, R.J.; Brunton, N.P. Phenolic content and antioxidant activity of fractions obtained from selected irish macroalgae species (laminaria digitata, fucus serratus, gracilaria gracilis and codium fragile). *J. Appl. Phycol.* **2015**, *27*, 519–530. [CrossRef]

76. Leyton, A.; Pezoa-Conte, R.; Barriga, A.; Buschmann, A.H.; Maki-Arvela, P.; Mikkola, J.P.; Lienqueo, M.E. Identification and efficient extraction method of phlorotannins from the brown seaweed macrocystis pyrifera using an orthogonal experimental design. *Algal Res.* **2016**, *16*, 201–208. [CrossRef]

77. Feng, Y.; Dunshea, F.R.; Suleria, H.A.R. Lc-esi-qtof/ms characterization of bioactive compounds from black spices and their potential antioxidant activities. *J. Food Sci. Technol.* **2020**. [CrossRef]

78. Rebaya, A.; Belghith, S.I.; Baghdikian, B.; Leddet, V.M.; Mabrouki, F.; Olivier, E.; Cherif, J.; Ayadi, M.T. Total phenolic, total flavonoid, tannin content, and antioxidant capacity of halimium halimifolium (cistaceae). *J. Appl. Pharm. Sci.* **2014**, *5*, 52–57.

79. Matanjun, P.; Mohamed, S.; Mustapha, N.M.; Muhammad, K.; Ming, C.H. Antioxidant activities and phenolics content of eight species of seaweeds from north borneo. *J. Appl. Phycol.* **2008**, *20*, 367. [CrossRef]

Article

Bioactive Polyphenols from Southern Chile Seaweed as Inhibitors of Enzymes for Starch Digestion

Luz Verónica Pacheco [1], Javier Parada [2,*], José Ricardo Pérez-Correa [3], María Salomé Mariotti-Celis [4], Fernanda Erpel [3], Angara Zambrano [5] and Mauricio Palacios [6,7,8]

[1] Graduate School, Faculty of Agricultural Sciences, Universidad Austral de Chile, Valdivia 5090000, Chile; luz_pacheco007@hotmail.com

[2] Institute of Food Science and Technology, Faculty of Agricultural Sciences, Universidad Austral de Chile, Valdivia 5090000, Chile

[3] Department of Chemical and Bioprocess Engineering, Pontificia Universidad Católica de Chile, Macul, Santiago 7810000, Chile; perez@ing.puc.cl (J.R.P.-C.); faerpel@uc.cl (F.E.)

[4] Programa Institucional de Fomento a la Investigación, Desarrollo e Innovación, Universidad Tecnológica Metropolitana, Santiago 8940577, Chile; mmariotti@utem.cl

[5] Instituto de Bioquímica y Microbiología, Facultad de Ciencias, Universidad Austral de Chile, Valdivia 5090000, Chile; angara.zambrano@uach.cl

[6] Centro FONDAP de Investigación Dinámica de Ecosistemas Marinos de Altas Latitudes (IDEAL), Valdivia 5090000, Chile; mauricio.palacios@alumnos.uach.cl

[7] Programa de Doctorado en Biología Marina, Facultad de Ciencias, Universidad Austral de Chile, Valdivia 5090000, Chile

[8] Facultad de Ciencias, Universidad de Magallanes, Punta Arenas 6200000, Chile

* Correspondence: javier.parada@uach.cl; Tel.: +56-63-222-1619

Received: 26 May 2020; Accepted: 7 July 2020; Published: 8 July 2020

Abstract: The increment of non-communicable chronic diseases is a constant concern worldwide, with type-2 diabetes mellitus being one of the most common illnesses. A mechanism to avoid diabetes-related hyperglycemia is to reduce food digestion/absorption by using anti-enzymatic (functional) ingredients. This research explored the potential of six common Chilean seaweeds to obtain anti-hyperglycemic polyphenol extracts, based on their capacity to inhibit key enzymes related with starch digestion. Ethanol/water hot pressurized liquid extraction (HPLE), which is an environmentally friendly method, was studied and compared to conventional extraction with acetone. Total polyphenols (TP), antioxidant activity, cytotoxicity and inhibition capacity on α-glucosidase and α-amylase were analyzed. Results showed that the *Durvillaea antarctica* (cochayuyo) acetone extract had the highest TP content (6.7 ± 0.7 mg gallic acid equivalents (GAE)/g dry seaweed), while its HPLE ethanol/water extract showed the highest antioxidant activity (680.1 ± 11.6 µmol E Trolox/g dry seaweed). No extract affected cell viability significantly. Only cochayuyo produced extracts having relevant anti-enzymatic capacity on both studied enzymes, showing a much stronger inhibition to α-glucosidase (even almost 100% at 1000 µg/mL) than to α-amylase. In conclusion, from the Chilean seaweeds considered in this study, cochayuyo is the most suitable for developing functional ingredients to moderate postprandial glycemic response (starchy foods), since it showed a clear enzymatic inhibition capacity and selectivity.

Keywords: seaweed polyphenols; hypoglycemic effect; starch digestion; enzyme inhibition; cochayuyo

---

1. Introduction

Seaweeds or macroalgae are the most important benthic organisms in coastal marine ecosystems. According to their specific pigments, algae are classified into three divisions. The first group is the so-called brown algae or Ochrophyta (Phaeophyceae class), whose pigmentation varies from brown yellow to dark brown and produces a large amount of protective mucus. Red algae or Rhodophyta is the second largest group of algae and is found in various media. Finally, green algae or Chlorophyta is less common than brown and red algae. Its pigmentation varies from greenish yellow to dark green [1]. Besides its chemical composition, seaweeds have been studied as sources of a variety of compounds with potential biological activities such as antitumoral, antidiabetic and antioxidant, among others. These bioactive compounds are synthesized according to the level of maturity and capacity of the plant to interact with environmental conditions such as radiation, water pressure and salinity, which make them particularly attractive [2]. For example, phlorotannins (the polyphenols found in brown algae) comprise oligomers or polymers of phloroglucinol (1,3,5-trihydroxybenzene) with different antioxidant activities [3–5]. Moreover, it has been shown that these polyphenols from algae have significant in vitro inhibitory activities against α-glucosidase and α-amylase. This might have a potential application to control type-2 diabetes, since the inhibition of these enzymes would reduce the severity of postprandial hyperglycemia by delaying starch hydrolysis [6,7]. As an example, *Sargassum patens*, a brown alga of the Noto peninsula in the Ishikawa prefecture of Japan, contains 2-(4-(3,5-dihydroxyphenoxy)-3,5-dihydroxyphenoxy) benzene-1,3,5-triol DDBT, a phlorotannin with an inhibiting effect on the enzymes that hydrolyze carbohydrates (IC$_{50}$ 3.2 μg/mL for α-amylase inhibition) [8]. The topography, waves, and exposure to wind of the benthic habitats of continental Chile favor the growth of seaweeds, where approximately 440 species have been identified [9]. Thus, Chilean seaweeds are an interesting source of new compounds with different applications.

Currently, the screening of the inhibitors of α-amylase and α-glucosidase of natural origin has received attention because this would avoid the side effects of commercial inhibitors for treating type-2 diabetes [10]. Extracts of natural plants with inhibitory α-glucosidase activity, such as tea and raspberry, are recommended as substitutes for synthetic drugs [11]. In addition to their anti-enzymatic activities, plant polyphenols are also capable of capturing free radicals and therefore act as antioxidants. It has been postulated that in diabetic patients, antioxidants help to prevent vascular diseases, the destruction of pancreatic cells and the formation of reactive oxygen species [12]. Seaweeds have recently gained significant interest as a sustainable source of various bioactive natural compounds, including polyphenols, carotene, lutein, astaxanthin, zeaxanthin, violaxanthin and fucoxanthin (pigments) [13]. Conventional processing technologies, based on organic solvent extraction at low pressures and temperatures, offer a simple approach for isolating such compounds; they have been used for a long time and are still widely applied. The main factors that should be considered to achieve an effective solid–liquid conventional extraction are the polarity of the compounds of interest, the characteristics of the solvent (toxicity, volatility, polarity, viscosity and purity) and the possible formation of other compounds during the extraction process. In addition, the yield, selectivity and contamination with undesired compounds are the usual performance indices used to compare different options. However, conventional techniques suffer from several limitations, such as being time-consuming and requiring large amounts of polluting or toxic solvents [14].

Alternative, faster and more efficient/selective techniques have been developed in the recent decades. Hot pressurized liquid extraction (HPLE), also known as accelerated solvent extraction, is a green alternative technique that has been widely applied to recover bioactive compounds from plant matrices. In this method, the extraction occurs at elevated temperatures and pressures (below the critical point of the solvent), normally in the ranges of 50–200 °C and 35–200 bar, respectively. Under these conditions, the viscosity and surface tension of the solvent are significantly reduced, whereas the solubility and mass transfer of the solute are greatly enhanced. Consequently, HPLE is fast, reduces the solvent consumption and allows for an efficient usage of green solvents such as water and ethanol for the extraction of a variety of compounds by changing their solvation dielectric constants (polarity) to values similar to those of organic solvents [15].

The objective of this research was to explore the potential of common seaweeds present in southern Chile to obtain anti-glycemic polyphenol rich extracts for functional food development. HPLE was applied and compared with an acetone conventional extraction method.

2. Results and Discussion

2.1. Cytotoxicity Assay

The growing interest in plant-derived compounds for commercial human applications requires the assessment of their possible toxicity. Cytotoxicity studies in cell lines can be considered as a first step in the development of pharmaceutical, cosmetic or food products; toxicity levels (minimal or no toxicity) should be verified [16]. In this test, HT-29 cells, the human colon adenocarcinoma cell line expressing the characteristics of mature intestinal cells, such as enterocytes or mucus-producing cells [17] were used. As seen in Figure 1, the incubation with extracts at low concentrations (1 and 10 µg/mL) did not reveal significant changes in the MTT conversion rates associated with cell viability. *Nothogenia* sp. and *M. laminarioides* extracts negatively affected the cellular metabolism of HT-29 cells at 24 and 48 h, respectively, at the maximum concentration evaluated (1000 µg/mL); however, no extract decreased the cell viability in the same way as the positive control (DMSO) that reduced viability to $17.1 \pm 1.6\%$ and $18.4 \pm 1.4\%$ at 24 and 48 h of exposure, respectively. In general, no statistically significant differences were observed between the type of extract (ethanol and acetone) or between the incubation times (24 and 48 h). According to Galindo et al. [18], a cytotoxic effect can be considered when the viability is less than 75%. None of the seaweed extracts reduced the cellular viability to that level, so they may be considered to not be cytotoxic.

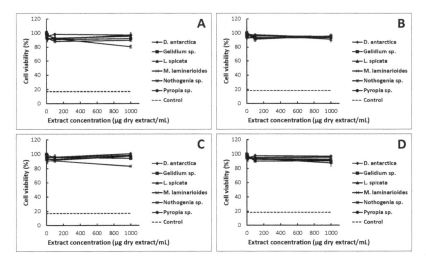

Figure 1. HT-29 cell viability at different dry extracts concentrations: (**A**) the ethanolic extract at 24 h incubation; (**B**) the ethanolic extract at 48 h incubation; (**C**) the acetone extract at 24 h incubation; (**D**) the acetone extract at 48 h incubation. Each point represents the mean of viable cells ± SD ($n = 2$). The dashed line is the cell viability using the positive control (DMSO 16.7%). The 100% viability was assigned to the cell culture without extracts.

The diversity of compounds derived from algae can generate different actions; these can stimulate growth or apoptosis. *Nothogenia* sp. and *Pyropia* sp. extracts slightly decreased cell viability at higher concentrations, therefore, they could be analyzed from a chemotherapeutic perspective. Seaweed extracts could also be used as antitumoral agents. Using different cell types can show different responses towards a specific compound or plant extract. Some authors indicate that polyphenols are toxic to certain types of cancer cells that proliferate rapidly but are nontoxic to others [19,20].

2.2. Total Polyphenol Content and Antioxidant Capacities

Both the species and extraction method significantly affect ($p < 0.05$) the total polyphenol content (TP) of the extracts (Table 1). The seaweed with the highest TP was *D. antarctica* (cochayuyo), whose acetone and ethanol extracts contained 7.4 ± 0.2 and 6.7 ± 0.7 mg gallic acid equivalents (GAE)/g dry seaweed, respectively, followed by *Pyropia* sp., whose acetone and ethanol extracts contained 6.2 ± 0.2 and 4.8 ± 0.3 mg GAE/g, respectively. In most cases, conventional extraction with acetone yielded higher values of TP (10%–20%) than HPLE with ethanol ($p < 0.05$). Acetone, being a dipolar compound, has an intermediate polarity and solubilizes solutes with a similar relative polarity. Wang et al. [21,22] found that acetone (70%) was the most suitable solvent to recover algae polyphenols. Nonetheless, acetone is a toxic solvent not permitted for human consumption. However, HPLE with ethanol (50%) allowed for seaweed extracts that are of food grade, which is mandatory for future applications such as nutraceuticals and functional food ingredients.

Table 1. Antioxidant activity of the seaweed extracts.

Species	Type	Total Polyphenolsmg GAE/g Dry Seaweed		DPPH μmol ET/g Dry Seaweed		ORAC μmol ET/g Dry Seaweed	
		Ethanol	Acetone	Ethanol	Acetone	Ethanol	Acetone
Durvillaea antarctica	Brown	7.4 ± 0.2 [b]	6.7 ± 0.7 [a]	48.5 ± 4.2 [a]	27.8 ± 2.2[a]	680.1 ± 11.6 [a]	64.7 ± 0.0 [a]
Gelidium sp.	Red	3.2 ± 0.3 [a]	3.4 ± 0.2 [a]	4.8 ± 0.4 [b]	4.7 ± 0.2 [c]	277.8 ± 15.5 [bc]	6.9 ± 0.5 [c]
Lessonia spicata	Brown	3.3 ± 0.2 [b]	3.8 ± 0.1 [a]	6.6 ± 0.7 [a]	10.7 ± 0.6 [a]	448.3 ± 33.4 [a]	21.3 ± 1.3 [b]
Nothogenia sp.	Red	4.8 ± 0.3 [b]	6.0 ± 0.3 [a]	5.4 ± 0.3 [ab]	6.9 ± 0.1 [bc]	371.6 ± 12.3 [ab]	18.1 ± 0.9 [b]
Mazzaella laminarioides	Red	1.9 ± 0.1 [a]	3.1 ± 0.1 [a]	2.2 ± 0.1 [c]	7.05 ± 0.8 [b]	208.1 ± 10.4 [c]	8.7 ± 0.6 [c]
Pyropia sp.	Red	2.2 ± 0.0 [b]	3.1 ± 0.1 [a]	6.5 ± 0.4 [ab]	10.6 ± 0.9 [a]	455.3 ± 3.4 [a]	30.7 ± 2.9 [a]

Different letters indicate the statistically significant differences for the Tukey multiple range test with 95% confidence, into each column.

Two complementary methods were applied to determine the antioxidant activity of the extracts: the oxygen radical absorbance capacity (ORAC) test, which corresponds to an electron transfer model, and the DPPH assay that determines the transfer of hydrogen atoms.

Statistically significant differences were observed among the species studied ($p < 0.05$), with brown seaweed extracts being the ones with the highest antioxidant activities in both extraction methods. The high antioxidant activity of brown species could be due mainly to their content of phlorotannins, whose phenolic ring can eliminate the reactive species of oxygen [23–26].

Ethanol *D. antarctica* extracts presented the highest antioxidant activity, both in the ORAC (680.1 ± 11.0 μmol ET/g dry seaweed) and DPPH (48.5 ± 4.2 μmol ET/g dry seaweed) assays (Table 1). Thus, HPLE with ethanol is a feasible food grade alternative to conventional extraction since it allows obtaining polyphenol extracts with higher antioxidant activity ($p < 0.05$). It is worth mentioning that the TP extraction yields differed only between 10–20%. Contrarily, for the antioxidant activity, the HPLE extracts presented two and 10 times higher DPPH and ORAC values, respectively, as compared to the extracts obtained using acetone. The advantage of using HPLE with ethanol is clear; the antioxidant capacity is associated with the potential benefic effect of polyphenols on human health, and above all because ethanol is a food grade solvent.

The major compounds contributing to the overall antioxidant activity in seaweeds are frequently phenolic compounds and polysaccharides; they may be found alone or associated with other components such as polyphenols, amino acid, protein, lipids, and sometimes polysaccharide conjugates [27] (Jacobsen et al., 2019). Since some seaweed polyphenol compounds may be conjugated with different types of sugars, the chemical structure of these conjugated compounds makes them more soluble in alcoholic solvents compared to acetone. This may explain the higher antioxidant activity of our extracts. The higher solubility of sugars can be associated with the dielectric constant of ethanol (24.55 at 25 °C), which is closer to water (78.54 at 25 °C) than acetone (20.7 at 25 °C) [28].

2.3. Anti-Enzymatic Activities

One of the strategies to manage type-2 diabetes is through the inhibition of enzymes such as α-amylase and α-glucosidase. The selective inhibition of these enzymes reduces the hydrolysis of carbohydrates and the rate of glucose uptake into the bloodstream [29]. The ability of the seaweed extracts to inhibit the activity of α-amylase and α-glucosidase at different concentrations was determined.

The α-amylase activity did not decrease significantly with any of the ethanol seaweed extracts within the concentration range studied (Figure 2). Acetone extracts of *D. antarctica* and *Gelidium* sp. inhibited the activity of this enzyme, although to a lesser extent than acarbose. The acetone extract of *D. antarctica* (2000 μg of dry extract/mL) decreased the α-amylase activity to 56.6 ± 2.0% and *Gelidium* sp. to 77.9 ± 2.1%. In contrast, acarbose (concentration ≥ 1000 μg/mL) decreased the enzyme activity to 37.5 ± 0.4%.

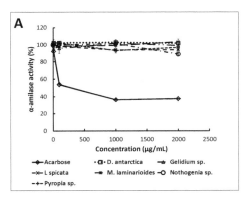 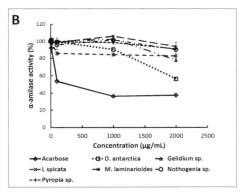

Figure 2. Percentages of the α-amylase activity under different concentrations of ethanol (**A**) and acetone (**B**) extracts (μg/mL). The points represent the average enzymatic activity (%) ± standard deviation (*n* = 3).

Like α-amylase inhibitors, α-glucosidase inhibitors can slow down the breakdown and absorption of dietary carbohydrates [30]. The acetone and ethanol extracts of *D. antarctica* and the acetone extract of *L. spicata* were the most effective inhibitors of α-glucosidase (Figure 3). The *D. antarctica* acetone extract almost completely inhibited the enzymatic activity (to 0.7 ± 0.3%), followed by the acetone extract of *L. spicata* (to 1.2 ± 0.3%) and by the ethanol extract of *D. antarctica* (to 3.1 ± 0.4%); all of them to a concentration of 1000 μg dry extract/mL. These extracts were much more effective than the commercial enzymatic inhibitor acarbose (1000 μg/mL), which reduced the enzymatic activity only to 40.4 ± 1.1%. The ethanol and acetone extracts of *Gelidium* sp., *M. laminarioides* and *Nothogenia* sp. did not show any significant inhibitory effect against α-glucosidase in the concentration range assessed (1, 10, 100, 1000 μg dry extract/mL). In general, the differences in the inhibition capacity among the extracts would be derived from their compositions, which in turn is the result of the seaweed species and extraction methods used.

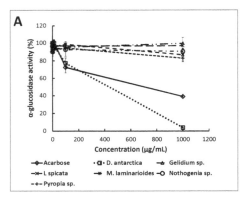 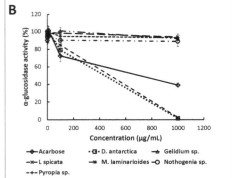

Figure 3. Percentages of α-glucosidase activity under different concentrations of ethanol (**A**) and acetone (**B**) extracts (μg/mL). Points represent the average enzymatic activity (%) ± standard deviation (*n* = 3).

The IC$_{50}$ values for the α-glucosidase (Table 2) of acetone extracts of *D. antarctica* and *L. spicata* (the only two seaweed species that generated inhibition higher than 50%) were 466 and 479.21 μg/mL, respectively, and both were lower than the IC$_{50}$ of acarbose (797.9 μg/mL). Hence, *D. antarctica* and *L. spicata* have an effective inhibitory effect on the activity of α-glucosidase. Brown algae were better inhibitors of α-glucosidase than red algae such as *Gelidium* sp., *M. laminarioides* and *Nothogenia* sp. Likewise, other authors found that the water extract of *Laminaria digitata*, a member of the Laminariaceae family and a brown seaweed, was effective in inhibiting α-glucosidase [11].

Table 2. The IC50 values (μg extract/mL) of brown seaweeds for α-glucosidase inhibition.

Species	Extract Type	
	Ethanol	Acetone
D. antarctica	473.4 ± 0.9 [b]	466 ± 1.3 [a]
L. spicate	5317.6 ± 0.75 [e]	479.2 ± 1.7 [c]
Acarbose	797.85 ± 1.1 [d]	

Different letters indicate the statistically significant differences for the Tukey multiple range test with 95% confidence.

The diversity and abundance of some compounds may hamper the interactions with the enzyme at a molecular level; however, it is possible that the purification of extracts may enhance the enzymatic inhibitory activity. Therefore, our crude algae extracts were not as effective in inhibiting these enzymes as the pure polyphenols of vegetable origin (IC$_{50}$ ~ 5 μg/mL) [14] or purified polyphenol extracts from pomegranate extracts (IC$_{50}$ of 278 μg/mL) or black tea (IC$_{50}$ of 64 μg/mL) [31].

Several reports have been published on the established enzyme inhibitors, such as acarbose, miglitol, voglibose and nojirimycin, and their favorable effects on blood glucose levels after food intake [32,33]. This fact is attributed to their capacity to diminish the assimilation of sugars, both monosaccharides and more complex carbohydrates. It seems appropriate to ingest the potent inhibitors of α-glucosidase and α-amylase to control the release of glucose from saccharides in the intestine. However, the complete inhibition of both enzymes could result in the poor absorption of nutrients. Inhibitors should be administered in doses that allow all carbohydrates to be digested. Otherwise, undigested carbohydrates will enter the colon and, as a result of bacterial fermentation, will lead to side effects such as flatulence. The most adequate inhibitors should delay the digestion and absorption of carbohydrates without overloading the colon with them [34]. It has been proposed that the administration of products presenting the moderate inhibition of α-amylase and strong inhibition of α-glucosidase is an effective therapeutic strategy that could decrease the availability of monosaccharides for absorption in the intestine [35–37]. Hence, the consumption of some of our seaweed extracts may be a preferred alternative for modulating the glycemic index of some food products.

The enzymatic inhibition capacity (α-amylase and α-glucosidase) of seaweed extracts followed the same trend as their antioxidant activities. The ethanol and acetone extracts of *D. antarctica*, the species with the highest average antioxidant activity (ORAC and DPPH), presented the highest α-glucosidase inhibition activity. Consequently, antioxidant activity indices can be used to optimize the design of the extraction process.

3. Materials and Methods

3.1. Chemicals and Cell Culture

All the chemicals and cell culture reagents were obtained from Sigma Chemical Co. (Saint Louis, MO, USA) unless stated otherwise. The cell lines were obtained from The Institute of Biochemistry and Microbiology UACh (Valdivia, Chile).

3.2. Seaweed Collection and Identification

Six species (4 red and 2 brown) were harvested in September 2017 from Corral port at "Region de los Rios" South of Chile. The red species collected were: *Gelidium* sp., *Mazzaella laminarioides* (Luga cuchara), *Nothogenia* sp. (Chascuo) and *Pyropia* sp. (Luche); the brown species were: *Durvillaea antarctica* (cochayuyo) and *Lessonia spicata* (huiro negro) (Figure 4). All of them were morphologically identified by taking into account taxonomical characteristics such as the shape of the thallus, the number of pyrenoids, the presence/absence of marginal teeth, and the thickness of the cross section. The six seaweeds were identified by Ph.D (c). M. Palacios (Algae Ecophysiology Laboratory of the Universidad Austral de Chile, Valdivia, Chile), an experienced researcher on Sub Antarctic macroalgae taxonomy.

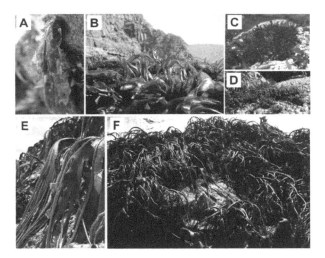

Figure 4. Macroalgae used in this study. The red macroalgae species (**A**) *Pyropia* sp., (**B**) *M. laminarioides*, (**C**) *Gelidium* sp., (**D**) *Nothogenia* sp.; and the brown macroalgae species (**E**) *D. antarctica* and (**F**) *L. spicata*. (Photograph by Mauricio Palacios- IDEAL Center).

3.3. Cytotoxicity Assays

To ensure the safety of the extracts as edible products, the cytotoxicity was assessed according to Lordan et al. [11]. The HT-29 cell line (ATTC number: HTB-3) from human colon carcinoma was maintained in Dulbecco's modified Eagle's medium (DMEM), containing 10% fetal bovine serum (FBS),

50 U/mL penicillin, 50 mg/mL streptomycin and 2 mM L-glutamine. The cells were cultured at 37 °C with 5% CO_2 in a humidified incubator. The cells in exponential growth were used.

The seaweed extracts were prepared like enzymatic assays and the stock solution was diluted with the culture medium to give the desired concentrations. For each assay, the HT-29 cells were seeded and acclimated for 24 h before treatment in a DMEM culture medium without phenol red, for which 100 μL of HT-29 cells were added per well (50,000 cells/well). Subsequently, 2 mL of increasing concentrations of algae extract were added (between 1 and 1000 mg seaweed extract/mL), done in triplicate; a positive control was used, that of a high concentration of DMSO (16.7% as a final concentration). Each microplate was incubated at 37 °C with 5% CO_2 for 24 and 48 h. The MTT (3(4,5-dimethyl-2-thiazoyl)-2,5-diphenyltetrazolic) from Sigma-Aldrich (cat # M2128; Saint Louis, MO, USA) was used to evaluate the effect of different concentrations of extracts on the cells. The MTT reagent (5 mg/mL) was added 4 h before the end of the treatment. Then, a lysis buffer (50% dimethylformamide and 20% SDS) was added, homogenized and finally incubated 10–15 min to subsequently record its absorbance at 545 nm. Cell survival was determined using the MTT (3(4,5-dimethyl-2-thiazoyl)-2,5-diphenyltetrazolic) assay, which measures the change in metabolic activity, proportional to the number of viable cells.

3.4. Polyphenol Extraction Methods

The collected seaweed samples were quickly washed in cold water to remove sand and other particles; they were immediately frozen and stored in vacuum-packed bags at −80 °C prior to freeze drying. Two methods of extraction for each of the seaweeds were employed: HPLE with ethanol/water (50%) and conventional extraction (at atmospheric conditions: 20 °C-14.69 psi) with acetone/water (60%) at atmospheric pressure. For both extraction methods, freeze-dried seaweed material was mixed using a 1:32 *w/v* ratio with the respective extraction solvent, ethanol (50% *v/v*) and acetone (60%). HPLE were carried out in an accelerated solvent extractor (Thermo Scientific™ Dionex™ ASE™ 150, Waltham, MA, USA) at 120 °C and 1500 psi. Each sample was subjected to one cycle of extraction of 20 min and using 150 mL of washing solvent. After extraction, water–ethanol (50%) extracts were transferred to an amber PET bottle and stored at −20 °C. Using a conventional water/acetone (60%) extraction, each seaweed sample was extracted for 1 h on a thermoregulatory rotary shaker at 100 rpm at 30 °C. Then, the mixture was centrifuged for 5 min at 6000 rpm and 20 °C, and the supernatant was transferred to a 50 mL flask wrapped in aluminum. A volume of water was added to the remaining solid, stirred manually and centrifuged. The washed supernatant was transferred to the previous flask and adjusted to 50 mL with a 60% water–acetone solution. Finally, the extract was transferred to an amber PET bottle and stored at −20 °C.

3.5. Total Polyphenol Content

The total polyphenol content of the extracts was determined using the Folin–Ciocalteu (FC) method, with gallic acid as the standard. In brief, 0.5 mL of the sample or solvent blank was diluted in 3.75 mL of distilled water. Subsequently, 0.25 mL of the FC reagent was added and homogenized. Then, 0.5 mL of the sodium carbonate solution (10% *w/v*) was added, and it was homogenized for 1 h at room temperature. The absorbance of the reaction product was measured at 765 nm (UV spectrophotometer 1240, Shimadzu, Kyoto, Japan). The total polyphenol content was calculated as a mg of gallic acid equivalents (mg GAE) per gram of dry seaweed, using an absolute standard curve of calibration in the range of 0.01–0.1 mg GAE/mL (r^2). Each extract was analyzed in duplicate.

3.6. Free Radical Scavenging Using DPPH Radical

The quantification of the anti-radical activity, 2,2-diphenyl-1-picrylhydracil (DPPH) of the extracts, was carried out by the method of Tierney et al. [38] with minor modifications. Prior to each batch of analysis, a working solution of DPPH (0.048 mg/mL) was prepared by diluting a stock of 0.238 mg/mL prepared in methanol. In addition, three serial dilutions of the samples were performed with the extraction solvent, at concentrations in the range of 0.025–15 mg/mL. For the analysis, 0.5 mL of DPPH solution was added to microtubes with 0.5 mL of the extract's dilutions. After homogenizing, the tubes reacted for 30 min at room temperature. Absorbance was measured at 520 nm on a UV 1240 spectrophotometer (Shimadzu, Kyoto, Japan). Likewise, the serial dilutions of Trolox were measured as the reference standard, and from which a standard curve was determined. The results were expressed in μmol equivalent of Trolox (ET)/g dry seaweed.

3.7. Free Radical Scavenging by Oxygen Radical Absorbance Capacity (ORAC) Assay

The ORAC method was carried out based on the procedure described by Caor and Prio [39], with minor modifications. The reaction was carried out in a 75 mM phosphate buffer (pH 7.4), in a 96-well microplate. Then, 45 μL of the sample and 175 μL of fluorescein were deposited at 108 nm. This mixture was incubated for 30 min at 37 °C; after that time, 50 μL of the AAPH solution was added to 108 mM. The microplate was immediately placed in the dual-scan microplate spectrofluorometer (Gemini XPS, San Jose, CA, USA) for 60 min; fluorescence readings were recorded every 3 min. The microplate was automatically shaken before and after each reading. For the calibration curve, Trolox was used at 6, 12, 18 and 24 μM. All reactions were carried out in triplicate. The area under the curve (AUC) was calculated for each sample by integrating the relative fluorescence curve ($r^2 > 0.99$). The net AUC of the sample was calculated by subtracting the AUC of the blank. The regression equation between the net AUC and Trolox concentration was determined, and the ORAC values were expressed as μmol Trolox equivalents/g of dry seaweed (ET/g) using the standard curve established previously.

3.8. Inhibition of α-Amylase Activity

The ability of each extract to inhibit α-amylase activity was measured using the method described by Nampoothiri et al. [40] and adapted by Lordan et al. [11]. For the preparation of the samples, each extract was dried by aeration (using an aeration pump). After solvent evaporation, each dry sample was re-suspended in DMSO and filtered, obtaining a stock solution of 10 mg of dry seaweed extract/mL. From this stock, the assay dilutions were made: 0.01–2 mg/mL phosphate buffer (pH 6.9). Then, 100 μL of each sample dilution and 1% starch solution in 20 mM sodium phosphate buffer (pH 6.9 with 6 mM sodium chloride) were incubated in microtubes at 25 °C for 10 min in a water bath. A volume of 100 μL of porcine pancreatic α-amylase (0.5 mg/mL) was added to each tube and the samples were incubated at 25 °C for another 10 min. Then, 200 μL of dinitrosalicylic acid reagent were added and the tubes incubated at 100 °C for 5 min in a water bath. Subsequently, 50 μL of each reaction mixture was transferred to wells of a 96-well microplate and diluted by adding 200 μL of water to each well and the absorbance was measured at 540 nm in a microplate reader. The enzymatic activity was determined as follows:

$$Enzymatic\ activity\ (\%) = \frac{Absorbance\ of\ extract}{Absorbance\ of\ control} \cdot 100 \tag{1}$$

where the control is the enzyme–substrate reaction in the absence of inhibitors. The effect of the pharmacological inhibitor, acarbose, was also determined.

3.9. Inhibitions of α-Glucosidase Activity

The ability of each extract to inhibit α-glucosidase activity was measured using the method described by Nampoothiri et al. [40] and adapted by Lordan et al. [11]. The samples were prepared with the same methodology as the α-amylase activity assay. The inhibitory effect of each extract was measured at concentrations from 0.1 to 1000 μg/mL in 100 mM sodium phosphate buffer (pH 6.9). A volume of 50 μL of the extract solution and 50 μL of the 5 mM p-nitrophenyl-α-ᴅ-glucopyranoside (PNPG) solution (in a phosphate buffer) was mixed in a 96-well microplate and incubated at 37 °C for 5 min. Then, a phosphate buffer (100 μL) containing 0.1 U/mL of μ-glucosidase (from *S. cerevisiae*) was added to each well. The absorbance at 405 nm was recorded for 15 min using a microplate reader at 37

Mar. Drugs **2020**, *18*, 353

°C. The effect of the commercial inhibitor on the α-glucoside activity was also determined, and the data were processed as in the previous assay.

The IC$_{50}$ value was also calculated, representing the concentration of the extract that caused 50% enzyme inhibition, which was calculated by linear regression analysis. That value was determined only when the inhibition was higher than 50%.

3.10. Statistics

All the data points were mean values of at least two or three independent experiments. Where appropriate, the data were analyzed by a one-way analysis of variance (ANOVA) followed by Tukey's Multiple Comparison test. The software employed for the statistical analysis was STATGRAPHICS Centurion XV.II (Old Tavern Rd, The Plains, VA, USA).

4. Conclusions

Extraction is the first step in studies regarding plant bioactive compounds and plays a significant role in the final outcome. Currently, there is growing demand for sustainable extraction methods of bioactive compounds from plant sources. HPLE is a green method that allows obtaining food-grade plant extracts that have specific healthy properties. Using this technology, we were able to produce extracts of *D. antarctica* and *L. spicata* that showed higher antioxidant capacities than the extracts from red and green algae. From the extracts that we obtained in this study, only three (of two species) strongly inhibited α-glucosidase and one appreciably inhibited α-amylase. From the six species under study, *D. antarctica* stood out as particularly useful for developing an antihyperglycemic agent given its enzyme inhibition profile: the strong inhibition of α-glucosidase and moderate inhibition of α-amylase. The seaweed considered in this study showed no cytotoxicity under the tested conditions.

Author Contributions: Conceptualization, J.P. and J.R.P.-C.; methodology, L.V.P. and M.S.M.-C.; formal analysis, L.V.P. and J.P.; investigation, L.V.P., F.E. and M.P.; resources, J.P., A.Z. and J.R.P.-C.; writing—original draft preparation, L.V.P.; writing—review and editing, J.P., J.R.P.-C. and M.S.M.-C.; supervision, J.P., A.Z. and J.R.P.-C.; project administration, J.P.; funding acquisition, J.P. and J.R.P.-C. All authors have read and agreed to the published version of the manuscript.

Funding: This research was funded by FONDECYT, grant number 1170594.

Acknowledgments: Authors greatly acknowledge to Jorge Rivas for helping us with the recollection of seaweeds, and Sandy Gonzalez for technical support. Lisa Gingles reviewed the text.

Conflicts of Interest: The authors declare no conflict of interest.

References

1. Swamy, M.L.A. Marine algal sources for treating bacterial diseases. In *Advances in Food and Nutrition Research*, 1st ed.; Academic Press: Waltham, MA, USA, 2011; Volume 64, pp. 71–84. [CrossRef]
2. Gutiérrez-Rodríguez, A.G.; Juárez-Portilla, C.; Olivares-Bañuelos, T.; Zepeda, R.C. Anticancer activity of seaweeds. *Drug Discov. Today* **2017**. [CrossRef] [PubMed]
3. Zaragoza, M.C.; Lopez, D.; Saiz, M.P.; Poquet, M.; Perez, J.; Puig-Parellada, P.; Màrmol, F.; Simonetti, P.; Gardana, C.; Lerat, Y.; et al. Toxicity and antioxidant activity in vitro and in vivo of two Fucus vesiculosus extracts. *J. Agric. Food Chem.* **2008**, *56*, 7773–7780. [CrossRef] [PubMed]
4. Li, Y.; Wijesekara, I.; Li, Y.; Kim, S. Phlorotannins as bioactive agents from brown algae. *Process Biochem.* **2011**, *46*, 2219–2224. [CrossRef]
5. Sathya, R.; Kanaga, N.; Sankar, P.; Jeeva, S. Antioxidant properties of phlorotannins from brown seaweed *Cystoseira trinodis* (Forsskål) C. Agardh. *Arab. J. Chem.* **2013**, *10*, S2608–S2614. [CrossRef]
6. Zhou, K.; Hogan, S.; Canning, C.; Sun, S. Inhibition of intestinal α-glucosidases and anti-postprandial hyperglycemic effect of grape seed extract. In *Emerging Trends in Dietary Components for Preventing and Combating Disease*; Patil, B.S., Jayaprakasha, G.K., Murthy, K.N., Seeram, N.P., Eds.; American Chemical Society: Washington, DC, USA, 2012; pp. 431–441. [CrossRef]

7. Jayaraj, S.; Suresh, S.; Kadeppagari, R.; Kadeppagari, R. Amylase inhibitors and their biomedical applications. *Starch* **2013**, *65*, 535–542. [CrossRef]
8. Kawamura-Konishi, Y.; Watanabe, N.; Saito, M.; Nakajima, N.; Sakaki, T.; Katayama, T.; Enomoto, T. Isolation of a new phlorotannin, a potent inhibitor of carbohydrate-hydrolyzing enzymes, from the brown alga sargassum patens. *J. Agric. Food Chem.* **2012**, *60*, 5565–5570. [CrossRef] [PubMed]
9. Ramírez, M.E. Algas Marinas Bentónicas. In *Biodiversidad de Chile, Patrimonio y Desafíos*, Ministerio del Medio Ambiente, 3rd ed.; Gobieron de Chile: Santiago, Chile, 2018; Volume I.
10. Naquvi, K.J.; Ahamad, J.; Mir, S.R.; Ali, M.; Shuaib, M. Review on role of natural alpha-glucosidase inhibitors for management of diabetes mellitus. *Int. J. Biomed. Res.* **2011**, *2*. [CrossRef]
11. Lordan, S.; Smyth, T.J.; Soler-Vila, A.; Stanton, C.; Ross, R.P. The α-amylase and α-glucosidase inhibitory effects of Irish seaweed extracts. *Food Chem.* **2013**, *141*, 2170–2176. [CrossRef]
12. Gulati, V.; Harding, I.H.; Palombo, E.A.; Palombo, E.A. Enzyme inhibitory and antioxidant activities of traditional medicinal plants: Potential application in the management of hyperglycemia. *BMC Complement. Altern. Med.* **2012**, *12*, 77. [CrossRef]
13. Khalid, S.; Abbas, M.; Saeed, F.; Bader-Ul-Ain, H.; Rasul, H. Therapeutic Potential of Seaweed Bioactive Compounds. In *Seaweed Biomaterials*; IntechOpen: London, UK, 2018; pp. 7–25. [CrossRef]
14. Grussu, D.; Stewart, D.; McDougall, G.J. Berry Polyphenols Inhibit α-Amylase in Vitro: Identifying active components in rowanberry and raspberry. *J. Agric. Food. Chem.* **2011**, *59*, 2324–2331. [CrossRef]
15. Poojary, M.M.; Barba, F.J.; Aliakbarian, B.; Donsì, F.; Pataro, G.; Dias, D.A.; Juliano, P. Innovative alternative technologies to extract carotenoids from microalgae and seaweeds. *Mar. Drugs* **2016**, *14*, 214. [CrossRef] [PubMed]
16. McGaw, L.J.; Elgorashi, E.E.; Eloff, J.N. Cytotoxicity of African medicinal plants against normal animal and human cells. In *Toxicological Survey of African Medicinal Plants*; Kuete, V., Ed.; Academic Press: Waltham, MA, USA, 2014. [CrossRef]
17. Martinez-Maqueda, D.; Millares, B.; Recio, I. HT29 Cell Line. In *The Impact of Food Bioactives on Health: In Vitro and Ex Vivo Models*; Verhoeckx, K., Cotter, P., López-Expósito, I., Kleiveland, C., Lea, T., Eds.; Springer: London, UK, 2015; pp. 113–130.
18. Galindo, C.V.; Castell, E.C.; Montell, L.T.; Segura, E.S.; Baeza, M.M.R.; Guilen, G. Evaluación de la citotoxicidad y bioseguridad de un extracto de polifenoles de huesos de aceitunas. *Nutr. Hosp.* **2014**, *29*, 1388–1393. [CrossRef]
19. Kovalchuk, A.; Aladedunye, F.; Rodriguez-Juarez, R.; Li, D.; Thomas, J.; Kovalchuk, O.; Przybylski, R. Novel antioxidants are not toxic to normal tissues but effectively kill cancer cells. *Cancer Biol. Ther.* **2013**, *14*, 907–915. [CrossRef] [PubMed]
20. Niedzwiecki, A.; Roomi, M.W.; Kalinovsky, T.; Rath, M. Anticancer efficacy of polyphenols and their combinations. *Nutrients* **2016**, *8*, 552. [CrossRef] [PubMed]
21. Wang, T.; Jónsdóttir, R.; Ólafsdóttir, G. Screening of antioxidant activity in Icelandic seaweed. Available online: http://www.avs.is/media/avs/Screening_of_antioxidant_activity_final_II.pdf (accessed on 6 June 2020).
22. Wang, T.; Jónsdóttir, R.; Ólafsdóttir, G. Total phenolic compounds, radical scavenging and metal chelation of extracts from Icelandic seaweeds. *Food Chem.* **2009**, *116*, 240–248. [CrossRef]
23. Shibata, T.; Ishimaru, K.; Kawaguchi, S.; Yoshikawa, H.; Hama, Y. Antioxidant activities of phlorotannins isolated from Japanese Laminariaceae. *J. Appl. Phycol.* **2008**, *20*, 705–711. [CrossRef]
24. Miranda-Delgado, A.; Montoya, M.J.; Paz-Araos, M.; Mellado, M.; Villena, J.; Arancibia, P.; Madrid, A.; Jara-Gutiérrez, C. Antioxidant and anti-cancer activities of brown and red seaweed extracts from Chilean coasts. *Lat. Am. J. Aquat. Res.* **2018**, *46*, 301–313. [CrossRef]
25. Sanz-Pintos, N.; Pérez-Jiménez, J.; Buschmann, A.H.; Vergara-Salinas, J.R.; Pérez-Correa, J.R.; Saura-Calixto, F. Macromolecular antioxidants and dietary fiber in edible seaweeds. *J. Food. Sci.* **2017**, *82*, 289–295. [CrossRef] [PubMed]
26. Jiménez-Escrig, A.; Gómez-Ordóñez, E.; Rupérez, P. Brown and red seaweeds as potential sources of antioxidant nutraceuticals. *J. Appl. Phycol.* **2012**, *24*, 1123–1132. [CrossRef]
27. Jacobsen, C.; Sørensen, A.M.; Holdt, S.L.; Akoh, C.C.; Hermund, D.B. Source, extraction, characterization, and applications of novel antioxidants from seaweed. *Annu. Rev. Food Sci. Technol.* **2019**, *10*, 541–568. [CrossRef] [PubMed]

28. Sepahpour, S.; Selamat, J.; Manap, M.Y.A.; Khatib, A.; Razis, A.F.A. Comparative analysis of chemical composition, antioxidant activity and quantitative characterization of some phenolic compounds in selected herbs and spices in different solvent extraction systems. *Molecules* **2018**, *23*, 402. [CrossRef] [PubMed]

29. McDougall, G.J.; Stewart, D. The inhibitory effects of berry polyphenols on digestive enzymes. *Biofactors* **2005**, *23*, 189–195. [CrossRef]

30. Proença, C.; Freitas, M.; Ribeiro, D.; Oliveira, E.F.T.; Sousa, J.L.C.; Tomé, S.M.; Ramos, M.J.; Silva, A.M.S.; Fernandes, P.A.; Fernandes, E. α-Glucosidase inhibition by flavonoids: An *in vitro* and *in silico* structure–activity relationship study. *Enzyme J. Enzyme Inhib. Med. Chem.* **2017**, *32*, 1216–1228. [CrossRef] [PubMed]

31. Bellesia, A.; Verzelloni, E.; Tagliazucchi, D. Pomegranate ellagitannins inhibit a -glucosidase activity in vitro and reduce starch digestibility under simulated gastro-intestinal conditions. *Food Sci. Nutr.* **2014**, *7486*, 1–8. [CrossRef]

32. Kim, Y.; Jeong, Y.; Wang, M.; Lee, W.; Rhee, H. Inhibitory effect of pine extract on a-glucosidase activity and postprandial hyperglycemia. *Nutrition* **2005**, *21*, 756–761. [CrossRef] [PubMed]

33. van de Laar, F.A. Alpha-glucosidase inhibitors in the early treatment of type 2 diabetes. *Vasc. Health Risk Manag.* **2008**, *4*, 1189–1195. [CrossRef]

34. Bischoff, H. Pharmacology of α-glucosidase inhibition. *Eur. J. Clin. Invest.* **1994**, *24*, 3–10. [CrossRef]

35. Oboh, G.; Akinyemi, A.J.; Ademiluyi, A.O.; Adefegha, S.A. Inhibitory effects of aqueous extract of two varieties of ginger on some key enzymes linked to type-2 diabetes in vitro. *Food Nutr. Res.* **2010**, *49*, 14–20.

36. Kajaria, D.; Ranjana, J.T.; Tripathi, Y.B.; Tiwari, S. In-vitro α amylase and glycosidase inhibitory effect of ethanolic extract of antiasthmatic drug—Shirishadi. *J. Adv. Pharm. Technol. Res.* **2013**, *4*, 2–5. [CrossRef]

37. Rasouli, H.; Hosseini-Ghazvini, S.M.; Adibi, H.; Khodarahmi, R. Diferential α-amylase/α-glucosidase inhibitory activities of plant-derived phenolic compounds: a virtual screening perspective for the treatment of obesity and diabetes. *Food Funct.* **2017**, *8*, 1942–1954. [CrossRef]

38. Tierney, M.S.; Smyth, T.J.; Hayes, M.; Soler-Vila, A.; Croft, A.K.; Brunton, N. Influence of pressurised liquid extraction and solid–liquid extraction methods on the phenolic content and antioxidant activities of Irish macroalgae. *Int. J. Food Sci. Technol.* **2013**, *48*, 860–869. [CrossRef]

39. Cao, G.; Prior, R. Measurement of Oxygen Radical Absorbance in Biological Samples. In *Methods in Enzymology*; Academic Press: Waltham, MA, USA, 1999; Volume 299, pp. 50–62. [CrossRef]

40. Nampoothiri, S.V.; Prathapan, A.; Cherian, O.L.; Raghu, K.G.; Venugopalan, V.V.; Sundaresan, A. *In vitro* antioxidant and inhibitory potential of *Terminalia bellerica* and *Emblica officinalis* fruits against LDL oxidation and key enzymes linked to type 2 diabetes. *Food Chem. Toxicol.* **2011**, *49*, 125–131. [CrossRef] [PubMed]

Review

Emerging Technologies for the Extraction of Marine Phenolics: Opportunities and Challenges

Adane Tilahun Getachew, Charlotte Jacobsen * and Susan Løvstad Holdt

National Food Institute, Technical University of Denmark, Kemitorvet Building 204, 2800 Kgs Lyngby, Denmark; atige@food.dtu.dk (A.T.G.); suho@food.dtu.dk (S.L.H.)
* Correspondence: chja@food.dtu.dk; Tel.: +45-23-279075

Received: 29 June 2020; Accepted: 23 July 2020; Published: 27 July 2020

Abstract: Natural phenolic compounds are important classes of plant, microorganism, and algal secondary metabolites. They have well-documented beneficial biological activities. The marine environment is less explored than other environments but have huge potential for the discovery of new unique compounds with potential applications in, e.g., food, cosmetics, and pharmaceutical industries. To survive in a very harsh and challenging environment, marine organisms like several seaweed (macroalgae) species produce and accumulate several secondary metabolites, including marine phenolics in the cells. Traditionally, these compounds were extracted from their sample matrix using organic solvents. This conventional extraction method had several drawbacks such as a long extraction time, low extraction yield, co-extraction of other compounds, and usage of a huge volume of one or more organic solvents, which consequently results in environmental pollution. To mitigate these drawbacks, newly emerging technologies, such as enzyme-assisted extraction (EAE), microwave-assisted extraction (MAE), ultrasound-assisted extraction (UAE), pressurized liquid extraction (PLE), and supercritical fluid extraction (SFE) have received huge interest from researchers around the world. Therefore, in this review, the most recent and emerging technologies are discussed for the extraction of marine phenolic compounds of interest for their antioxidant and other bioactivity in, e.g., cosmetic and food industry. Moreover, the opportunities and the bottleneck for upscaling of these technologies are also presented.

Keywords: marine phenolics; emerging technologies; extraction

1. Introduction

Phenolic compounds are secondary metabolites produced by plants, microorganisms, and algae. The marine environment is an excellent source of bioactive compounds, and this includes phenolic compounds. There are many theories about why marine plant (seaweeds) tissues produce and accumulate phenolic compounds, and one of the most widely accepted theories states that phenolic compounds are produced as a defense mechanism to protect from biotic and abiotic factors. To adapt and survive in a very competitive and challenging marine environment, seaweeds (macroalgae) need to produce defense mechanisms from their metabolic pathways that protect them against the harsh environment and biological agents like UV protectant, anti-herbivory, and antioxidant [1]. The most pronounced environmental stress is that seaweeds often grow in intertidal locations where they are exposed to a frequently changing environment with high fluctuation of sunlight and oxygen, demanding from them a strong antioxidative defense system [1,2]. Nevertheless, the type and level of compounds vary from species to species, e.g., the geographic location, the growing season, and the environmental stress [3,4]. For instance, phlorotannins are phenolic compounds that are exclusively found in brown seaweeds [5], which grow in the mentioned harsh environments, and their concentration is highly variable across the different seasons [6,7].

Several types of phenolic compounds were isolated and characterized from brown, green, and red seaweed. Both the type and composition of the phenolic compounds vary across the different genera of seaweed [8]. The most commonly reported phenolic compounds are simple phenolic compounds like gallic acid, epicatechin, epigallocatechin, catechin gallate, protocatechuic acid, hydroxybenzoic acid, chlorogenic acid, caffeic acid, and vanillic acid [9,10]. Several flavonoids like quercetine, hesperidin, myricetin, and rutin were also reported in marine seaweeds [10–13] (Figure 1). Phlorotannins are the most commonly reported complex phenolic compounds in several brown seaweed species. Phlorotannins are oligomers of a phloroglucinol (1,3,5-trihydroxy benzene) monomer unit, with a large molecular size ranging from 126 Da to several thousand Da. Depending on the nature of the linkage between the aromatic units, there are four different classes of phlorotannins. When the linkage between the phloroglucinol unit is aryl-ether (C-O-C), they are called phloroethols and fuhalols; those with phenyl linkage (C-C) are fucols; fucophloroethols have both phenyl and aryl-ether linkages, and ecols with a benzodioxin linkage [14] (Figure 2). As secondary metabolites, phlorotannins play a series of recognized roles in the cell, such as anti-herbivory defense, antifouling activity, and UV protectants [15].

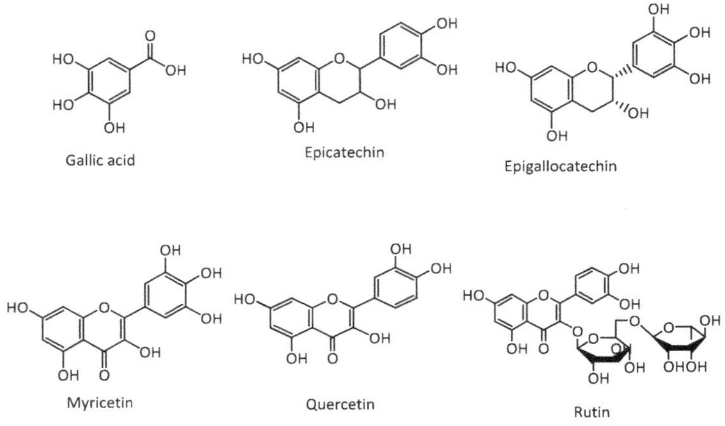

Figure 1. Chemical structures of some commonly reported seaweed phenolic and flavonoid compounds.

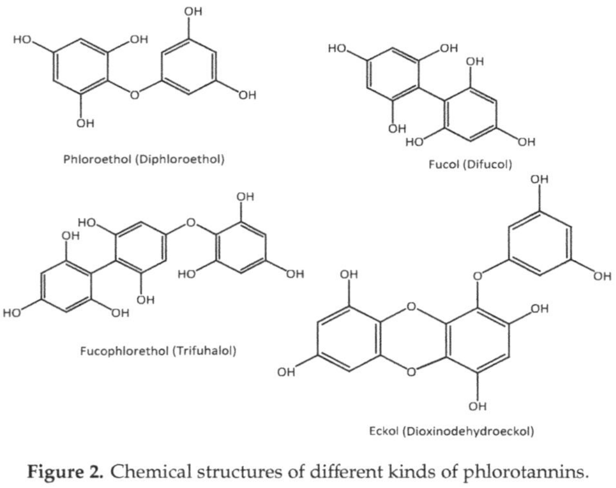

Figure 2. Chemical structures of different kinds of phlorotannins.

Marine phenolics show several biological activities [16], such as antioxidants [10,17], anti-inflammation [18–20], antimicrobial [21–23], anticoagulant [24], plant growth promoters [25], anticancer agents [26,27], angiotensin I-converting enzyme (ACE) inhibiting activity [28], antidiabetic [29], and antiproliferative [30]. To get all these benefits of the marine phenolics and use them in different delivery systems, such as food, pharmaceuticals, and cosmetics, the compounds need to be extracted from the sample matrix, characterized, and purified. Traditionally, marine phenolic compounds were extracted from marine resources using organic solvents. However, this extraction method had several drawbacks, including the use of a large volume of solvents, long extraction time, low extraction yield, degradation of extracted compounds, and difficulty of separating the extract from the solvent. In order to mitigate these challenges, several newly emerging technologies were used to extract phenolics from marine resources. Therefore, this review aimed to summarize the most frequently used emerging and sustainable technologies in the extraction of marine phenolics from seaweed. Moreover, the opportunities and challenges are discussed for the upscaling of these technologies.

2. Traditional Extraction of Marine Phenolics

Traditional extraction, commonly known as solvent extraction or solid–liquid extraction (SLE), is the most commonly and frequently used extraction technique. SLE can be done in several ways, such as; refluxing using soxhlet; boiling the sample and solvent with or without stirring for a certain duration; or maceration with continuous stirring (often for long duration, and sometimes even overnight). Several kinds of solvents and a mixture of solvents with wide polarity ranges are used to extract marine phenolics. The solvents include methanol, ethanol, acetone, ethyl acetate, trichloromethane, a mixture of water and organic solvents like ethanol, acetone, and acetonitrile, and methanol, at different mixing ratios [31–36]. When a mixture of solvents like water and ethanol is used, the extraction yield of phenolic compounds is reported to be higher than each of the solvents used individually, due to different optimums of extractability of compounds of different polarities, for the different solvents in a mix [37]. However, conventional extraction methods usually involve the use of a large volume of solvents, longer extraction time, and high temperature. Such harsh extraction conditions lead to the possibility of oxidation and hydrolysis of the phenolic compounds. Moreover, the upscaling of this technology at an industrial level would be difficult, owing to practicality, energy, economic, and environmental considerations [38]. Thus, to overcome the challenges associated with the method, several newly emerging extraction technologies were introduced.

3. Emerging Technologies for Extraction of Marine Phenolics

The newly emerging technologies could be based on the energy/mechanism they use for the extraction, such as ultrasound-assisted extraction (UAE); microwave-assisted extraction (MAE); based on the use of biological agents such as enzymes, enzyme-assisted extraction (EAE); based on the use of new type of solvents like pressurized solvent extraction (PLE); and supercritical fluid extraction (SFE). There are also less frequently used technologies, such as pulsed electric field-assisted extraction (PEF) and ohmic heating, where an electrical current is passed through the material and generates heat. Moreover, in recent years, there are new types of solvents, which are increasingly used for the extraction of phenolic compounds, and these solvents include, ionic liquids and deep eutectic solvents utilizing the different melting points of the constituents. Additionally, two or more of these methods were also used in combination [39,40]. Recent advances in the use of these extraction methods to extract marine phenolics from seaweeds are discussed below.

3.1. Enzyme-Assisted Extraction (EAE)

EAE offers several advantages, as compared to conventional extraction methods, including low operation temperature and the use of environment-friendly solvent. EAE is not a new extraction method to recover bioactives from a marine organism, however, ever-increasing research in the extraction process optimization and intensification coupled with a constant study in the discovery of new robust enzymes makes it an emerging technology.

Phenolic compounds can exist in several forms in the plant, microorganism, and algal material, such as in free soluble form or as insoluble complex forms [41]. Most marine phenolic compounds either make a complex with other macromolecules like protein and carbohydrate, or inside the cell, surrounded by a thick cell wall polysaccharide like alginate. Some phenolics are, however, easily accessible when located in the outer cell walls in physodes [8,42]. The discovery and application of cell wall degrading enzymes are based on their ability to degrade polysaccharides like alginate. For instance, phlorotannins, the most abundant form of phenolic compounds in brown seaweeds, usually form a covalent bond with proteins to produce a protein–polyphenol complex. To release these bound phenolic compounds, and make them available for extraction, an enzyme that is capable of hydrolyzing the protein–polyphenol bond is required [41]. In recent years, several novel enzymes were discovered from marine organisms. Most of the enzymes are cell wall polysaccharides degrading enzymes like alginate-lyases, glucuronan-lyases, carrageenases, laminarinases, and agarases [43]. When such structural and cell wall polysaccharides are degraded, the contents of the cell, including the phenolic compounds are released into the extraction medium, and thus increases the extraction yield. Some recently discovered enzymes had a higher activity, as compared to the commercially available enzymes. Ihua et al. [44] used extracts from enzyme-active bacterial isolate, which was obtained from decaying *Ascophyllum nodosum* for EAE of phenolic compounds from *Fucus vesiculosus*. They reported that the yield of phenolic compounds extracted, even at a low temperature of 28 °C, was higher than all three types of commercial enzymes (cellulase, proteases, and xylanase) used in the study. From the three commercial enzymes studied, xylanase yielded the best phenolic compounds (35.7 mg PGE/g dry weight (DW)), whereas enzyme active bacterial isolate yielded 44.8 mg PGE/g DW. However, most of such enzymes are isolated and tested at laboratory scale and are not currently available for industrial applications [45]. Thus, this research indicated that new types of robust and efficient enzymes are yet to be discovered from the marine environment and more integrated research is needed to accomplish this.

Although there is a problem in attacking specific cell wall polysaccharides, nonspecific enzymes that can degrade the polysaccharide and protein are available in large quantities for use in labs, pilots, and even at an industrial scale. Several researchers reported the use of such enzymes to recover marine phenolics from seaweed (Table 1). As shown in Table 1, the yield of phenolic compounds was affected by different factors, including the type of enzyme used, the enzyme concentration, temperature, the pH of the media, and the treatment duration. The type of enzymes used for the extraction process is very important. Generally, when comparing the yield of phenolic compounds produced by carbohydrases and proteases, most reports showed that those treated with carbohydrases had higher phenolic compounds [39,46,47]. The reason could be the differences in their attacking targets. Carbohydrases attack the cell wall polysaccharides and release the cell contents into the extraction medium to increase the extraction yield of the phenolic compounds. In contrast, proteases attack the protein and increase the concentration of protein in the extraction medium, which might subsequently react with phenolic compounds to form a protein–polyphenol complex in the extraction medium. The produced complex eventually makes aggregates and precipitates, and thereby results in a possible reduction of the phenolic compounds extraction yield [48]. Therefore, a careful selection of enzymes and optimizing the extraction parameters is a fundamental part of EAE. EAE is reported by itself or with the combination of other types of extraction technologies like UAE, MAE, PLE, and SFE. When EAE is combined with PLE and SFE, the pretreatment of biomass, together with the enzyme degrades the cell wall, and subsequently, the phenolic compounds are easily available for extraction by the solvents [48].

Table 1. Enzyme-assisted extraction (EAE) of marine phenolics.

Seaweed Type	State of the Seaweed (Wet/Dry/Particle Size)	Type of Enzyme Used	Extraction Conditions Enzyme Conc./Temperature (°C)/Time (min)/pH	Yield (mg GAE/g DW)	Application of the Extract	Reference
Sargassum baceanum, Sargassum angustifolium, Padina gymnospora, Canistrocarpus cervicornis, Colpomenia sinuosa, Iyengaria stellata, Feldmannia irregularis	Freeze dried/powdered	Viscozyme AMG 300 L Celluclast Termamyl Ultraflo L Flavourzyme Alcalase Neutrase	0.1%/50/1200/4.5 0.1%/60/1200/4.5 0.1%/50/1200/4.5 0.1%/60/1200/6 0.1%/40/1200/6 0.1%/50/1200/7 0.1%/50/1200/8 0.1%/50/1200/8	32.4–74.8 * 50.0–84.0 22.3–63.8 26.1–43.0 16.7–32.0 9.5–38.4 28.2–82.5	Antioxidant Antimicrobial	[46]
Lessonia nigrescens, Macrocystis pyrifera, Durvillaea antarctica	Air dried/powdered (100 μm)	Cellulase α-Amylase	10%/50/1020/4.5	17.38–19.31 21.3 ~13	Angiotensin I-converting enzyme (ACE) activity	[28]
Enteromorpha prolifera	Dried/pulverized	AMG 300 L Celluclast Dextrozyme, Maltogenase, Promozyme, Viscozyme Termamyl Alcalase Flavourzyme Neutrase Protamex	2% (w/w, DW)/-/480/-	2.47 1.5 2.02 1.24 1.83 2.53 2.02 8.44 6.43 6.98 1.83	Antioxidant Anti-acetylcholinesterase Anti-Inflammatory	[47]
Sargassum muticum, Osmundea pinnatifida, Codium tomentosum	Oven dried (60 °C)/Powdered (<1.0 mm)	Alcalase Flavourzyme Cellulase Viscozyme L	5% (w/w, DW)/50/8.0 5% (w/w, DW)/50/7.0 5% (w/w, DW)/50/4.5 5% (w/w, DW)/50/4.5	0.2–0.3 mg CE/g LE 0.1–0.12 mg CE/g LE 0.11–0.16 mg CE/g LE	Antioxidant Antidiabetic	[39]
Ulva armoricana	Wet/grounded	Neutral endo-protease A mix of neutral and alkaline endo-proteases A multiple-mix of carbohydrases Mix of endo-1,4-β-xylanase/endo-1,3(4)-β-glucanase Cellulase Exo-β-1,3(4)-glucanase	6% (w/w, DW)/50/240/6.2	9 11 7 6 4 7	Antioxidant and antiviral	[45]

* When multiple samples are treated with multiple enzymes, the yield of phenolic contents is described in ranges of minimum to maximum values. Further detailed values can be found in the respective references; CE—catechetol equivalent, and LE—lyophilized extracts.

3.2. Ultrasound-Assisted Extraction (UAE)

Ultrasound-assisted extraction is an emerging potential technology that can accelerate heat and mass transfer and was successively used in extraction. Ultrasound waves alter the physical and chemical properties after interaction with the exposed material. The cavitational effect of the ultrasound waves facilitates the release of extractable compounds, and furthermore, enhances the mass transfer by disrupting the plant cell walls. UAE is a clean method that avoids the use of a large quantity of solvent, along with cutting down in the working time. Ultrasounds are successfully employed in the extraction field [49,50], and is well-known to have a significant effect on the rate of various processes in the chemical and food industry. Much attention is given to the application of ultrasound for the extraction of natural products that typically need hours or days to reach completion with conventional methods. With the use of ultrasound, full extractions can now be completed in minutes, with high reproducibility, reducing the consumption of solvent, simplifying manipulation and work-up, and yielding higher purity of the final product. This also eliminates post-treatment of wastewater and consumes only a fraction of the fossil energy normally needed for a conventional extraction method like the Soxhlet extraction, maceration, or steam distillation. Several classes of food components like aromas, pigments, polyphenols, and other organic and mineral compounds were extracted and analyzed efficiently from a variety of matrices, such as oils from garlic, phenolics from citrus peel and wheat bran, aromas from tea, and lycopene/pigment from tomatoes [49].

In more detail, the UAE is based on cavitation, i.e., the creation, growth, and sudden implosion (collapse) of bubbles with tremendous energy release from each of them, sometimes referred to as hot spots [51,52]. In the bulk solvent, the bubble collapse is symmetrical, but near to the surface, it is asymmetrical and generates a high-speed jet of liquids. In the case of cavitation in an extraction system, the jet hits the surface of the sample matrix and provides a continuous circulation of new solvents at the surface, producing deep penetration of solvent into the sample particle, continuous solvent mixing, and sometimes particle size reduction. The deep penetration of the solvent coupled with size reduction, enhances the extraction efficiency of the phenolic compounds [53]. UAE is used to extract phenolic compounds from marine resources. Several researchers extracted phenolics from brown algae (*Phaeophyceae*); *Ascophyllum nodosum* [54], *Laminaria japonica*, *Fucus vesiculosus* [55], green algae (*Chlorophyceae*); *Codium tomentosum* [39], and the red alga (*Rhodophyceae*); *Laurencia obtusa* [56] (Table 2). The extraction yield of phenolic compounds using UAE could be significantly affected by several experimental parameters. The parameters include the ultrasound energy/power, the sample-to-solvent ratio, a combined ratio of different solvents, the extraction temperature, time, and the particle size of the sample. To investigate the effect of each parameter, and the combination of two or more parameters on the extraction efficiency of marine phenolics using UAE, several statistical models were evaluated.

In their optimization study for extraction of phenolic compounds from red seaweed *L. obtusa* using UAE, Topuz et al. [56] determined the optimum extraction conditions of solvent—seaweed ratio, 24.3:1; extraction temperature, 45.3 °C; and the extraction time of 58 min, to recover a maximum yield of TPC 26.2 mg GAE/g seaweed. Dang et al. [57] also optimized the UAE of phenolic compounds from brown seaweed *Hormosira banksii*, using response surface methodology, and reported that the optimum conditions for maximum phenolic content were temperature, 30 °C; time, 60 min; and power, 150 W. According to this study, the application of UAE at the optimal conditions increased the yield of TPC by 143%, as compared to the conventional extraction method. In another study, Vázquez-Rodríguez et al. [58] optimized the extraction parameters, namely, extraction temperature (50–65 °C), ultrasound power density (1.2–3.8 W cL^{-1}), solvent/seaweed ratio (10–30 mL g^{-1}), and ethanol concentration (25–100% ethanol in water), to recover phlorotannins from brown seaweed *Silvetia compressa*. They reported a maximum of 7.3 mg PGE/g dry seaweed at (X1 = 50 °C, X2 = 3.8 W cL^{-1}, X3 = 30 mL g^{-1} seaweed meal, and X4 = 32.3%). Moreover, the study showed that the ultrasound power density was the most influential parameter for the extraction of the phlorotannins.

3.3. Microwave-Assisted Extraction (MAE)

Microwave heating or microwave-assisted extraction (MAE) is used extensively to extract valuable materials from plant and animal resources. In details, microwave heating is generated by dipole rotation of a polar solvent and ionic condition of dissolved ions and this rapid volumetric heating leads to effective cell rupture, releasing the compounds into the solvent. MAE is used to extract phenolic compounds from several seaweed species (Table 3). Yuan et al. [59] applied microwave irradiation at 110 °C for 15 min, to extract phenolic compounds from four economically important brown seaweed species *Ascophyllum nodosum*, *Laminaria japonica*, *Lessonia trabeculate*, and *Lessonia nigrecens*. They also conducted the conventional method, agitation at room temperature for 4 h, for comparison. They found a TPC concentration of 139.8, 73.1, 74.1, and 107.1, GAE mg/100 g dry weight of each seaweed, respectively, which were much higher than the TPC obtained through the conventional extraction technique of 51.5, 38.5, 49.8, and 78.1 GAE mg/100 g DW, respectively, for the same set of seaweed samples.

The advantage of MAE over the conventional method, is getting a higher amount of TPC at a very short time, at a temperature, as high as 110 °C, which normally degrades and reduces the yield of phenolic compounds. Manusson et al. [60] conducted a comparison study of MAE extraction with conventional solid–liquid extraction of phloroglucinol from several brown seaweed species, and reported that the yield of polyphenols using MAE increased up to 70%, as compared to SLE. According to this study, the high amount of phenolic compounds in MAE, as compared to SLE is due to the accessibility of up to 40% of cell-wall bound polyphenols by MAE, which was not possible with SLE. Li et al. [61] optimized MAE of phenolic compounds from green seaweed species *Caulerpa racemose*, using an $L_{18}(3)^5$ orthogonal experimental array. In this study, a maximum of 67.9 mg GAE/100 g dried sample was obtained at optimum conditions of microwave power, 200 W; ethanol concentration, 60%; extraction time, 40 min; extraction temperature, 50 °C; and solvent-to-material ratio, 40 mL/g. In general, the efficiency of MAE was affected by several factors, including the microwave power, the type of solvents, the composition of solvents used, extraction time, temperature, and the sample-to-solvent ratio. Microwave power was found to be effective in increasing the extraction yield. However, too much power, on the contrary, would result in overheating of the system and lead to the degradation of heat-labile compounds such as polyphenols. The effect of temperature was also observed to be similar to microwave power. Another important factor affecting the yield of MAE is the type of solvent used in the extraction process. Water is a good solvent to absorb the microwave energy, and distribute the energy evenly in the extraction medium. However, the solubility of phenolic compounds in water is low, as compared to other organic solvents like ethanol. To overcome such problems, several researchers tested a mixture of water and ethanol in different mixing ratios. Several studies optimized the extraction parameters using statistical models, so that they could get the maximum polyphenol compounds with a possible combination of extraction parameters [26,27,61–64].

Table 2. Ultrasound-assisted extraction (UAE) of marine phenolics.

Seaweed Type	State of the Seaweed (Wet/Dry/Particle Size)	Extraction Conditions Power (W)/Temperature (°C)/Time (min)	Solvent Used	Yield (TPC mg GAE/g DW)	Application of the Extract	Reference
Ascophyllum nodosum, Laminaria hyperborea	Freeze dried/powdered	750/-/15	Distilled water	0.365 mg PGE/g DW 0.156 mg PGE/g DW	Antioxidant	[65]
Ecklonia cava	Far infrared radiation dried (40 °C)/grounded (300 μm)	200/30/720	Water Methanol: water 50:50 Methanol	47.7 63.5 57.9	Antioxidant	[66]
Laurencia obtuse	Oven dried (50 °C)/Powdered (1.55 mm)	250/30–50/30–60	95% ethanol	26.23	Antioxidant	[56]
Hormosira banksii	Freeze dried/powdered (≤0.6 mm)	150–200/30–50/20–60	70% (v/v) Ethanol	23.12	Antioxidant	[57]
Sargassum muticum, Osmundea pinnatifida, Codium tomentosum	Oven dried (60 °C)/powdered (<1.0 mm)	400/50/60	Deionized water	235.0 ± 5.57 μg CE/g LE 103.7 ± 1.67 μg CE/g LE 141.1 ± 9.79 μg CE/g LE	antioxidant	[39]

Note: CE—catechetol equivalent, LE—lyophilized extracts; PEG—phloroglucinol equivalent.

3.4. Pressurized Liquid Extraction (PLE)

Pressurized liquid extraction (PLE) is also known as accelerated solvent extraction (ASE) or subcritical water extraction (SWE), when the solvent used is only water in a temperature range of boiling point of water (100 °C) and critical temperature of water (374 °C). This is a newly emerging technology for the extraction of marine phenolics. This technology has several advantages over many other extraction technologies. This extraction takes a very short time, uses a relatively lower amount of solvents, and hence requires minimum consumption of solvent, and since the extraction is conducted in the absence of light and oxygen, the degradation of phenolic compounds is very low [67]. For SWE, we can easily modify the polarity of water by tuning the extraction temperature. For instance, increasing the temperature of water from room temperature to 200–250 °C, reduces the dielectric constant of water, which is a measure of the polarity of water from 80 to 30–25. This value is close to the dielectric constant of those of organic solvents like ethanol and methanol. These properties of water, at subcritical conditions, would enable water to extract some of the organic compounds that are otherwise not extractable by water at normal conditions. Thus, this could make SWE, the greenest process of all PLEs [68].

In recent years, several researchers reported the extraction of phenolic compounds from a variety of seaweed species, using SWE (Table 5). Different solvents such as ethanol, hexane, ethyl acetate, acetone, and a mixture of different solvents, like water and ethanol at different ratios, were used to extract marine phenolics from seaweed [69]. The extraction yield of phenolic compounds using PLE is affected by numerous factors, including temperature, pressure, the type of the solvent, the ratio between the sample and solvent, the extraction time, and the particle size of the sample. Among these factors, the temperature is by far the most influential one determining the extraction yield. Increasing the temperature reduces the viscosity and surface tension, and increases the diffusivity of the solvent, which consequently increases the mass transfer of the solvent into the sample matrix, to enhance the extraction yield. However, high extraction temperature might not always favor the extraction yield. High temperatures could potentially lead to decomposition of heat-sensitive compounds like phenolics and hence reduce the yield [70]. Pangestuti et al. [71] investigated the effect of temperature (120–270 °C), solid-to-liquid ratio (1:150 to 1:50), and static extraction time of 10 min, on total phenolic content of tropical red seaweed *Hypnea musciformis* extracts, obtained using SWE. In this study, the authors observed that increasing the solid-to-liquid ratio, increased the TPC at all extraction temperatures. Similarly, increasing temperature, also linearly increased the TPC content until it reached 210 °C, and after 210 °C, the TPC started to decline. The possible reason mentioned in this study for the reduction of TPC at higher temperatures was the thermal degradation of the phenolic compound beyond 210 °C. The maximum TPC reported in this study was 39.75 ± 0.2 mg GAE/g dried seaweed and this was recorded at solid-to-liquid ratio of 1:50 and a temperature of 210 °C. In another study, Gereniu et al. [72] studied pressurized hot water extraction (PHWE) of *Kappaphycus alvarezii* at different temperatures (150 to 300 °C) and pressure (1 MPa to 10 MPa), and reported the maximum TPC content at 270 °C and 8 MPa.

The type of solvent or mixture of solvents and their polarity is another type of very important parameter that significantly affect the yield of phenolic compounds. Otero, López-Martínez, and García-Risco [69] applied four different solvents that had wide polarity ranges, like hexane, ethyl acetate, ethanol, and ethanol:water (1:1), to extract phenolic compounds from brown alga *Laminaria ochroleuca*, using PLE. They reported that PLE extract with ethanol:water (1:1, *v/v*) showed the highest 173.7 mg GAE/g extract TPC, whereas the hexane extract showed the least TPC, with only 6 mg GAE/g extract. This indicated that careful selection of solvent is very important in PLE to get the best possible extraction yield and quality of phenolic compounds.

Table 3. Microwave-assisted extraction (MAE) of marine phenolics.

Seaweed Type	State of the Seaweed (Wet/Dry/Particle Size)	Extraction Conditions Power (W) Temperature (°C)/Time (min)	Solvent Used/Solid:Solvent Ratio (g/mL)	Yield * (TPC mg GAE/g DW)	Application of the Extract	Reference
Sargassum vestitum	Freeze dried/powdered (≤600 μm)	720–1200/-/0.42–1.25	Ethanol: water (30–70%)/1:50	58.2	Antioxidant	[62]
Cystoseira sedoides	Shade dried/Powdered (200–500 μm)	-/-/0.17–3	Ethanol: water (0–100%)/1:10–1:60	0.38 mg PGE/g DW	Anticancer Activity	[27]
Ascophyllum nodosum	Oven dried/Powdered (1 mm)	250,600,1000/-/2–5	0.1 M HCl/1:10	17.9	Antioxidant	[40]
Chaetomorpha sp.	Shade dried/powdered (60μm)	200–600/-/4–12	Acetone: water (0–100%)/1:20	0.98 mg TAE/g DW		[64]
Enteromorpha prolifera	Shade dried (40 °C)/powdered	300–700/-/5–40 (1–4 cycles)	Ethanol: water (10–60%)/1:10–1:35	0.923	Antioxidant	[73]
Saccharina japonica	Dried/powdered (40 μm)	400–600/45–65/5–25	Ethanol: water (50–70%)/1:8–1:12	0.644 mg PGE/g DW	Inhibitory effects on HepG2 cancer cells	[26]
Caulerpa racemose	Oven dried (35 ± 2 °C)/Powdered	100–600/20–70/5–60	Ethanol: water (20–100%)/1:10–1:50	6.8	Antioxidant	[61]

* Some of the extraction processes were optimized to get the maximum possible phenolic compounds yield. In such cases, the yield indicated here is the maximum yield. For detailed processes, the readers of this paper are advised to refer to the respective references. PGE—Phloroglucinol equivalent; TAE—Tannic acid equivalent.

3.5. Supercritical Fluid Extraction (SFE)

A supercritical fluid is any substance that is kept at a temperature above its critical point (Tc) and critical pressure (Pc). When a substance is kept at such conditions, it has different physical and thermodynamic properties. The physical properties like viscosity, diffusibility, dielectric constant, density, and surface tension, all significantly change, compared to the substance at its standard atmospheric conditions. Moreover, we can tune these physical properties by changing the temperature and pressure. The most commonly used supercritical fluid in the food, pharmaceutical, and cosmetic industries is carbon dioxide ($ScCO_2$). This is because carbon dioxide has a low critical temperature (31.4 °C) and pressure (73.8 bar), is not toxic, is easily available at high purity, and it is very easy to separate the gas from the extract [74]. $ScCO_2$ can be used as the only solvent to extract marine phenolics or in a combination of organic solvents as entrainer. Several researchers reported the extraction of phenolic compounds from marine resources using $ScCO_2$ [75–79]. Different extraction conditions were reported like temperature (30–60 °C) [76], pressure (10–37.9 MPa) [79,80], extraction time, (60–240 min) [77,79], CO_2 flow rate (6.7–56.7 g/min) [77,80], and the use of different types of co-solvents in several compositions with respect to CO_2 (0.5–12%, *w/w*) [79,81] (Table 4).

In SFE, controlling the above extraction parameters is very important to maximize the extraction yield and minimize the operation cost. Understanding the interaction of temperature and pressure is a fundamental part of the SFE, as both affect the physical properties, like viscosity, diffusivity, and density of the solvent, $ScCO_2$. For instance, increasing extraction pressure, increases the density of the solvent and the solvating power of the $ScCO_2$, which then easily penetrates the sample matrix to facilitate the extraction rate. However, too high a pressure is not good, because it might result in compacting the extraction bed, and it could restrict the flow of CO_2, reduce the diffusivity of the solvent, and create channels within the extraction bed, which subsequently reduce the extraction yield [82]. The effect of temperature is inconsistent, especially when we are dealing with the solute that is in the form of a single compound. In such cases, increasing temperature increases the volume of the solvent, and decreases the density and solvating power of the solvent, which subsequently reduces the extraction yield. On the contrary, low extraction temperature reduces the vapor pressure of the solute and volume of the solvent, while increasing the density and solvating power, which results in a higher extraction yield. This unique phenomenon is called the "crossover effect". Depending on the nature of the solute, the crossover region of the compound in the extraction curve varies [83]. This phenomenon might not be applicable for plant material, where the solute is in the form of a crude extract with several kinds of phenolic compounds. Therefore, conducting several extraction experiments is required to understand the effect of the temperature and pressure, and their interaction effect on the overall extraction yield.

Increasing the flow rate of CO_2, enhances the mass transfer of the solvent into the sample matrix to facilitate the extraction of the phenolic compound. The increase in mass transfer can consequently shorten the time needed for the extraction. However, a very high flow rate could negatively affect extraction efficiency by flowing around the sample matrix and limiting the mass transfer [84]. Most of the time, SFE of phenolic compounds is conducted in the presence of co-solvents like ethanol, because the solubility of phenolic compounds in $ScCO_2$ is low, compared to nonpolar compounds. Enma et al. [79] observed an increase in the yield of phenolic compound extraction from *Sargassum muticum* by 1.5 times, as compared to pure $ScCO_2$, when they used ethanol as a co-solvent at 10% (*w/w*). They also investigated other solvent flow rates (ranging from 0.5–10%), but only 10% of ethanol showed the highest result. Therefore, careful selection and optimization of the solvent flow is important.

Table 4. Supercritical CO$_2$ extraction (SFE) of marine phenolics.

Seaweed Type	State of the Seaweed (wet/dry/particle size)	Co-Solvent Used/Co-Solvent Flow Rate (mL/min/%CO$_2$ flow rate)	Extraction Conditions Temperature (°C)/Pressure (MP)/CO$_2$ Flow Rate (g/min)/Time	Yield (mg GAE/g DW)	Application of the Extract	Reference
Gracilaria mammillaris	Vacuum oven dried (45 °C)/Powdered (0.15–0.6 mm)	Ethanol (2–8%, *w/w*)	40–60/15–30/6.7/240	3.79	Antioxidant	[77]
Sargassum muticum	Freeze dried/powdered (250 μm)	Ethanol (12% *w/w*)	60/15.2/-/90	34.5 mg PGE/g DE	Antioxidant	[81]
Sargassum muticum	Freeze dried/Powdered (<0.5 mm)	Ethanol (0.5–10%, *w/w*)	30–50/10–30/25/60	-	Antioxidant	[79]
Undaria pinnatifida	Freeze dried/Powdered (500 μm)	Ethanol/2	30–60/10–30/28.17/60	-	-	[76]
Laminaria digitata *Undaria pinnatifida* *Porphyra umbilicalis* *Eucheuma denticulatum* *Gelidium pusillum*	Dried/Powdered	-	50/37.9/56.7/120	23 mg GAE/g DE 4 3 2 15	Antifungal	[80]

PGE—phloroglucinol equivalent; and DE—dry extract.

SFE is used less frequently for the extraction of marine phenolics from seaweeds, compared to other emerging technologies reviewed in this paper, and the number of publications is very limited. However, the technology is used extensively for the extraction of phenolic compounds from plant materials. More details and a comprehensive review on the application of supercritical fluid for extraction of phenolic compounds from several types of terrestrial plant materials, were presented by Katarzyna at al. [89]. Some information in these studies could be extrapolated for optimizing the extraction of marine phenolics.

3.6. Other Emerging Technologies

The need for an efficient, green, and sustainable method to recover plant and marine bioactives has encouraged researchers to develop several new techniques, with some of them still in development stages. These techniques are based on the energy sources they use for extraction, such as pulsed electric field extraction (PEF), ohmic heating [90], and the use of a centrifugal field for the case of centrifugal partition extraction (CPE) [81]. Based on the use of surfactants to facilitate the extraction (SME) and based on the type of extraction medium they use, such as extraction using a new type of "designer solvents"; including, ionic liquids (IL), deep eutectic solvents (DES), and natural deep eutectic solvents (NDES).

In PEF applications, high voltages (kV range) are applied in pulses of short duration (nano or micro-seconds) with the main objective of causing electro-permeabilization and destroying the cell membranes to accelerate the extraction rate [91]. PEF is used extensively for the extraction of phenolic compounds from terrestrial plants [92–94]. There is very limited information on the application of PEF marine resources, however, since the method is proven to be effective for the extraction of phenolic compounds from land plants, it should also be replicated for the extraction of marine phenolics, including seaweeds.

CPE is a multi-stage liquid–liquid extraction technique conducted under a centrifugal field. The extraction of the specific components is based on its partition coefficients between the two liquid phases [95]. This technique is used most frequently for the purification of phenolic compounds from marine resources [96–98]. However, Anaëlle et al. [81] used CPE to extract bioactive phenolic compounds from brown seaweed *Sargassum muticum*, and compared the yield and activity of the extracts with two green techniques, SFE and PLE. The result of their study showed that the total phenolic content of the CPE extract was higher than both the PLE and SFE extracts. The concentration of the total phenolic compounds in CPE was twice that of PLE, which gave the second-highest concentration. This study indicated that more studies are needed to explore such kind of alternative techniques, optimize the operation conditions, and apply to other seaweed species.

The use of surfactants in surfactant-mediated extraction (SME) is also a promising, newly emerging technology for the extraction of phenolic compounds [99]. Surfactants can form monomolecular layers on the surface of a liquid, decreasing the interfacial tension between two liquids, allowing the miscibility of two liquid. This could enable SME to be used in the isolation of compounds with a wide range of polarities and complex chemical structures [100]. Yılmaz et al. [100] have conducted a comparative study of SME with EAE and PLE, for isolation of total phenolic compounds and phlorotannins from brown seaweed *Lobophora variegata*. They reported that the yield of both total phenolics and phlorotannins were higher for SME, than EAE and PLE.

Ionic liquids (ILs) are types of simple molten salts, containing a relatively large organic cation and an inorganic anion, which are liquid at or near room temperature. Compared to common organic solutions, ionic liquids have potential advantages like a low melting point, broad liquid temperature, negligible vapor pressure, and extended, specific, solvent properties [101]. In recent years, ILs-based extraction techniques were used for the extraction of phenolic compounds from plant materials [102–104]. However, the use of ILs-based extraction of phenolic compounds from the marine resources like seaweeds is scarce [86,105].

Table 5. Pressurized liquid extraction (PLE) of marine phenolics.

Seaweed Type	State of the Seaweed (Wet/Dry/Particle Size)	Extraction Solvent	Extraction Temperature (°C)/Pressure (MPa)/Time (min)	Solid: Liquid Ratio (g/mL)	Yield (mg GAE/g DW)	Application of the Extract	Reference
Sargassum muticum	Freeze dried/powdered (250 μm)	Ethanol: water (25:75, and 75:25)	120/10.3/20	1:5	101.8	Antioxidant	[81]
Gracilaria chilensis	Oven dried (50 °C)/ Powdered (0.5 mm)	Water	100/-/5 (3 extraction cycles) 150/-/5 (3 extraction cycles) 200/-/30 (3 extraction cycles)	-	2.06 0.78 10.17	Antioxidant	[85]
Saccharina japonica	Freeze dried/Powdered (710 μm)	Water + 0.25 M 1-Butyl-3-methylimidazolium tetrafluoroborate	175/5/5	1:32	58.92 mg PGE/g DW	Antioxidant	[86]
Laminaria ochroleuca	Freeze dried/Powdered (<500 μm)	Hexane Ethyl Acetate Ethanol Ethanol: Water (1:1)	80,120,160/10/10	1:20	6 - 83 173.65	Bioactive	[69]
Fucus serratus,	Freeze dried/powdered	Ethanol: water (80:20) Methanol: water (70:30)	100/6.9/25 90/6.9/25		75.96 80.70		
Laminaria digitata,		Ethanol: water (80:20) Methanol: water (70:30)	100/6.9/25 90/6.9/25	-	1.39 2.93	Antioxidant	[87]
Gracilaria gracilis,		Ethanol: water (80:20) Methanol: water (70:30)	100/6.9/25 90/6.9/25		2.40 0.93		
Codium fragile		Ethanol: water (80:20) Methanol: water (70:30)	100/6.9/25 90/6.9/25		4.76 5.36		
Ascophyllum nodosum,	Freeze dried/powdered	Water Ethanol: water (80:20) Acetone: water (80:20)	120/10.3/60 120/10.3/60 60/6.9/60		70.4 mg PGE/g DW 66.26 155.95		
Pelvetia canaliculata,		Water Ethanol: water (80:20) Acetone: water (80:20)	120/10.3/60 120/10.3/60 60/6.9/60	-	41.13 40.07 168.82	Antioxidant	[88]
Fucus spiralis,		Water Ethanol: water (80:20) Acetone: water (80:20)	120/10.3/60 120/10.3/60 60/6.9/60		90.79 124.30 204.40		
Ulva intestinalis		Water Ethanol: water (80:20) Acetone: water (80:20)	120/10.3/60 120/10.3/60 60/6.9/60		33.75 20.95 48.56		

Deep eutectic solvents (DESs) as a system formed from a mixture of two or more Lewis acids and bases or Brønsted-Lowry acids and bases that has the lowest freezing point, compared to its starting constituents. The formation of DES results from the complexation of a halide salt, which acts as hydrogen-bond acceptor and a hydrogen-bond donor (HBD). The physical structure of some DESs is thought to be similar to that of the ILs. However, generally, DESs are different in terms of the source of the starting ingredients and the chemical formation process. Hence, the applications of their chemical characteristics are different in many ways [106]. DESs were used extensively for the extraction of phenolic compounds from terrestrial plants [107–110]. They have already found an application in the extraction of hydrocolloids from the red seaweed *Kappaphycus alvarezii* [111]. However, unlike ILs, to the best of our knowledge, there is no single published article on the use of DESs for the extraction of marine phenolics. Therefore, as these solvents are considered "green", more research is needed to utilize the potentials of their unique properties for the extraction of phenolic compounds from marine resources for a sustainable future.

3.7. Combination of Different Emerging Technologies

A combination of two or more emerging technologies was applied to enhance the extractability of marine phenolics from seaweeds. Thus far, several researchers reported the use of a combination of the emerging technologies, MAE with PLE, UAE with EAE, MAE with EAE, EAE with SFE, and a combination of UAE with MAE. In a recent study, Garcia-Vaquero et al. [40], combined UAE with MAE, for the extraction of bioactive compounds from *Ascophyllum nodosum*. They reported a 30.3% and 10.2% increase of phenolic compounds yield, when a combination of the two methods was used at the same time, as compared to UAE and EAE alone, respectively.

A combination of two different extraction methods might not always result in the high content of phenolic compounds. Sánchez-Camargo et al. [48] pretreated *Sargassum muticum* with different carbohydrases and proteases, to evaluate the effect of phlorotannins extraction, using PLE. The total phenolic contents of all PLE extracts, which were pretreated with enzymes, were lower than the extracts obtained with PLE alone. As described by the authors, the possible reason for the lower result in the total phenolic content of enzyme-pretreated samples is that when the algal cell wall is degraded by proteases or carbohydrases, intracellular contents like proteins are released in the extraction medium. The released proteins bind to phenolic compounds to form complexes, i.e., polyphenols–protein, which further leads to aggregation and eventually precipitation. Therefore, careful selection of the extraction methods to use them in a combined extraction systems is crucial to minimize the cost of extraction and increase the extraction yield and quality of the extracts.

4. Opportunities and Challenges with Emerging Technologies

Although the newly emerging technologies show promising opportunities in terms of increasing the extraction efficiency of phenolic compounds, minimizing the extraction time, improving the quality of the extracts, and minimization of the generation of hazardous waste and its associated impact on the environment, the new technologies also have some challenges, especially in terms of industry applications. If we consider EAE, the most commonly available enzymes are nonspecific enzymes like carbohydrases and proteases, which could result in the co-extraction of other components. This could be solved by using specific enzymes like alginate–lyases, glucuronan–lyases, carrageenases, laminarinases, and agarases, specifically targeting the cell wall polysaccharides; alginate, glucuronic acid, laminarian, and agars, respectively. However, most of these enzymes are under laboratory research stage, produced in lower quantity, and have high prices. Thus, further biotechnological and bioengineering techniques, like cloning, are required to produce these enzymes in large quantity, and at a reasonable price, to use them at an industrial level. Technologies like PLE and SFE require state-of-the art equipment like high-pressure pumps, stainless steel extraction vessels, pipes, fittings, valves, pressure regulators, condensers, and process control systems. Therefore, upscaling these technologies would require very high equipment fixed capital and make the technologies expensive to implement at an industrial level. However, once the equipment is installed, the use of cheap and widely available solvents, like CO_2 and water, and the possibility of recycling the solvents would make the operational costs relatively low. Moreover, the sustainable nature of these processes creates more opportunities to use them at the industrial level.

Similarly, UAE and MAE also have challenges and opportunists, when we consider them for upscaling, and there are some success as well as failure stories on trials for upscaling for extraction of phenolic compounds from other kinds of biomasses. These success stories can be reproduced here for upscaling of phenolics from marine biomass. The authors refer to the work by Belwal et al. [112], which recently published a comprehensive review of scholarly articles on the scaling up of emerging technologies for extraction process, such as EAE, MAE, PEE, PLE SFE, and UAE. In this review work, several lessons learned from the successes and failures stories were critically evaluated and presented to show the opportunities and possible challenges associated with the non-conventional extraction methods. Nevertheless, the opportunities associated with these technologies far outweighed the challenges, especially considering when their impact on the environment, and this makes them preferable and a more applicable technology for a sustainable future.

5. Conclusions

Phenolics are a diverse group of compounds found in seaweeds with several biological activities that attract interest from the food, pharmaceutical, and cosmetic industries. To benefit from the marine organisms and more specifically the seaweeds as a potential source of unique and diverse phenolic compounds, extraction plays a crucial role, especially when it is green. Traditional extraction involves the use of organic solvents and requires a longer extraction time. The newly emerging technologies avoid the challenges associated with conventional extraction methods and are considered to be green. They are not only green, but perform better in terms of maximizing the extraction yield, as compared to the conventional methods. A combination of these methods further increase the yield. These techniques act differently to extract the phenolic compounds, depending on their energy source and the extraction mechanism. Some generate heat during the extraction process, which might degrade the phenolic compounds, some others require further purification and cleaning steps. Some of the emerging technologies were tested and upscaled to extract interesting compounds from terrestrial plants, however, more attention is needed to study the phenolics of seaweeds. In most of the extracts reviewed in this study, the phenolics are reported as the total phenolic content. Information regarding the individual phenolics content of seaweed extracts obtained using emerging extraction technologies is very scarce. Such information is important to understand the influence of each emerging extraction technologies on the individual phenolic contents. Thus, in future, researchers working this area should consider including this important information.

Author Contributions: Conceptualization, A.T.G. and C.J. A.T.G. writing—original draft preparation; writing—review and editing, C.J. and S.L.H. All authors have read and agreed to the published version of the manuscript.

Funding: This work has received funding from the European Union's Horizon 2020 research and innovation program under the Marie Sklodowska-Curie grant agreement no. 713683 (COFUNDfellowsDTU).

Acknowledgments: A.T.G. is grateful for the grant he has received for his post-doctoral study.

Conflicts of Interest: The authors declare no conflict of interest.

References

1. Jiménez-Escrig, A.; Gómez-Ordóñez, E.; Rupérez, P. Brown and red seaweeds as potential sources of antioxidant nutraceuticals. *J. Appl. Phycol.* **2012**, *24*, 1123–1132. [CrossRef]
2. Blanchette, C.A. Size and survival of intertidal plants in response to wave action: A case study with fucus gardneri. *Ecology* **1997**, *78*, 1563–1578. [CrossRef]
3. Parys, S.; Rosenbaum, A.; Kehraus, S.; Reher, G.; Glombitza, K.W.; König, G.M. Evaluation of quantitative methods for the determination of polyphenols in algal extracts. *J. Nat. Prod.* **2007**, *70*, 1865–1870. [CrossRef]

4. Marinho, G.S.; Sørensen, A.D.M.; Safafar, H.; Pedersen, A.H.; Holdt, S.L. Antioxidant content and activity of the seaweed Saccharina latissima: A seasonal perspective. *J. Appl. Phycol.* **2019**, *31*, 1343–1354. [CrossRef]
5. Steevensz, A.J.; MacKinnon, S.L.; Hankinson, R.; Craft, C.; Connan, S.; Stengel, D.B.; Melanson, J.E. Profiling phlorotannins in brown macroalgae by liquid chromatography-high resolution mass spectrometry. *Phytochem. Anal.* **2012**, *23*, 547–553. [CrossRef]
6. Gager, L.; Connan, S.; Molla, M.; Couteau, C.; Arbona, J.F.; Coiffard, L.; Cérantola, S.; Stiger-Pouvreau, V. Active phlorotannins from seven brown seaweeds commercially harvested in Brittany (France) detected by 1H NMR and in vitro assays: Temporal variation and potential valorization in cosmetic applications. *J. Appl. Phycol.* **2020**, 1–12. [CrossRef]
7. Kirke, D.A.; Rai, D.K.; Smyth, T.J.; Stengel, D.B. An assessment of temporal variation in the low molecular weight phlorotannin profiles in four intertidal brown macroalgae. *Algal Res.* **2019**, *41*, 101550. [CrossRef]
8. Jacobsen, C.; Sørensen, A.-D.M.; Holdt, S.L.; Akoh, C.C.; Hermund, D.B. Source, extraction, characterization, and applications of novel antioxidants from seaweed. *Annu. Rev. Food Sci. Technol.* **2019**, *10*, 541–568. [CrossRef]
9. Wang, T.; Jónsdóttir, R.; Ólafsdóttir, G. Total phenolic compounds, radical scavenging and metal chelation of extracts from Icelandic seaweeds. *Food Chem.* **2009**, *116*, 240–248. [CrossRef]
10. Sabeena Farvin, K.H.; Jacobsen, C. Phenolic compounds and antioxidant activities of selected species of seaweeds from Danish coast. *Food Chem.* **2013**, *138*, 1670–1681. [CrossRef]
11. Machu, L.; Misurcova, L.; Vavra Ambrozova, J.; Orsavova, J.; Mlcek, J.; Sochor, J.; Jurikova, T. Phenolic content and antioxidant capacity in algal food products. *Molecules* **2015**, *20*, 1118–1133. [CrossRef]
12. Rajauria, G.; Foley, B.; Abu-Ghannam, N. Identification and characterization of phenolic antioxidant compounds from brown Irish seaweed Himanthalia elongata using LC-DAD–ESI-MS/MS. *Innov. Food Sci. Emerg. Technol.* **2016**, *37*, 261–268. [CrossRef]
13. Yoshie, Y.; Wang, W.; Petillo, D.; Suzuki, T. Distribution of catechins in Japanese seaweeds. *Fish. Sci.* **2000**, *66*, 998–1000. [CrossRef]
14. Singh, I.P.; Sidana, J. Phlorotannins. In *Functional Ingredients from Algae for Foods and Nutraceuticals*; Elsevier: Amsterdam, The Netherland, 2013; Volume 256, pp. 181–204. ISBN 9780857095121.
15. Gómez, I.; Huovinen, P. Brown algal phlorotannins: An overview of their functional roles. In *Antarctic Seaweeds*; Springer International Publishing: Cham, Switzerland, 2020; pp. 365–388.
16. Holdt, S.L.; Kraan, S. Bioactive compounds in seaweed: Functional food applications and legislation. *J. Appl. Phycol.* **2011**, *23*, 543–597. [CrossRef]
17. Pinteus, S.; Silva, J.; Alves, C.; Horta, A.; Fino, N.; Rodrigues, A.I.; Mendes, S.; Pedrosa, R. Cytoprotective effect of seaweeds with high antioxidant activity from the Peniche coast (Portugal). *Food Chem.* **2017**, *218*, 591–599. [CrossRef]
18. Liu, N.; Fu, X.; Duan, D.; Xu, J.; Gao, X.; Zhao, L. Evaluation of bioactivity of phenolic compounds from the brown seaweed of Sargassum fusiforme and development of their stable emulsion. *J. Appl. Phycol.* **2018**, *30*, 1955–1970. [CrossRef]
19. Abdelhamid, A.; Jouini, M.; Bel Haj Amor, H.; Mzoughi, Z.; Dridi, M.; Ben Said, R.; Bouraoui, A. Phytochemical analysis and evaluation of the antioxidant, anti-inflammatory, and antinociceptive potential of phlorotannin-rich fractions from three mediterranean brown seaweeds. *Mar. Biotechnol.* **2018**, *20*, 60–74. [CrossRef]
20. Tenorio-Rodríguez, P.A.; Esquivel-Solis, H.; Murillo-Álvarez, J.I.; Ascencio, F.; Campa-Córdova, Á.I.; Angulo, C. Biosprospecting potential of kelp (Laminariales, Phaeophyceae) from Baja California Peninsula: Phenolic content, antioxidant properties, anti-inflammatory, and cell viability. *J. Appl. Phycol.* **2019**, *31*, 3115–3129. [CrossRef]
21. Devi, K.P.; Suganthy, N.; Kesika, P.; Pandian, S.K. Bioprotective properties of seaweeds: In vitro evaluation of antioxidant activity and antimicrobial activity against food borne bacteria in relation to polyphenolic content. *BMC Complement. Altern. Med.* **2008**, *8*, 1–11. [CrossRef]
22. Moubayed, N.M.S.; Al Houri, H.J.; Al Khulaifi, M.M.; Al Farraj, D.A. Antimicrobial, antioxidant properties and chemical composition of seaweeds collected from Saudi Arabia (Red Sea and Arabian Gulf). *Saudi J. Biol. Sci.* **2017**, *24*, 162–169. [CrossRef]
23. Zouaoui, B.; Ghalem, B.R. The phenolic contents and antimicrobial activities of some marine algae from the mediterranean sea (Algeria). *Russ. J. Mar. Biol.* **2017**, *43*, 491–495. [CrossRef]

24. Karthik, R.; Manigandan, V.; Sheeba, R.; Saravanan, R.; Rajesh, P.R. Structural characterization and comparative biomedical properties of phloroglucinol from Indian brown seaweeds. *J. Appl. Phycol.* **2016**, *28*, 3561–3573. [CrossRef]

25. Michalak, I.; Górka, B.; Wieczorek, P.P.; Rój, E.; Lipok, J.; Łęska, B.; Messyasz, B.; Wilk, R.; Schroeder, G.; Dobrzyńska-Inger, A.; et al. Supercritical fluid extraction of algae enhances levels of biologically active compounds promoting plant growth. *Eur. J. Phycol.* **2016**, *51*, 243–252. [CrossRef]

26. He, Z.; Chen, Y.; Chen, Y.; Liu, H.; Yuan, G.; Fan, Y.; Chen, K. Optimization of the microwave-assisted extraction of phlorotannins from Saccharina japonica Aresch and evaluation of the inhibitory effects of phlorotannin-containing extracts on HepG2 cancer cells. *Chin. J. Oceanol. Limnol.* **2013**, *31*, 1045–1054. [CrossRef]

27. Abdelhamid, A.; Lajili, S.; Elkaibi, M.A.; Ben Salem, Y.; Abdelhamid, A.; Muller, C.D.; Majdoub, H.; Kraiem, J.; Bouraoui, A. Optimized extraction, preliminary characterization and evaluation of the in vitro anticancer activity of phlorotannin-rich fraction from the Brown Seaweed, *Cystoseira sedoides*. *J. Aquat. Food Prod. Technol.* **2019**, *28*, 892–909. [CrossRef]

28. Olivares-Molina, A.; Fernández, K. Comparison of different extraction techniques for obtaining extracts from brown seaweeds and their potential effects as angiotensin I-converting enzyme (ACE) inhibitors. *J. Appl. Phycol.* **2016**, *28*, 1295–1302. [CrossRef]

29. Zhang, J.; Tiller, C.; Shen, J.; Wang, C.; Girouard, G.S.; Dennis, D.; Barrow, C.J.; Miao, M.; Ewart, H.S. Antidiabetic properties of polysaccharide- and polyphenolic-enriched fractions from the brown seaweed *Ascophyllum nodosum*. *Can. J. Physiol. Pharmacol.* **2007**, *85*, 1116–1123. [CrossRef]

30. Yuan, Y.V.; Walsh, N.A. Antioxidant and antiproliferative activities of extracts from a variety of edible seaweeds. *Food Chem. Toxicol.* **2006**, *44*, 1144–1150. [CrossRef]

31. Pantidos, N.; Boath, A.; Lund, V.; Conner, S.; McDougall, G.J. Phenolic-Rich extracts from the edible seaweed, ascophyllum nodosum, inhibit α-amylase and α-glucosidase: Potential anti-hyperglycemic effects. *J. Funct. Foods* **2014**, *10*, 201–209. [CrossRef]

32. Chowdhury, M.T.H.; Bangoura, I.; Kang, J.Y.; Cho, J.Y.; Joo, J.; Choi, Y.S.; Hwang, D.S.; Hong, Y.K. Comparison of ecklonia cava, ecklonia stolonifera and eisenia bicyclis for phlorotannin extraction. *J. Environ. Biol.* **2014**, *35*, 713–719.

33. Lopes, G.; Sousa, C.; Silva, L.R.; Pinto, E.; Andrade, P.B.; Bernardo, J.; Mouga, T.; Valentão, P. Can phlorotannins purified extracts constitute a novel pharmacological alternative for microbial infections with associated inflammatory conditions? *PLoS ONE* **2012**, *7*. [CrossRef]

34. Parys, S.; Kehraus, S.; Krick, A.; Glombitza, K.W.; Carmeli, S.; Klimo, K.; Gerhäuser, C.; König, G.M. In vitro chemopreventive potential of fucophlorethols from the brown alga Fucus vesiculosus L. by anti-oxidant activity and inhibition of selected cytochrome P450 enzymes. *Phytochemistry* **2010**, *71*, 221–229. [CrossRef]

35. Vijayan, R.; Chitra, L.; Penislusshiyan, S.; Palvannan, T. Exploring bioactive fraction of *Sargassum wightii*: *In vitro* elucidation of angiotensin-I-converting enzyme inhibition and antioxidant potential. *Int. J. Food Prop.* **2018**, *21*, 674–684. [CrossRef]

36. Catarino, M.; Silva, A.; Mateus, N.; Cardoso, S. Optimization of phlorotannins extraction from fucus vesiculosus and evaluation of their potential to prevent metabolic disorders. *Mar. Drugs* **2019**, *17*, 162. [CrossRef]

37. Tierney, M.S.; Smyth, T.J.; Rai, D.K.; Soler-Vila, A.; Croft, A.K.; Brunton, N. Enrichment of polyphenol contents and antioxidant activities of Irish brown macroalgae using food-friendly techniques based on polarity and molecular size. *Food Chem.* **2013**, *139*, 753–761. [CrossRef]

38. Ojha, K.S.; Aznar, R.; O'Donnell, C.; Tiwari, B.K. Ultrasound technology for the extraction of biologically active molecules from plant, animal and marine sources. *TrAC Trends Anal. Chem.* **2020**, *122*, 115663. [CrossRef]

39. Rodrigues, D.; Sousa, S.; Silva, A.; Amorim, M.; Pereira, L.; Rocha-Santos, T.A.P.; Gomes, A.M.P.; Duarte, A.C.; Freitas, A.C. Impact of enzyme- and ultrasound-assisted extraction methods on biological properties of red, brown, and green seaweeds from the Central West Coast of Portugal. *J. Agric. Food Chem.* **2015**, *63*, 3177–3188. [CrossRef]

40. Garcia-Vaquero, M.; Ummat, V.; Tiwari, B.; Rajauria, G. Exploring ultrasound, microwave and ultrasound–microwave assisted extraction technologies to increase the extraction of bioactive compounds and antioxidants from brown macroalgae. *Mar. Drugs* **2020**, *18*, 172. [CrossRef] [PubMed]

41. Gligor, O.; Mocan, A.; Moldovan, C.; Locatelli, M.; Crișan, G.; Ferreira, I.C.F.R. Enzyme-Assisted extractions of polyphenols—A comprehensive review. *Trends Food Sci. Technol.* **2019**, *88*, 302–315. [CrossRef]

42. Deniaud-Bouët, E.; Kervarec, N.; Michel, G.; Tonon, T.; Kloareg, B.; Hervé, C. Chemical and enzymatic fractionation of cell walls from Fucales: Insights into the structure of the extracellular matrix of brown algae. *Ann. Bot.* **2014**, *114*, 1203–1216. [CrossRef] [PubMed]

43. Michel, G.; Czjzek, M. Polysaccharide-Degrading enzymes from marine bacteria. In *Marine Enzymes for Biocatalysis: Sources, Biocatalytic Characteristics and Bioprocesses of Marine Enzymes*; Elsevier Ltd.: Amsterdam, The Netherland, 2013; pp. 429–464. ISBN 9781907568800.

44. Ihua, M.W.; Guihéneuf, F.; Mohammed, H.; Margassery, L.M.; Jackson, S.A.; Stengel, D.B.; Clarke, D.J.; Dobson, A.D.W. Microbial population changes in decaying ascophyllum nodosum result in macroalgal-polysaccharide-degrading bacteria with potential applicability in enzyme-assisted extraction technologies. *Mar. Drugs* **2019**, *17*, 200. [CrossRef]

45. Hardouin, K.; Bedoux, G.; Burlot, A.S.; Donnay-Moreno, C.; Bergé, J.P.; Nyvall-Collén, P.; Bourgougnon, N. Enzyme-Assisted extraction (EAE) for the production of antiviral and antioxidant extracts from the green seaweed Ulva armoricana (Ulvales, Ulvophyceae). *Algal Res.* **2016**, *16*, 233–239. [CrossRef]

46. Sabeena, S.F.; Alagarsamy, S.; Sattari, Z.; Al-Haddad, S.; Fakhraldeen, S.; Al-Ghunaim, A.; Al-Yamani, F. Enzyme-Assisted extraction of bioactive compounds from brown seaweeds and characterization. *J. Appl. Phycol.* **2020**, *32*, 615–629.

47. Ahn, C.-B.; Park, P.-J.; Je, J.-Y. Preparation and biological evaluation of enzyme-assisted extracts from edible seaweed (*Enteromorpha prolifera*) as antioxidant, anti-acetylcholinesterase and inhibition of lipopolysaccharide-induced nitric oxide production in murine macrophages. *Int. J. Food Sci. Nutr.* **2012**, *63*, 187–193. [CrossRef]

48. Sánchez-Camargo, A.D.P.; Montero, L.; Stiger-Pouvreau, V.; Tanniou, A.; Cifuentes, A.; Herrero, M.; Ibáñez, E. Considerations on the use of enzyme-assisted extraction in combination with pressurized liquids to recover bioactive compounds from algae. *Food Chem.* **2016**, *192*, 67–74. [CrossRef]

49. Chemat, F.; Zill-e-Huma; Khan, M.K. Applications of ultrasound in food technology: Processing, preservation and extraction. *Ultrason. Sonochem.* **2011**, *18*, 813–835. [CrossRef]

50. Luque-García, J.L.; Luque De Castro, M.D. Ultrasound: A powerful tool for leaching. *TrAC Trends Anal. Chem.* **2003**, *22*, 41–47. [CrossRef]

51. Romdhane, M.; Gourdon, C. Investigation in solid-liquid extraction: Influence of ultrasound. *Chem. Eng. J.* **2002**, *87*, 11–19. [CrossRef]

52. Saleh, I.A.; Vinatoru, M.; Mason, T.J.; Abdel-Azim, N.S.; Aboutabl, E.A.; Hammouda, F.M. A possible general mechanism for ultrasound-assisted extraction (UAE) suggested from the results of UAE of chlorogenic acid from Cynara scolymus L. (artichoke) leaves. *Ultrason. Sonochem.* **2016**, *31*, 330–336. [CrossRef]

53. Vinatoru, M. An overview of the ultrasonically assisted extraction of bioactive principles from herbs. *Ultrason. Sonochem.* **2001**, *8*, 303–313. [CrossRef]

54. Moreira, R.; Sineiro, J.; Chenlo, F.; Arufe, S.; Díaz-Varela, D. Aqueous extracts of Ascophyllum nodosum obtained by ultrasound-assisted extraction: Effects of drying temperature of seaweed on the properties of extracts. *J. Appl. Phycol.* **2017**, *29*, 3191–3200. [CrossRef]

55. Moreira, R.; Chenlo, F.; Sineiro, J.; Arufe, S.; Sexto, S. Drying temperature effect on powder physical properties and aqueous extract characteristics of Fucus vesiculosus. *J. Appl. Phycol.* **2016**, *28*, 2485–2494. [CrossRef]

56. Topuz, O.K.; Gokoglu, N.; Yerlikaya, P.; Ucak, I.; Gumus, B. Optimization of antioxidant activity and phenolic compound extraction conditions from red seaweed (*Laurencia obtuse*). *J. Aquat. Food Prod. Technol.* **2016**, *25*, 414–422. [CrossRef]

57. Dang, T.T.; Van Vuong, Q.; Schreider, M.J.; Bowyer, M.C.; Van Altena, I.A.; Scarlett, C.J. Optimisation of ultrasound-assisted extraction conditions for phenolic content and antioxidant activities of the alga Hormosira banksii using response surface methodology. *J. Appl. Phycol.* **2017**, *29*, 3161–3173. [CrossRef]

58. Vázquez-Rodríguez, B.; Gutiérrez-Uribe, J.A.; Antunes-Ricardo, M.; Santos-Zea, L.; Cruz-Suárez, L.E. Ultrasound-Assisted extraction of phlorotannins and polysaccharides from Silvetia compressa (Phaeophyceae). *J. Appl. Phycol.* **2020**, *32*, 1441–1453. [CrossRef]

59. Yuan, Y.; Zhang, J.; Fan, J.; Clark, J.; Shen, P.; Li, Y.; Zhang, C. Microwave assisted extraction of phenolic compounds from four economic brown macroalgae species and evaluation of their antioxidant activities and inhibitory effects on α-amylase, α-glucosidase, pancreatic lipase and tyrosinase. *Food Res. Int.* **2018**, *113*, 288–297. [CrossRef]

60. Magnusson, M.; Yuen, A.K.L.; Zhang, R.; Wright, J.T.; Taylor, R.B.; Maschmeyer, T.; de Nys, R. A comparative assessment of microwave assisted (MAE) and conventional solid-liquid (SLE) techniques for the extraction of phloroglucinol from brown seaweed. *Algal Res.* **2017**, *23*, 28–36. [CrossRef]

61. Li, Z.; Wang, B.; Zhang, Q.; Qu, Y.; Xu, H.; Li, G. Preparation and antioxidant property of extract and semipurified fractions of Caulerpa racemosa. *J. Appl. Phycol.* **2012**, *24*, 1527–1536. [CrossRef]

62. Dang, T.T.; Bowyer, M.C.; Van Altena, I.A.; Scarlett, C.J. Optimum conditions of microwave-assisted extraction for phenolic compounds and antioxidant capacity of the brown alga Sargassum vestitum. *Sep. Sci. Technol.* **2018**, *53*, 1711–1723. [CrossRef]

63. Poole, J.; Diop, A.; Rainville, L.-C.; Barnabé, S. Bioextracting polyphenols from the brown seaweed *Ascophyllum nodosum* from Québec's North Shore Coastline. *Ind. Biotechnol.* **2019**, *15*, 212–218. [CrossRef]

64. Safari, P.; Rezaei, M.; Shaviklo, A.R. The optimum conditions for the extraction of antioxidant compounds from the Persian gulf green algae (*Chaetomorpha* sp.) using response surface methodology. *J. Food Sci. Technol.* **2015**, *52*, 2974–2981. [CrossRef]

65. Kadam, S.; O'Donnell, C.; Rai, D.; Hossain, M.; Burgess, C.; Walsh, D.; Tiwari, B. Laminarin from Irish Brown Seaweeds Ascophyllum nodosum and Laminaria hyperborea: Ultrasound assisted extraction, characterization and bioactivity. *Mar. Drugs* **2015**, *13*, 4270–4280. [CrossRef]

66. Lee, S.H.; Kang, M.C.; Moon, S.H.; Jeon, B.T.; Jeon, Y.J. Potential use of ultrasound in antioxidant extraction from Ecklonia cava. *Algae* **2013**, *28*, 371–378. [CrossRef]

67. Getachew, A.T.; Chun, B.S. Influence of hydrothermal process on bioactive compounds extraction from green coffee bean. *Innov. Food Sci. Emerg. Technol.* **2016**, *38*, 24–31. [CrossRef]

68. Herrero, M.; Sánchez-Camargo, A.d.P.; Cifuentes, A.; Ibáñez, E. Plants, seaweeds, microalgae and food by-products as natural sources of functional ingredients obtained using pressurized liquid extraction and supercritical fluid extraction. *TrAC Trends Anal. Chem.* **2015**, *71*, 26–38. [CrossRef]

69. Otero, P.; López-Martínez, M.I.; García-Risco, M.R. Application of pressurized liquid extraction (PLE) to obtain bioactive fatty acids and phenols from Laminaria ochroleuca collected in Galicia (NW Spain). *J. Pharm. Biomed. Anal.* **2019**, *164*, 86–92. [CrossRef]

70. Plaza, M.; Marina, M.L. Pressurized hot water extraction of bioactives. *TrAC Trends Anal. Chem.* **2019**, *116*, 236–247. [CrossRef]

71. Pangestuti, R.; Getachew, A.T.; Siahaan, E.A.; Chun, B.S. Characterization of functional materials derived from tropical red seaweed Hypnea musciformis produced by subcritical water extraction systems. *J. Appl. Phycol.* **2019**, *31*, 2517–2528. [CrossRef]

72. Gereniu, C.R.N.; Saravana, P.S.; Getachew, A.T.; Chun, B.-S. Characteristics of functional materials recovered from Solomon Islands red seaweed (*Kappaphycus alvarezii*) using pressurized hot water extraction. *J. Appl. Phycol.* **2017**, *29*. [CrossRef]

73. Luo, H.; Wang, B.; Yu, C.; Xu, Y. Optimization of microwave-assisted extraction of polyphenols from enteromorpha prolifra by orthogonal test. *Chin. Herb. Med.* **2010**, *2*, 321–325.

74. Liu, J.; Liu, J.; Lin, S.; Wang, Z.; Wang, C.; Wang, E.; Zhang, Y. Supercritical fluid extraction of flavonoids from Maydis stigma and its nitrite-scavenging ability. *Food Bioprod. Process.* **2011**, *89*, 333–339. [CrossRef]

75. Messyasz, B.; Michalak, I.; Łęska, B.; Schroeder, G.; Górka, B.; Korzeniowska, K.; Lipok, J.; Wieczorek, P.; Rój, E.; Wilk, R.; et al. Valuable natural products from marine and freshwater macroalgae obtained from supercritical fluid extracts. *J. Appl. Phycol.* **2018**, *30*, 591–603. [CrossRef]

76. Roh, M.K.; Uddin, M.S.; Chun, B.S. Extraction of fucoxanthin and polyphenol from Undaria pinnatifida using supercritical carbon dioxide with co-solvent. *Biotechnol. Bioprocess Eng.* **2008**, *13*, 724–729. [CrossRef]

77. Ospina, M.; Castro-Vargas, H.I.; Parada-Alfonso, F. Antioxidant capacity of Colombian seaweeds: 1. Extracts obtained from Gracilaria mammillaris by means of supercritical fluid extraction. *J. Supercrit. Fluids* **2017**, *128*, 314–322. [CrossRef]

78. Klejdus, B.; Lojková, L.; Vlcek, J. Hyphenated solid phase extraction/supercritical fluid extraction methods for extraction of phenolic compounds from algae. *Curr. Anal. Chem.* **2014**, *10*, 86–98. [CrossRef]

79. Conde, E.; Moure, A.; Domínguez, H. Supercritical CO_2 extraction of fatty acids, phenolics and fucoxanthin from freeze-dried Sargassum muticum. *J. Appl. Phycol.* **2015**, *27*, 957–964. [CrossRef]

80. De Corato, U.; Salimbeni, R.; De Pretis, A.; Avella, N.; Patruno, G. Antifungal activity of crude extracts from brown and red seaweeds by a supercritical carbon dioxide technique against fruit postharvest fungal diseases. *Postharvest Biol. Technol.* **2017**, *131*, 16–30. [CrossRef]

81. Anaëlle, T.; Serrano Leon, E.; Laurent, V.; Elena, I.; Mendiola, J.A.; Stéphane, C.; Nelly, K.; Stéphane, L.B.; Luc, M.; Valérie, S.P. Green improved processes to extract bioactive phenolic compounds from brown macroalgae using Sargassum muticum as model. *Talanta* **2013**, *104*, 44–52. [CrossRef]

82. Moon, J.N.; Getachew, A.T.; Haque, A.S.M.T.; Saravana, P.S.; Cho, Y.J.; Nkurunziza, D.; Chun, B.S. Physicochemical characterization and deodorant activity of essential oil recovered from Asiasarum heterotropoides using supercritical carbon dioxide and organic solvents. *J. Ind. Eng. Chem.* **2019**, *69*, 217–224. [CrossRef]

83. Ghosh, S.; Chatterjee, D.; Das, S.; Bhattacharjee, P. Supercritical carbon dioxide extraction of eugenol-rich fraction from Ocimum sanctum Linn and a comparative evaluation with other extraction techniques: Process optimization and phytochemical characterization. *Ind. Crops Prod.* **2013**, *47*, 78–85. [CrossRef]

84. Gallego, R.; Bueno, M.; Herrero, M. Sub- and supercritical fluid extraction of bioactive compounds from plants, food-by-products, seaweeds and microalgae—An update. *TrAC Trends Anal. Chem.* **2019**, *116*, 198–213. [CrossRef]

85. Sanz-Pintos, N.; Pérez-Jiménez, J.; Buschmann, A.H.; Vergara-Salinas, J.R.; Pérez-Correa, J.R.; Saura-Calixto, F. Macromolecular antioxidants and dietary fiber in edible seaweeds. *J. Food Sci.* **2017**, *82*, 289–295. [CrossRef]

86. Vo Dinh, T.; Saravana, P.S.; Woo, H.C.; Chun, B.S. Ionic liquid-assisted subcritical water enhances the extraction of phenolics from brown seaweed and its antioxidant activity. *Sep. Purif. Technol.* **2018**, *196*, 287–299. [CrossRef]

87. Heffernan, N.; Smyth, T.J.; FitzGerald, R.J.; Soler-Vila, A.; Brunton, N. Antioxidant activity and phenolic content of pressurised liquid and solid-liquid extracts from four Irish origin macroalgae. *Int. J. Food Sci. Technol.* **2014**, *49*, 1765–1772. [CrossRef]

88. Tierney, M.S.; Smyth, T.J.; Hayes, M.; Soler-Vila, A.; Croft, A.K.; Brunton, N. Influence of pressurised liquid extraction and solid-liquid extraction methods on the phenolic content and antioxidant activities of Irish macroalgae. *Int. J. Food Sci. Technol.* **2013**, *48*, 860–869. [CrossRef]

89. Tyśkiewicz, K.; Konkol, M.; Rój, E. The application of supercritical fluid extraction in phenolic compounds isolation from natural plant materials. *Molecules* **2018**, *23*, 2625. [CrossRef]

90. Zhang, R.; Lebovka, N.; Marchal, L.; Vorobiev, E.; Grimi, N. Pulsed electric energy and ultrasonication assisted green solvent extraction of bio-molecules from different microalgal species. *Innov. Food Sci. Emerg. Technol.* **2020**, *62*, 102358. [CrossRef]

91. Kotnik, T.; Frey, W.; Sack, M.; Haberl Meglič, S.; Peterka, M.; Miklavčič, D. Electroporation-Based applications in biotechnology. *Trends Biotechnol.* **2015**, *33*, 480–488. [CrossRef]

92. Rocha, C.M.R.; Genisheva, Z.; Ferreira-Santos, P.; Rodrigues, R.; Vicente, A.A.; Teixeira, J.A.; Pereira, R.N. Electric field-based technologies for valorization of bioresources. *Bioresour. Technol.* **2018**, *254*, 325–339. [CrossRef]

93. Ameer, K.; Shahbaz, H.M.; Kwon, J.-H. Green extraction methods for polyphenols from plant matrices and their byproducts: A review. *Compr. Rev. Food Sci. Food Saf.* **2017**, *16*, 295–315. [CrossRef]

94. Yan, L.-G.; He, L.; Xi, J. High intensity pulsed electric field as an innovative technique for extraction of bioactive compounds—A review. *Crit. Rev. Food Sci. Nutr.* **2017**, *57*, 2877–2888. [CrossRef]

95. Nagaosa, Y.; Wang, T. High performance centrifugal partition chromatographic separation of alkaline earth metal ions with bis-2-ethylhexylphosphinic acid. *J. Sep. Sci.* **2003**, *26*, 953–956. [CrossRef]

96. Lee, J.H.; Ko, J.Y.; Oh, J.Y.; Kim, C.Y.; Lee, H.J.; Kim, J.; Jeon, Y.J. Preparative isolation and purification of phlorotannins from Ecklonia cava using centrifugal partition chromatography by one-step. *Food Chem.* **2014**, *158*, 433–437. [CrossRef]

97. Lee, J.H.; Ko, J.Y.; Samarakoon, K.; Oh, J.Y.; Heo, S.J.; Kim, C.Y.; Nah, J.W.; Jang, M.K.; Lee, J.S.; Jeon, Y.J. Preparative isolation of sargachromanol E from Sargassum siliquastrum by centrifugal partition chromatography and its anti-inflammatory activity. *Food Chem. Toxicol.* **2013**, *62*, 54–60. [CrossRef]

98. Lee, J.H.; Ko, J.Y.; Kim, H.H.; Kim, C.Y.; Jang, J.H.; Nah, J.W.; Jeon, Y.J. Efficient approach to purification of octaphlorethol A from brown seaweed, Ishige foliacea by centrifugal partition chromatography. *Algal Res.* **2017**, *22*, 87–92. [CrossRef]

99. Sharma, S.; Kori, S.; Parmar, A. Surfactant mediated extraction of total phenolic contents (TPC) and antioxidants from fruits juices. *Food Chem.* **2015**, *185*, 284–288. [CrossRef]

100. Gümüş Yılmaz, G.; Gómez Pinchetti, J.L.; Cifuentes, A.; Herrero, M.; Ibáñez, E. Comparison of extraction techniques and surfactants for the isolation of total polyphenols and phlorotannins from the brown algae lobophora variegata. *Anal. Lett.* **2019**, *52*, 2724–2740. [CrossRef]

101. Niu, J.; Qiu, H.; Li, J.; Liu, X.; Jiang, S. 1-Hexadecyl-3-methylimidazolium ionic liquid as a new cationic surfactant for separation of phenolic compounds by MEKC. *Chromatographia* **2009**, *69*, 1093–1096. [CrossRef]
102. Cláudio, A.F.M.; Ferreira, A.M.; Freire, C.S.R.; Silvestre, A.J.D.; Freire, M.G.; Coutinho, J.A.P. Optimization of the gallic acid extraction using ionic-liquid-based aqueous two-phase systems. *Sep. Purif. Technol.* **2012**, *97*, 142–149. [CrossRef]
103. Du, F.Y.; Xiao, X.H.; Luo, X.J.; Li, G.K. Application of ionic liquids in the microwave-assisted extraction of polyphenolic compounds from medicinal plants. *Talanta* **2009**, *78*, 1177–1184. [CrossRef]
104. Lou, Z.; Wang, H.; Zhu, S.; Chen, S.; Zhang, M.; Wang, Z. Ionic liquids based simultaneous ultrasonic and microwave assisted extraction of phenolic compounds from burdock leaves. *Anal. Chim. Acta* **2012**, *716*, 28–33. [CrossRef]
105. Han, D.; Zhu, T.; Row, K.H. Ultrasonic extraction of phenolic compounds from laminaria japonica aresch using ionic liquid as extraction solvent. *Bull. Korean Chem. Soc.* **2011**, *32*, 2212–2216. [CrossRef]
106. Zainal-Abidin, M.H.; Hayyan, M.; Hayyan, A.; Jayakumar, N.S. New horizons in the extraction of bioactive compounds using deep eutectic solvents: A review. *Anal. Chim. Acta* **2017**, *979*, 1–23. [CrossRef]
107. Wei, Z.F.; Wang, X.Q.; Peng, X.; Wang, W.; Zhao, C.J.; Zu, Y.G.; Fu, Y.J. Fast and green extraction and separation of main bioactive flavonoids from Radix Scutellariae. *Ind. Crops Prod.* **2015**, *63*, 175–181. [CrossRef]
108. Dai, Y.; Witkamp, G.J.; Verpoorte, R.; Choi, Y.H. Natural deep eutectic solvents as a new extraction media for phenolic metabolites in carthamus tinctorius L. *Anal. Chem.* **2013**, *85*, 6272–6278. [CrossRef]
109. Wei, Z.; Qi, X.; Li, T.; Luo, M.; Wang, W.; Zu, Y.; Fu, Y. Application of natural deep eutectic solvents for extraction and determination of phenolics in Cajanus cajan leaves by ultra performance liquid chromatography. *Sep. Purif. Technol.* **2015**, *149*, 237–244. [CrossRef]
110. Ruesgas-Ramón, M.; Figueroa-Espinoza, M.C.; Durand, E. Application of deep eutectic solvents (DES) for phenolic compounds extraction: Overview, challenges, and opportunities. *J. Agric. Food Chem.* **2017**, *65*, 3591–3601. [CrossRef]
111. Das, A.K.; Sharma, M.; Mondal, D.; Prasad, K. Deep eutectic solvents as efficient solvent system for the extraction of κ-carrageenan from *Kappaphycus alvarezii. Carbohydr. Polym.* **2016**, *136*, 930–935. [CrossRef]
112. Belwal, T.; Chemat, F.; Venskutonis, P.R.; Cravotto, G.; Jaiswal, D.K.; Bhatt, I.D.; Devkota, H.P.; Luo, Z. Recent advances in scaling-up of non-conventional extraction techniques: Learning from successes and failures. *TrAC Trends Anal. Chem.* **2020**, *127*, 115895. [CrossRef]

Article

Extraction, Enrichment, and LC-MS^n-Based Characterization of Phlorotannins and Related Phenolics from the Brown Seaweed, *Ascophyllum nodosum*

J. William Allwood [1], Huw Evans [2], Ceri Austin [1] and Gordon J. McDougall [1,*]

[1] Plant Biochemistry and Food Quality Group, Environmental and Biochemical Sciences Department, The James Hutton Institute, Dundee DD2 5DA, UK; will.allwood@hutton.ac.uk (J.W.A.); Ceri.Austin@hutton.ac.uk (C.A.)

[2] Byotrol Ltd., Thornton Science Park, Chester CH2 4NU, UK; hevans@byotrol.com

* Correspondence: gordon.mcdougall@hutton.ac.uk; Tel.: +44-1382-568782

Received: 30 July 2020; Accepted: 24 August 2020; Published: 27 August 2020

Abstract: Phenolic components from the edible brown seaweed, *Ascophyllum nodosum*, have been associated with considerable antioxidant activity but also bioactivities related to human health. This study aims to select and identify the main phlorotannin components from this seaweed which have been previously associated with potential health benefits. Methods to enrich phenolic components then further select phlorotannin components from ethanolic extracts of *Ascophyllum nodosum* were applied. The composition and phenolic diversity of these extracts were defined using data dependent liquid chromatography mass spectroscopic (LC-MS^n) techniques. A series of phlorotannin oligomers with apparent degree of polymerization (DP) from 10 to 31 were enriched by solid phase extraction and could be selected by fractionation on Sephadex LH-20. Evidence was also obtained for the presence of dibenzodioxin linked phlorotannins as well as sulphated phlorotannins and phenolic acids. As well as diversity in molecular size, there was evidence for potential isomers at each DP. MS^2 fragmentation analyses strongly suggested that the phlorotannins contained ether linked phloroglucinol units and were most likely fucophlorethols and MS^3 data suggested that the isomers may result from branching within the chain. Therefore, application of these LC-MS^n techniques provided further information on the structural diversity of the phlorotannins from *Ascophyllum*, which could be correlated against their reported bioactivities and could be further applied to phlorotannins from different seaweed species.

Keywords: phlorotannins; *Ascophyllum*; seaweed; health benefits; isomers; LC-MS^n; diversity; phenolics

---

1. Introduction

Phlorotannins are dehydro-polymers of phloroglucinol units particularly associated with brown seaweeds (Phaeophyceae) [1–3]. They appear to play a defensive role in the seaweeds, protecting against herbivory [4–6] and UV-B radiation e.g., [7]. Their levels vary with season, developmental stages and abiotic stresses [8–11] and it has been suggested that there is balance between a structural role in the algal cell wall and these protective roles [12]. They exist in structurally different forms depending on how the phloroglucinol units are interlinked [2]. If the phloroglucinol units are only linked by phenyl -C-C bonds, they are termed fucols; if they are only linked by -C-O-C- aryl ether bonds, they are termed phlorethols and if both linkages are present they are termed fucaphlorethols. Some phlorotannins also have dibenzodioxin linkages and these are generally termed as eckols. The type of inter-linkages of phlorotannins vary notably between species as is suggested by their names, *Fucus* species are rich in fucols and fucophlorethols and *Ecklonia* species notably contain eckols.

Molecular size or the degree of polymerization of phlorotannins also varies greatly between species and may also be affected by biotic and abiotic stresses [2].

The phlorotannins have considerable antioxidant activity e.g., [13–15] and have been mooted as food-grade additives to prevent spoilage [16]. They have also been associated with specific health beneficial effects in related to disease states such as inflammation [17], cancers [18], diabetes [19] and hyperlipidemia associated with cardiovascular issues [20], which may be independent of their antioxidant activity. They have also been suggested to have valuable antimicrobial effects [21], with some efficacy against viruses e.g., [22]. The link between phlorotannin structure and activity has not particularly been well defined as many studies use extracts with varying extents of enrichment.

Their potential bioactivities are particularly relevant as brown seaweeds are generally edible and have formed part of the food culture of humans across the world, but notably in the Far East [23]. The edible brown seaweed *Ascophyllum nodosum* is common around the coasts of the UK and reaches high abundance around Scotland. Our previous work indicated that phlorotannins from *Ascophyllum* can inhibit digestive enzymes and thereby could modulate glycemic responses or reduce calorie intake from fats [18,24,25]. In this paper, we report on the enrichment and fractionation of phlorotannins and related phenolics from *Ascophyllum nodosum* and apply a series of liquid chromatography mass spectrometric (LC-MS^n) methods to define their structural diversity.

2. Results and Discussion

2.1. Fractionation of Phenolic Material Using Solid Phase Extraction (SPE) and Sephadex LH-20

The total phenol content (TPC) of the *Ascophyllum* extract was mainly retained in the SPE-bound fraction with an overall recovery of ~90% of applied material in the fractionation (see Supplementary Data; Figure S1). The reasonably high TPC in the unbound material may be due to non-phenolic material that cross-reacts with the non-specific Folin reagent [26]. In the Sephadex LH-20 fractionation, the majority of the phenolic material was recovered in the first 80% acetone fraction and >80% of the total in the fractions released by acetone. The requirement for higher concentrations of acetone for release of bound material has been noted before [24]. Once again, the TPC in the unbound material may be due to non-phenolic material that cross-reacts with the non-specific Folin reagent.

2.2. Liquid Chromatography Mass Spectrometric (LC-MS^n) Analysis

Using the LC-MS data acquired on the LCQ Fleet MS system, the SPE bound sample was enriched in later eluting UV-absorbing peaks (Figure 1A) whereas the unbound fraction was essentially devoid of these peaks. The MS spectrum across the retention time of the UV peaks from 12–21 min (Figure 1B) gave a set of *m/z* signals in negative mode which were characteristic of phlorotannins. These could be thought of as two series of *m/z* values that differ by 124 amu, an extension unit mass equivalent to phloroglucinol minus 2 H atoms, which have been noted previously in our laboratory [18,24,25], or indeed as a single series that differ by 62 amu (124 divided by 2). If the *m/z* values from 621, 745, 869, 993, 1117, 1241, 1365 and 1489 were single charged then they could arise from successive phloroglucinol additions to a phloroglucinol dimer with *m/z* [M − H]- of 249 (such as diphlorethol) and working on this basis, the signals at *m/z* 621 and 745 could be pentaphloroethol and hexaphloroethol structures and the major signal noted at *m/z* 1117 could be a nonaphloroethol derivative (see Figure 2) [2]. However, if we assume singly charged ions then the possible nature of the *m/z* series at 683, 807, 931, 1055, 1179, 1303 etc. is not apparent unless all the signals were from double charged species. However, it was not possible to confirm the charge status of the ions as the resolution of the Fleet LCQ MS cannot discern this in normal mode function.

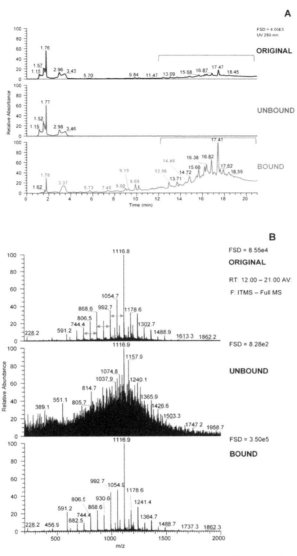

Figure 1. Profiles of the fractions from the SPE procedure. (**A**) UV profiles, (**B**) MS spectra from 12–21 min for each sample. Blue bracket in (**A**) represents area for MS spectra in (**B**). Arrows show 124 amu differences between peaks. FSD = full scale deflection.

The phlorotannin series outlined above is based on a phlorethol structure, oligomers of phloroglucinol attached through aryl ether (-C-O-C) bonds but the addition of phloroglucinol units (124 amu) to achieve this series could also be achieved through phenyl (-C-C-) linkages (see Figure 3). In fact, this series of *m/z* values could arise from phloroglucinol units linked by all phenyl linkages (called fucols), all aryl ether linkages (called phlorethols) or a mixture of both (called fucophlorethols) [2,15].

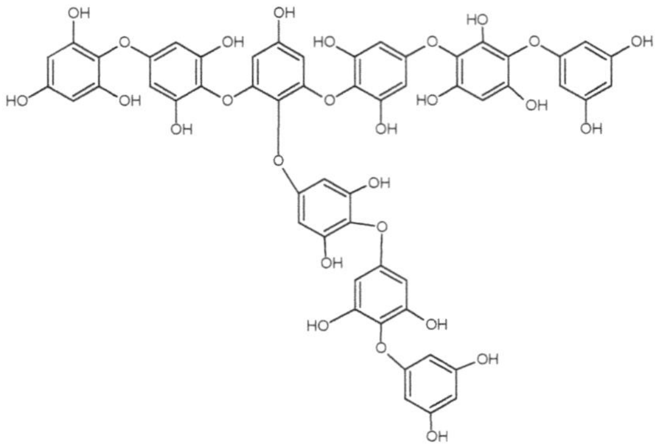

Figure 2. Possible nonaphorethol structure.

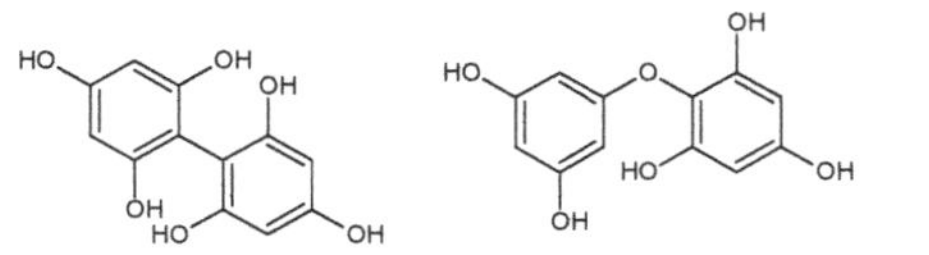

Difucol; $C_{12}H_{10}O_6$, MW = 250   Diphorethol; $C_{12}H_{10}O_6$, MW = 250

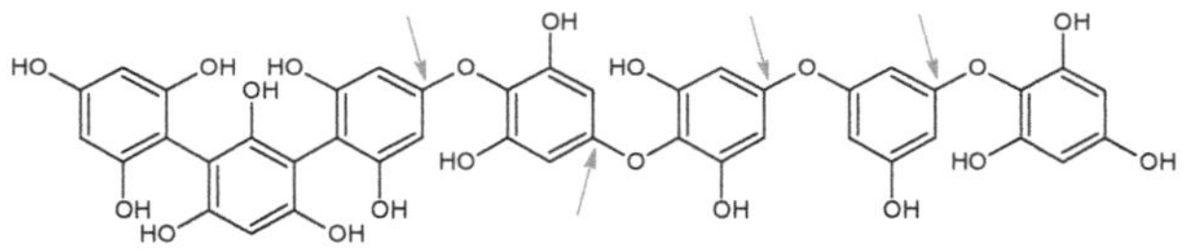

Trifucotetraphlorethol; $C_{42}H_{30}O_{21}$, MW = 994

Figure 3. Structures of difucol, diphlorethol and trifucotetraphlorethol. These structures are discussed in the text. Fragmentation at the positions noted with blue arrows would produce THB containing fragments with neutral losses of 142 amu = THB, 266 amu = PG-THB, 390 amu = 2PG-THB and 514 amu = 3PG-THB respectively from right to left.

It was notable that these "phlorotannin" MS signals were approximately 4-fold enriched in the bound fraction obtained after SPE (see FSD in Figure 1B). Also, the bound acetone fractions from the Sephadex LH-20 separation also showed this enrichment in UV-absorbing peaks between 12 and 21 min (Figure 4A). However, in this case, the enrichment was less apparent as the phlorotannins were spread over the three acetone fractions and there was a dilution factor inherent in the procedure. However, these UV peaks contain the same set of *m/z* signals characteristic of phlorotannins (Figure 4B) and these are sufficient at this stage to follow the enrichment of these phlorotannin species.

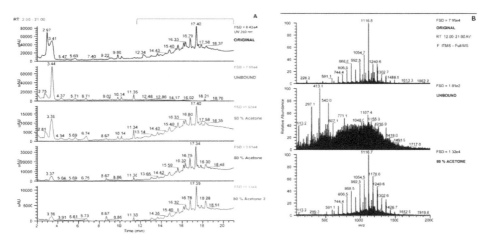

Figure 4. Profiles of the fractions from the Sephadex LH-20 procedure. (**A**) UV profiles, (**B**) MS spectra from 12–21 min for each sample. FSD = full scale deflection.

2.3. Differences between Positve and Negative Mode MS Data

Using the Fleet LCQ MS system, the MS profiles of the phlorotannin peaks in the SPE bound sample were different in negative and positive mode (Figure 5A, panels A–C). In general, intensities were lower in positive mode but there were qualitative differences in the ionization of different peaks. For example, the major peak in negative mode is at 17.46 min and the MS profile largely matches the UV profile (compare panel A and B). In positive mode, it is notable that the later eluting major UV peaks (e.g., RT = 16.8 and 17.4) have lower relative MS intensity then the negative mode data. The negative mode MS data across RT 10–21 min showed the series of major peaks that differ by 62 amu (panel D). The positive mode MS data over the same retention range showed a series beginning at m/z 747 that increased by 124 amu through 871, 995, 1119, etc. However, there was also a minor series of m/z signals that began at 1180 (in grey) and also increased by 124 amu. When the MS data under specific phlorotannin peaks was examined, the situation became clearer. The spectra under the MS peaks shown in bold in Figure 5A are shown in negative mode then positive mode in Figure 5B. The MS peak at ~14.8 in positive mode had three main signals at m/z 1243, 1491 and 1739. In negative mode, the same peak gave the corresponding m/z $[M - H]^-$ signals at 1241, 1489 and 1737 but also had strong signals at 620, 746 and 868 which were effectively half the values of the other signals. The other major peak at RT 15.6 showed a similar picture. Therefore, in both examples, the negative mode MS spectra contained m/z signals that are effectively half the m/z value of the positive mode signals. This disparity can be explained if the phlorotannins ionise mainly as singly charged ions in positive mode (i.e., as m/z $[M + H]^+$ ions) but they have a propensity to ionise as doubly charged ions (i.e., as $[M - 2H]^{2-}$ ions) in negative mode. In fact, this can be neatly illustrated by the co-chromatography of peaks with base peaks at m/z $[M - H]^{2-}$ values against $[M + H]^+$ ions in the SPE bound sample (see example in Figure S2).

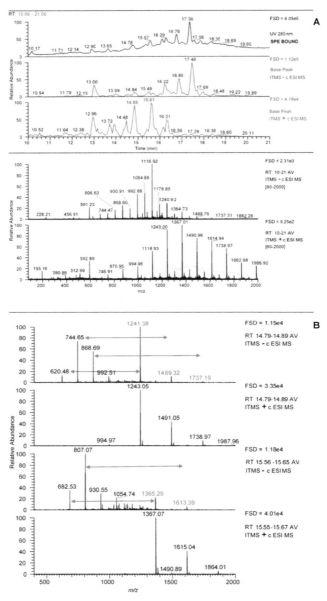

Figure 5. (**A**) MS properties of phlorotannin peaks in negative and positive mode. The top panel shows the UV spectra of the SPE bound sample from 10–21 min, the next shows the MS spectra in negative mode from 10–21 min and the third panel shows the MS spectra in positive mode from 10–21 min. The fourth panel shows the MS spectrum over 10–21 min in negative mode and the last panel shows the same in positive mode. FSD = full scale deflection. The peaks denoted in bold green were examined further (Figure 5B). (**B**). MS spectra of selected phlorotannin peaks in negative and positive mode. Examples of negative and positive MS spectra of specific peaks labelled in bold in Figure 5A. Green arrows denote m/z signals discussed in the text.

As noted above, operating in normal full scan mode, the Fleet MS system cannot discriminate between single and double charged ions. Therefore, we re-examined the samples using the FT-MS detector of an LTQ Orbitrap XL system that has sufficient resolution to differentiate 0.5 amu isotope spacing, as observed in double charged ions.

2.4. Re-Examination of Data Using the LTQ-Orbitrap XL FT-MS System

When analyzed on the LTQ-Orbitrap XL system in ESI negative mode, the major peak at RT 17.8 showed revealed ions that differed by 0.5 amu, indicating that they are double charged ions (i.e., m/z [M − 2H]$^{2-}$, Figure 6). As a double charged ion, the true mass would be twice as high at 2234, and this suggests that the main UV peak at RT 17.8 actually contained an oligomer of 18 phloroglucinol (PG) units rather than the nonaphloroethol suggested before. In fact, all of the m/z values in the negative mode series from m/z 621 upwards were doubly double charged which indicates that they are all double the MW than first suggested by the original LCQ-Fleet MS data. It also confirms that the original observed peak spacing of 62 amu observed upon the LCQ-Fleet MS system, when accounting for the ions being double charged, in fact represents a single series of ions with peak spacing of 124 amu, corresponding to phloroglucinol minus 2 H atoms. Peaks for all the ion species with apparent double charged [M − 2H]$^{2-}$ ions from m/z 621 through to m/z 1429 were identified in the MS profiles of the SPE bound samples (see Figure S3). At least two main peaks were apparent for each m/z species which suggests the presence of isomers of the different oligomers of each DP.

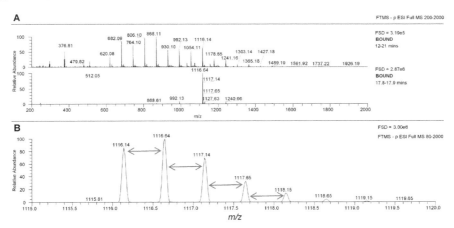

Figure 6. The phlorotannin peaks yield doubly charged MS signals in negative mode. The top panel (**A**) shows the MS spectra across 12–21 min of the SPE bound sample and the second panel shows the MS spectra of the main UV peak. The third panel (**B**) shows the zoom spectra across the m/z 1116 peak.

For quantification, we summed the areas for the major putative isomer peaks but ignored other small peaks which may be due to in-source fragmentation within the MS spectra of other oligomers. The relative abundance of these putative phlorotannin oligomers from DP 10 to 24 are shown in Figure 7. Plotted against DP, the oligomers of DP 11–18 were the most abundant with a drop-off at DP > 18. In fact, although some peaks areas for m/z values for phlorotannin oligomers >24 DP could be discerned, they were not significantly above baseline values. However, there were indications that phlorotannins of larger DP were present in the region between RT 19–21 min where triple charged m/z signals could be discerned (see Supplementary Data, Figure S4) that suggested the presence of oligomers of up to DP 31 (e.g., a triple charged ion at 1281.42 yields an estimated MW of 3846). However, these were in much lower abundance and did not yield MS^2 data. Previously, the presence of triple charged phlorotannin species has been indicated [27] and the distribution and range of DP noted in our work fits in the range noted by this group.

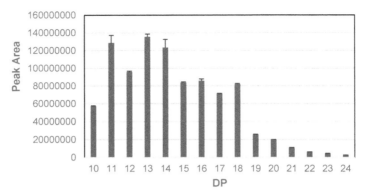

Figure 7. Relative abundance of phlorotannin oligomers in the SPE bound fraction.

It is important to note that as well as having more intense signals, the negative mode data with its double and triple charged ions actually allowed the detection of molecular species with MWs > 2000 amu that would not have been detected in positive mode. The positive mode analysis would have been limited to the detection and quantification of phlorotannins with DP = 16 PG units (m/z [M + H]$^+$ = 1987). This also explains why the later eluting phlorotannins noted above had such poor MS spectra in positive mode as their signals were effectively outside the detectable MS range of 100–2000 m/z. Indeed, Tierney et al. [28] reported that phlorotannins from *Ascophyllum* had a lower DP range, with a maximum of 16 PG units. Although they applied a dialysis step which may have altered the MW range, they crucially used an MS detector which limited detection at DP 16 (i.e., a m/z value = 1987). However, a previous paper by this group [29] detected phlorotannins of up to DP 20 in low molecular weight extracts of *Ascophyllum nodosum* by using direct infusion of SPE-purified extracts into a Q-Tof Premier mass spectrometer with detector mass range of m/z 100 to m/z 3000.

The MS data obtained on the LTQ-Orbitrap XL system also provided evidence for the presence of other phlorotannin components. For example, there was a peak at RT = 12.96 (see Figure 1B, bound sample, bold blue peaks), which preceded the cluster of phlorotannin oligomers. This gave m/z [M − H]$^-$ = 591.2 with fragmentation yielding MS^2 fragments of 511 & 385 and an accurate mass of 591.0075 derived a molecular formula of $C_{24}H_{15}O_{16}S$ at < 1 ppm error (Table 1). The neutral loss of 80 amu can be assigned to loss of a sulphate group (SO$_3$) to a compound with formula m/z [M − H]$^-$ of $C_{24}H_{15}O_{13}$, which matches with diphlorethohydroxycarmalol (Pub Chem 16075395), a phlorotannin component noted in the brown seaweed, *Ishige okamurae* [30]. The presence of this diphlorethohydroxycarmalol derivative was suggested but not confirmed in our previous studies of *Ascophyllum* [23,24]. In fact, a peak attributable to diphlorethohydroxycarmalol itself (RT = 14.48; m/z [M − H]$^-$ = 511.0506, predicted formula of $C_{24}H_{15}O_{13}$ and sole MS^2 fragment at 385; Table 1). was also present later in the separation. There was also evidence for the presence of a component with m/z = 246.9914, MS^2 = 203, 121, with a predicted formula $C_{12}H_7O_6$, that matched with the dimer, dibenzodioxin-1,3,6,8-tetraol, which has been reported previously in *Fucus* species e.g., [27]. These three phlorotannin components contain a dibenzodioxin-ring structure which has not been noted previously in *Ascophyllum*. There were also other components which appeared to be sulphated phenolic acids and another whose MS and MS^2 properties matched with a DOPA-sulphate-like component. However, confirmation of the identity of these components requires further isolation and characterization by techniques such as 2D NMR. It was notable that all these components were enriched by the SPE procedure but were greatly reduced by the selection of the phlorotannin oligomers on Sephadex LH-20.

Table 1. MS properties of phenolic components from *Ascophyllum nodosum*.

RT	m/z [M − H]$^-$	MS^2	Exact Mass Formulae as [M − H]$^-$	Putative Identity
1.78	191.0195	173, **111** *	$C_6H_7O_7$	Citric acid
3.37	276.0184	259, 231, **215**, **196**, 179, 150, 135	$C_9H_{10}O_7NS$	Dihydroxyphenylalanine (DOPA)-sulphate (PC 178810)
5.73	246.9914	203, **121**	$C_{12}H_7O_6$	Dibenzodioxin-1,3,6,8-tetraol (PC 14309078)
7.45	277.0924	**185**, 167, 141, 97	$C_{11}H_{17}O_8$	Unknown
9.08	318.0284	300, 276, **238**, 192	$C_{11}H_{12}NSO_8$	Unknown
8.93	216.9809	173, **137**, 97	$C_7H_5O_6S$	Hydroxybenzoic acid sulphate
9.89	230.9967	187, **151**	$C_8H_7O_6S$	Phenolic acid sulphate
12.96	591.0075	511, **385**	$C_{24}H_{15}O_{16}S$	Diphlorethohydroxycarmalol sulphate
14.48	511.0506	**385**	$C_{24}H_{15}O_{13}$	Diphlorethohydroxycarmalol (PC 16075395)

RT = retention time. All formulae were derived at <2 ppm error. * major fragments are in bold. PC = Pub Chem reference number.

2.5. Structural Information from MS^2 Fragmentation Data

Different collision energies were assessed to maximize the yield of MS^2 fragments and to increase the likelihood of providing useful structural data for the phlorotannin structures. The NCE of 45% applied in our standard MS^2 method often only provided weak fragmentation spectra and overall, better fragmentation required an NCE of 65% which was adopted for further analyses. The higher energies required for effective fragmentation probably reflect the stable nature of the phlorotannin oligomers. Fragmentation data was available for all peaks corresponding to phlorotannin oligomers up to DP = 23 at m/z [M − 2H]$^{2-}$ = 1427, which did not provide MS^2 data as it was below the MS^n minimum trigger intensity (Table 2). The fragmentation data for the phlorotannins was characterized by several common factors. Firstly, as all the target m/z values were double charged, there were fragments greater than the m/z [M − 2H]$^{2-}$ value. Secondly, often the major fragment resulted from the m/z [M − 2H]$^{2-}$ value minus H_2O (neutral loss of 18 amu). Overall, the MS^2 data gave fragments and neutral losses previously noted in reports of phlorotannin structures [31–36]. Notably, the neutral losses obtained through fragmentation showed patterns between the different phlorotannin structures with repeated neutral losses being observed (Table 2), which could largely be assigned to the loss of a fixed numbers of phloroglucinol (PG) groups or PG groups + H_2O. It was also notable that major fragments noted for different phlorotannin components that differed by 124 amu in full scan MS, also generated fragment ions that differed by 124 amu, suggesting that a consistent fragmentation mechanism was occurring. Other undefined fragmentations in Table 1 could arise through ring fission events within PG units. It was also notable that major fragments noted for different phlorotannin components that differed by 124 amu also differed by 124 amu, which suggests that a consistent fragmentation mechanism was occurring. Other undefined fragmentations in Table 2 could arise through ring fission events within PG units [28,29,34].

Table 2. MS² fragmentation properties for the phlorotannins of DP 10–23.

Left portion:

m/z [M−2H]²⁻	Calc m/z	DP	MS²	NL	ΔM−2H²⁻	Neutral Loss
621	1241	10	1097	144		PG+H₂O
			993	248		2PG*
			831	410		UK
			603	638	18	4PG-THB
			495	**746**	125	6PG
			477	764	144	6PG+H₂O
			247	994	373	8PG
			229	1012	391	8PG+H₂O
745	1489	12	1345	144		PG+H₂O
			1223	**266**		PG-THB
			975	514		3PG-THB
			727	**762**	18	5PG-THB
			672	817	72	?
			477	1012	267	8PG+H₂O
			229	1260	515	10PG+H₂O
869	1737	14	1471	266		PG-THB
			1223	514		3PG-THB
			975	762		5PG-THB
			851	**887**	18	7PG+H₂O*
			787	950	81	?
			477	1260	391	10PG+H₂O
			353	1384	515	11PG+H₂O
993	1985	16	1595	390		2PG-THB
			1223	762		5PG-THB
			974	**1011**	18	8PG+H₂O*
			911	1074	81	?
			848	1137	144	9PG+H₂O
			477	1508	515	12PG+H₂O
			353	1632	639	13PG+H₂O
1117	2233	18	1985	248		2PG*
			1719	514		3PG-THB
			1451	782		?
			1223	1010		7PG-THB
			1098	**1135**	18	9PG+H₂O*
			1089	1144	28	?
			1035	1198	81	?
			709	1524	407	?
			477	1756	639	14PG+H₂O
			337	1896	779	?
1241	2481	20	1967	514		3PG-THB
			1719	762		5PG-THB
			1489	992		8PG
			1222	**1259**	18	10PG+H₂O*
			1159	1322	81	?
			955	1526	285	?
			727	1754	513	13PG-THB
			477	2004	763	16PG+H₂O
			355	2126	885	16PG-THB
1365	2729	22	1967	762		5PG-THB
			1719	1010		7PG-THB
			1451	1278		?
			1337	**1392**	27	?
			1283	**1446**	81	?
			955	1774	409	?
			727	2002	637	15PG-THB
			477	2252	887	18PG+H₂O

Right portion:

m/z [M−2H]²⁻	Calc m/z	DP	MS²	NL	ΔM−2H²⁻	Neutral Loss
683	1365	11	**1117**	248		2PG*
			975	390		2PG-THB
			851	514		3PG-THB
			664	**701**	18	?
			610	755	72	?
			477	888	205	7PG+H₂O
			371	994	311	8PG
			355	1010	327	7PG-THB
			229	**1136**	453	9PG+H₂O
807	1613	13	1469	144		PG+H₂O
			1223	390		2PG-THB
			955	658		?
			788	**825**	18	?
			725	888	81	7PG+H₂O
			477	1136	329	9PG+H₂O
			353	1260	453	10PG+H₂O
931	1861	15	1700	161		?
			1595	266		PG-THB
			1347	514		3PG-THB
			1099	762		5PG-THB
			912	**949**	18	?
			849	1012	81	8PG+H₂O
			477	1384	453	11PG+H₂O
			355	1506	575	11PG-THB
1055	2109	17	1929	180		?
			1719	390		2PG-THB
			1469	640		5PG+H₂O
			1223	886		6PG-THB
			1036	**1073**	18	?
			973	1136	81	9PG+H₂O
			831	1278	223	10PG+2H₂O
			477	1632	577	13PG+H₂O
			353	1756	701	14PG+H₂O
1179	2357	19	1825	532		?
			1595	762		6PG+H₂O
			1223	1134		8PG-THB
			1160	**1197**		?
			1151	1206	27	?
			973	1384	205	11PG+H₂O
			601	1756	577	14PG+H₂O
			479	1878	699	14PG-THB
			355	2002	823	15PG-THB
1303	2605	21	1805	800		?
			1595	1010		7PG-THB
			1284	**1321**	18	?
			1275	1330	27	?
			1221	**1384**	81	11PG+H₂O
			955	1650	347	?
			831	1774	471	14PG+2H₂O
			479	2126	823	16PG-THB
			365	2240	937	?
1427	2853	23	None			

The MS signals in bold are the predominant fragments and those underlined as the next more abundant, other MS² signals are more minor. NL = neutral loss. ?s denote undefined neutral losses that probably arise by cross ring fragmentations as noted previously [27]. (-THB) denotes the presence of a tetrahydroxybenzene unit in the neutral loss. * denotes that the NL differs by one amu from expected value.

Another important finding was the commonality of fragmentations that could be explained by the presence of a tetrahydroxylbenzene structure in the neutral loss (also called O-phloroglucinol moieties previously e.g., [34]. For example, a neutral loss of 514 amu was noted in the MS² spectra of many of the phlorotannin species (Table 2) and this could arise if the phlorotannin molecule fragmented at an ether bond and formed a phloroglucinol with an extra hydroxyl group linked to a trimer of PG units (i.e., 3PG + THB = 374 + 140; see Figure 3). In fact, this neutral loss (514 amu) is the same as the molecular weight of tetrafuhalol, a trimer of phloroglucinol units linked to a tetrahydroxybenzene unit [2], an example of another type of phlorotannin found in other brown seaweeds. Indeed, neutral losses consistent with one PG unit attached to one THB unit (i.e., 266 amu) up to 16 PG units attached to a THB group (i.e., 2126 amu) were indicated in the MS^2 fragments of the phlorotannins (Table 2). The only fragments missing from this series were those which represent 9, 10 and 12 PG units linked to a THB group and a THB group itself. THB-containing neutral losses were present in the MS^2 of every phlorotannin at each DP examined (Table 2) and since they can only occur at -C-O-C- ether linkages [31–34], this strongly suggests that the phlorotannins in *Ascophyllum* are fucophlorethol-type components. Certainly, evidence of at least one aryl ether bond in the oligomers means that they cannot be fucol oligomers, which are composed of only phenyl linked PG units. The possibility that cleavage of C-C phenyl bonds requires more energy than cleavage of C-O-C ether bonds has been discussed previously [2,34] and this may explain the preponderance of cleavages at ether bonds.

2.6. Structural Information from MS^3 Fragmentation Data

Further information was sought by acquiring MS^3 fragmentation data from the MS^2 fragments for each phlorotannin species. Once again, application of 45% NCE as defined in our original method did not provide strong and consistent compound fragmentation, therefore the NCE was increased to 65% for each MS^2 target. Using the $[M - 2H]^{2-}$ ion at *m/z* 1116.6 as an example (Figure 8), there was a major and a minor peak at *m/z* 1116.6 at RT 17.7 and RT 16.3 respectively which may represent isomers (Figure 8A). Indeed, the presence of 2 or 3 isomers for each phlorotannin signal was noted previously (Figure S3) and the different chromatographic behaviour of these apparent isomers suggests structural differences. The different peaks gave different MS^2 patterns. The more abundant isomer gave *m/z* 1098 as the predominant fragment whereas the less abundant isomer gave mainly *m/z* 1098 and 1089 but with other MS^2 products in more equal amounts. Once again, both isomers gave MS^2 products greater than the original target *m/z* value (e.g., *m/z* values of 1223, 1719 and 1949) due to their double charged nature. Similarly, the major and minor isomers of the other phlorotannin peaks gave different MS^2 patterns (results not shown).

The MS^3 fragments obtained by fragmentation of the dominant MS^2 signal at *m/z* 1098 for the two *m/z* 1117 isomers are shown in Figure 8B. As expected, many of these MS^3 fragments were also found as minor fragments in the MS^2 spectra. It was noticeable that the MS^3 products of the *m/z* 1098 MS^2 fragment for the different isomers were also different. The more minor isomer at RT 16.3 gave a simpler MS^3 pattern than the isomer at RT 17.8. The major isomer gave MS^3 products for *m/z* 1098 which were spread across apparent losses of 1, 2, 3, 4 and 5 PG units whereas the MS^3 products from the *m/z* 1098 MS^2 fragment from the minor isomer gave a major neutral loss of H_2O to *m/z* 1080 with smaller amounts of fragments resulting from losses of apparent 2, 4 and 5 PG units (see arrows, Figure 8B). The different pattern of losses at the MS^3 level suggest that these apparent phlorotannin isomers differ in their inter-linkages. For example, the major isomer seems to break into smaller fragments which could mean a more branched structure, perhaps with more ether bonds. However, confirmation of such differences would require purification of the isomers and further studies, perhaps using 2D-NMR approaches.

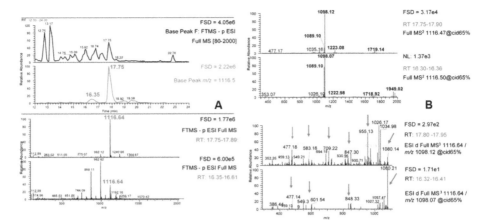

Figure 8. MS^2 and MS^3 data of major phlorotannin isomers at m/z 1116.6. (**A**) The top panel shows the MS profile between RT = 21–21 min, the second panel shows the profile for base peak at m/z 1116.6 and the bottom two panels show the MS spectra at the two peaks at 17.75 min and 16.35 min respectively. (**B**) The top panel shows the MS^3 fragments from the main MS^2 peak (m/z 1098) derived from the major m/z 1116.6 peak at RT = 17.75. The bottom panel shows the MS^3 fragments from the main MS^2 peak (m/z 1098) derived from the minor m/z 1116.6 peak at RT = 16.35.

MS^3 fragmentation data from the other notable MS^2 fragments from the m/z 1116.6 isomers could yield further information, but due to their lower ion intensities, MS^3 fragmentation data was not obtainable. The MS^2 product of m/z 1719 gave a single product ion at m/z 892 which suggests a clean fragmentation with a neutral loss of 827. However, the MS^3 DDA of the MS^2 product at m/z 1223 gave multiple MS^3 ions. Given that the intensities of these MS^2 products varied in abundance between the two isomers, this further suggests differences in structure. Overall, the MS^2 evidence (Table 2) strongly suggested that the phlorotannin oligomers were fucophlorethols as they contained aryl ether linkages [2]. Due to the three-way symmetry of the phloroglucinol molecule, -C-C- phenyl linkages are always effectively meta-orientated [15]. However, PG units attached through aryl ether bridges can occur at hydroxyls at ortho, meta or para positions on the rings so the possibilities for different isomeric structures are large especially as the DP increases. For example, extending from a diphorethol structure to a triphlorethol structure can produce structurally discernible triphlorethol A and B isomers [2] depending on the positioning of the "new" aryl ether bond. Also, branching can also occur within fucol type phlorotannins composed of only phenyl linkages if one of the PG units has 3 rather than 2 linkages to other PGs [2,15]. Indeed, other studies have reported multiple apparent isomers of phlorotannin oligomers from *Ascophyllum* [28,29,31] and other brown seaweeds [37]. In fact, the presence of only two (perhaps three maximum) major isomers as noted in this paper seems unusual compared to the 70 apparent isomers for a DP 10 oligomer reported in one study [29]. However, it is also possible that some of the isomers noted result from in-source fragments of larger phlorotannins, or from a greater number of different charge states being generated in ESI and higher MS scan ranges, which might be more apparent under the different MS conditions used in each study. Indeed, it is also possible that our chromatographic procedure did not separate all these different putative isomers, but the chromatographic conditions used similar C18 reverse phase conditions. It also seems particularly unlikely that our extraction procedure would have so drastically influenced isomer diversity.

3. Conclusions

The biosynthetic pathway by which brown seaweeds produce phloroglucinol via the acetate-malonate pathway, also known as the polyketide pathway, is understood, in a process which involves a specific polyketide synthase-type enzyme [37]. However, the mechanisms by which phloroglucinol groups are combined in such a controlled manner, especially to yield the diversity of phlorotannin structures found in different species [2], and the extent that the oligomers extend are less well understood. The LC-MS^n techniques used here have yielded useful information of the possible structural diversity of the phlorotannin oligomers in *Ascophyllum*, and further advances could be made, especially if ion-tree based fragmentation methods [38] could also be employed. However, the level of complexity for the major phlorotannin component (m/z [M − 2H]$^{2-}$ = 1117; apparent MW 2234; possible fucaphlorethol of DP 18) highlighted in this paper illustrates the daunting extent of data interpretation required to make inferences of overall structure. Also, MS-based analyses are inherently limited by their mass range and cannot deal with the high molecular weight phlorotannin species known to be present in brown seaweeds [2,15]. However, the molecular range could be extended by using other MS systems with higher MW ranges such as MALDI-TOF, as described previously e.g., [32]. Further work could focus effort on the main phlorotannin oligomers as these are most likely to be those most associated with specific bioactivities noted for the phlorotannin samples. In addition, future work could use these techniques to examine phlorotannins from different species of brown seaweed and begin to correlate their structural diversity with potential biological effectiveness.

4. Materials and Methods

4.1. Materials and General Methods

Dried and milled *Ascophyllum nodosum* powder was obtained from Hebridean Seaweeds in 2017 and stored at −20 °C until use. The ASCO extract was prepared from this dried powder using a propriety hydro-ethanolic extraction procedure carried out by Byotrol plc. The extract was filtered then reduced in volume by rotary evaporation until it could be freeze dried to a powder. The freeze-dried extract was soluble at 10% (*w/v*) in ultra-pure water.

Total phenol content (TPC) was assessed using the Folin-Ciocalteu method [24]. The TPC of the original extract was 0.394 ± 0.18 g GAE/g DW. It should be noted that the Folin method is not totally specific and these values probably over-estimate the yield of phenolic components.

4.2. Solid Phase Extraction

The method used was developed and scaled up from that reported previously [18]. In brief, a SPE unit (Strata C18-E GIGA tube, 10 g capacity & 60 mL volume; Phenomenex Ltd., Macclesfield, UK) was washed with 2 × 50 mL volumes of acetonitrile (ACN) containing 0.1% formic acid (FA) then equilibrated with 3 × 50 mL of UPW containing 0.1% FA.

The FD extract was dissolved at 5% (*w/v*) in UPW containing 0.1% FA and applied to the SPE unit, the unbound fraction was recovered, then the SPE unit was washed with 2 × volumes of UPW + 0.1% (*v/v*) FA, collected as the wash fraction. The bound fraction was obtained by eluting the unit with 2 volumes of 80% ACN + 0.1% (*v/v*) FA. The unit was then re-equilibrated for further use by washing with UPW + 0.1% (*v/v*) FA. At this stage, it was noted that the eluted fraction was cloudy so this bound-RW (rewash) fraction was also collected. The fractions were tested for total phenol content. Aliquots of each fraction were completely dried by speed vacuum concentration for LC-MS^n analysis.

4.3. Fractionation on Sephadex LH-20

A portion of the *Ascophyllum* extract was fractionated using Sephadex LH-20 applying a technique [23] well-known to select for phlorotannin-like components (https://www.users.miamioh.edu/hagermae/). In brief, a 25 mg/mL solution of the FD material in UPW was produced and then 5 mL was added to 5 mL ethanol and mixed well. This solution was added to 5 mL of a slurry of Sephadex LH-20 in 50% ethanol and mixed well for 10 min at room temperature. After centrifugation at 2500× *g* for 5 min at 5 °C, the unbound fraction was removed, and 5 mL of 50% ethanol added. The centrifugation procedure was repeated to give the wash fraction then similarly with 50% acetone and then two washes with 80% acetone to provide the bound fractions. The total phenol contents were measured as before, and aliquots of each fraction completely dried by centrifugal evaporation in a Speed Vac prior to LC-MS^n analysis.

4.4. Liquid Chromatography-Mass Spectrometry (LC-MS^n) Analysis

The samples were first analyzed on an LCQ Fleet Ion Trap mass spectrometer (Thermo Scientific Ltd., Hemel Hempstead, UK) attached to an HPLC system consisting of an Accela 600 quaternary pump and Accela photodiode array detector (PDAD) and autosampler. Spectra were collected in wavelength/absorbance mode 200–600 nm (1nm filter bandwidth and wavelength step, 1 s filter rise time, 10 Hz sample rate). Additionally, three UV channel set points were employed (A: 280 nm, B: 365 nm, C: 520 nm, 9 nm bandwidth, 10 Hz sample rate). Samples (20 µL injection volumes) were eluted on a Synergi Hydro C18 2.0 mm × 150 mm, 4 µm particle size column (Phenomenex Ltd., Macclesfield, UK) applying mobile phase A, HPLC grade water + 0.1% FA, and mobile phase B, HPLC grade Acetonitrile + 0.1% FA, at a flow-rate of 0.3 mL/min. The gradient was as follows: 0–2 min hold 2% B, 2–5 min 2–5% B, 5–25 min 5–45% B; 25–26 min 45–100% B, 26–29 min hold 100% B, 29–30 min 100–2% B, 30–35 min hold 2% B for HPLC equilibration. Mass spectra were collected with a primary full scan event (*m/z* 80–2000, profile mode) and a secondary data-dependent analysis (DDA) MS/MS scan (centroid mode) for the top three most intense ions. Helium was applied as a collision gas for collision-induced dissociation at a normalized collision energy (NCE) of 45%, a trapping window width of 2 (+/−1) *m/z* was applied, an activation time of 30 ms and activation Q of 0.25 were applied, only singly charged ions were selected for DDA, isotopic ions were excluded. The Automatic Gain Control was set to 1 × 104, scan speed to 0.1 s, the following settings were applied to ESI: Spray voltage −3.5 kV (ESI−) and +4.0 kV (ESI+); Sheath gas 60; Auxiliary gas 30; Capillary voltage at −35 V (ESI−) +35 V (ESI+); Tube lens voltage −100 V (ESI−) and +100 V (ESI+); Capillary temperature 280 °C; ESI probe temperature 100 °C.

Selected samples were separated using the same chromatographic conditions and PDA set points, but with a Thermo Dionex U3000 UHPLC-PDA (Thermo Fisher Scientific UK), coupled to a Thermo LTQ-Orbitrap XL mass spectrometry system capable of full scan accurate mass FT-MS (30,000 FWHM resolution defined at *m/z* 400), as well as DDA at MS^2 and MS^3 levels. The Orbitrap XL applied identical settings as the LCQ-Fleet, but with a scan speed of 0.1 s and 0.4 s and AGC of 1 × 10^5 and 5 × 10^5 for the LTQ-IT and FT-MS respectively. For selected samples, in addition to full scan accurate mass FT-MS and LTQ-IT MS^2, LTQ-IT MS^3 data were also collected in DDA mode for the top three most intense ions detected in each MS^2 scan. MS^2 and MS^3 data were collected at collision energies of 45% and 65% NCE.

Supplementary Materials: The following are available online at http://www.mdpi.com/1660-3397/18/9/448/s1, Figure S1. Recovery of TPC after fractionation by Solid Phase Extraction and Sephadex LH-20; Figure S2. Co-chromatography of *m/z* [M − H]$^-$ peaks at 745 and *m/z* [M + H]$^+$ peaks at 1491 in SPE bound samples. Figure S3. Peak areas of even DP phlorotannin oligomers from DP10-24. Figure S4. Evidence for triple charged phlorotannin species.

Author Contributions: Conceptualization, H.E., J.W.A. and G.J.M.; methodology, J.W.A. and G.J.M.; software, J.W.A.; validation, J.W.A. and G.J.M.; formal analysis, C.A. and J.W.A.; investigation, C.A., J.W.A. and G.J.M.; resources, G.J.M., J.W.A. and H.E.; data curation, C.A., J.W.A. and G.J.M.; writing—original draft preparation, C.A., H.E., J.W.A. and G.J.M.; writing—review and editing, H.E., J.W.A. and G.J.M.; visualization, J.W.A.; supervision, H.E., J.W.A. and G.J.M.; project administration, G.J.M. and H.E.; funding acquisition, G.J.M. All authors have read and agreed to the published version of the manuscript.

Funding: This research was funded by Byotrol PLC. The James Hutton Institute is grateful for support through the Strategic Research Portfolio from the Rural Affairs Food and Environment Strategic Research Group of the Scottish Government.

Acknowledgments: The authors acknowledge the support of the Environmental and Biochemical Sciences group at the James Hutton Institute.

Conflicts of Interest: The authors declare no conflict of interest.

References

1. Ragan, M.A.; Glombitza, K.W. Phlorotannins, brown algal polyphenols. *Prog. Phycol. Res.* **1986**, *4*, 129–241.
2. Martínez, J.H.; Castaneda, H.G. Preparation and chromatographic analysis of phlorotannins. *J. Chromat. Sci.* **2013**, *51*, 825–838. [CrossRef]
3. Li, S.-X.; Wijesekara, I.; Li, Y.; Kim, S.-K. Phlorotannins as bioactive agents from brown algae. *Process Biochem.* **2011**, *46*, 2219–2224. [CrossRef]
4. Jormalainen, V.; Honkanen, T.; Koivikko, R.; Eränen, J. Induction of phlorotannin production in a brown alga: Defense or resource dynamics? *Oikos* **2003**, *103*, 640–650. [CrossRef]
5. Amsler, C.D.; Fairhead, V.A. Defensive and sensory chemical ecology of brown algae. *Adv. Bot. Res.* **2006**, *43*, 1–91.
6. Svensson, C.J.; Pavia, H.; Toth, G.B. Do plant density, nutrient availability, and herbivore grazing interact to affect phlorotannin plasticity in the brown seaweed *Ascophyllum nodosum*. *Marine Biol.* **2007**, *151*, 2177–2181. [CrossRef]
7. Creis, E.; Delage, L.; Charton, S.; Goulitquer, S.; Leblanc, C.; Potin, P.; Ar Gall, E. Constitutive or inducible protective mechanisms against UV-B radiation in the brown alga *Fucus vesiculosus*? A study of gene expression and phlorotannin content responses. *PLoS ONE* **2015**, *10*, e0128003. [CrossRef]
8. Steinberg, P.D. Seasonal variation in the relationship between growth rate and phlorotannin. production in the kelp *Ecklonia radiata*. *Oecologia* **1995**, *102*, 169–173. [CrossRef]
9. Ragan, M.A.; Jensen, A. Quantitative studies on brown algal phenols. II. Seasonal variation in polyphenol content of *Ascophyllum nodosum* (L.) Le Jol. and *Fucus vesiculosus* (L.). *J. Exp. Mar. Biol. Ecol.* **1978**, *34*, 245–258. [CrossRef]
10. Pavia, H.; Toth, G.B. Influence of light and nitrogen on the phlorotannin content of the brown seaweeds *Ascophyllum nodosum* and *Fucus vesiculosus*. *Hydrobiology* **2000**, *440*, 299–307. [CrossRef]
11. Parys, S.; Kehraus, S.; Pete, R.; Küpper, F.C.; Glombitza, K.-W.; König, G.M. Seasonal variation of polyphenolics in *Ascophyllum nodosum* (Phaeophyceae). *Eur. J. Phycol.* **2009**, *44*, 331–338. [CrossRef]
12. Arnold, T.M.; Targett, N.M. To grow and defend: Lack of tradeoffs for brown algal phlorotannins. *Oikos* **2003**, *100*, 406–408. [CrossRef]
13. Nakamura, T.; Nagayama, K.; Uchida, K.; Tanaka, R. Antioxidant activity of phlorotannins isolated from the brown alga *Eisenia bicyclis*. *Fisheries Sci.* **1996**, *62*, 923–926. [CrossRef]
14. Farvin, S.K.H.; Jacobsen, C. Phenolic compounds and antioxidant activities of selected species of seaweeds from Danish coast. *Food Chem.* **2013**, *138*, 1670–1681. [CrossRef] [PubMed]
15. Erpel, F.; Mateos, R.; Pérez-Jiménez, J.; Pérez-Correa, J.R. Phlorotannins: From isolation and structural characterization, to the evaluation of their antidiabetic and anticancer potential. *Food Res. Int.* **2020**, *137*, 109589. [CrossRef]
16. Kirke, D.A.; Smyth, T.J.; Rai, D.K.; Kenny, O.; Stengel, D.B. The chemical and antioxidant stability of isolated low molecular weight phlorotannins. *Food Chem.* **2017**, *221*, 1104–1112. [CrossRef]
17. Yang, Y.I.; Woo, J.H.; Seo, Y.J.; Lee, K.T.; Lim, Y.; Choi, J.H. Protective effect of brown alga phlorotannins against hyper-inflammatory responses in lipopolysaccharide-induced sepsis models. *J. Agric. Food Chem.* **2016**, *64*, 570–578. [CrossRef]
18. Lopes, G.; Andrade, P.B.; Valentão, P. Phlorotannins: Towards new pharmacological interventions for Diabetes Mellitus Type 2. *Molecules* **2016**, *22*, 56. [CrossRef]
19. Cardoso, S.M.; Pereira, O.R.; Seca, A.M.; Pinto, D.C.; Silva, A.M. Seaweeds as preventive agents for cardiovascular diseases: From nutrients to functional foods. *Mar. Drugs* **2015**, *13*, 6838–6865. [CrossRef]
20. Eom, S.-H.; Kim, Y.-M.; Kim, S.-K. Antimicrobial effect of phlorotannins from marine brown algae. *Food Chem. Toxicol.* **2012**, *50*, 3251–3255. [CrossRef]
21. Vo, T.S.; Kim, S.K. Potential anti-HIV agents from marine resources: An overview. *Mar. Drugs* **2010**, *8*, 2871–2892. [CrossRef]
22. Cherry, P.; O'Hara, C.; Magee, P.J.; McSorley, E.M.; Allsopp, P.J. Risks and benefits of consuming edible seaweeds. *Nutr. Rev.* **2019**, *77*, 307–329. [CrossRef]
23. Nwosu, F.; Morris, J.; Lund, V.A.; Stewart, D.; Ross, H.A.; McDougall, G.J. Anti-proliferative and potential anti-diabetic effects of phenolic-rich extracts from edible marine algae. *Food Chem.* **2011**, *126*, 1006–1012. [CrossRef]

24. Pantidos, N.; Boath, A.; Lund, V.; Conner, S.; McDougall, G.J. Phenolic-rich extracts from the edible seaweed, *Ascophyllum nodosum*, inhibit α-amylase and α-glucosidase: Potential anti-hyperglycemic effects. *J. Funct. Foods* **2014**, *10*, 201–209. [CrossRef]
25. Austin, C.; Stewart, D.; Allwood, J.W.; McDougall, G.J. Extracts from the edible seaweed, *Ascophyllum nodosum*, inhibit lipase activity in vitro: Contributions of phenolic and polysaccharide components. *Food Funct.* **2018**, *24*, 502–510. [CrossRef]
26. Georgé, S.; Brat, P.; Alter, P.; Amiot, M.J. Rapid Determination of polyphenols and vitamin c in plant-derived products. *J. Agric. Food Chem.* **2005**, *53*, 1370–1373. [CrossRef]
27. Catarino, M.D.; Silva, A.M.S.; Mateus, N.; Cardoso, S.M. Optimization of phlorotannins extraction from *Fucus vesiculosus* and evaluation of their potential to prevent metabolic disorders. *Mar. Drugs* **2019**, *17*, 162–170. [CrossRef]
28. Tierney, M.S.; Soler-Vila, A.; Rai, D.K.; Croft, A.K.; Brunton, N.P.; Smyth, T.J. UPLC-MS profiling of low molecular weight phlorotannin polymers in *Ascophyllum nodosum*, *Pelvetia canaliculata* and *Fucus spiralis*. *Metabolomics* **2014**, *10*, 524–535. [CrossRef]
29. Tierney, M.S.; Smyth, T.J.; Hayes, M.; Soler-Vila, A.; Croft, A.K.; Brunton, N. Influence of pressurized liquid extraction and solid–liquid extraction methods on the phenolic content and antioxidant activities of Irish macroalgae. *Int. J. Food Sci. Technol.* **2013**, *48*, 860–869. [CrossRef]
30. Heo, S.-J.; Hwang, J.-Y.; Choi, J.-I.; Han, J.-S.; Kim, H.-J.; Jeon, Y.-J. Diphlorethohydroxycarmalol isolated from *Ishige okamurae*, a brown algae, a potent α-glucosidase and α-amylase inhibitor, alleviates postprandial hyperglycemia in diabetic mice. *Eur. J. Pharmacol.* **2009**, *615*, 252–256. [CrossRef]
31. Steevensz, A.J.; Mackinnon, S.L.; Hankinson, R.; Craft, C.; Connan, S.; Stengel, D.B.; Melanson, J.E. Profiling phlorotannins in brown macroalgae by liquid chromatography-high resolution mass spectrometry. *Phytochem. Anal.* **2012**, *23*, 547–553. [CrossRef]
32. Vissers, A.M.; Caligiani, A.; Sforza, S.; Vincken, J.-P.; Gruppen, H. Phlorotannin composition of *Laminaria digitata*. *Phytochem. Anal.* **2017**, *28*, 487–495. [CrossRef]
33. Li, Y.; Fu, X.; Duan, D.; Liu, X.; Xu, J.; Gao, X. Extraction and identification of phlorotannins from the brown alga, *Sargassum fusiforme* (Harvey) Setchell. *Mar. Drugs* **2017**, *15*, 49. [CrossRef]
34. Lopes, G.; Barbosa, M.; Vallejo, F.; Gil-Izquierdo, A.; Valentão, P.; Pereira, D.M.; Ferreres, F. Profiling phlorotannins from *Fucus* spp. of the Northern Portuguese coastline: Chemical approach by HPLC-DAD-ESI/MS and UPLC-ESI-QTOF/MSn. *Algal Res.* **2018**, *29*, 113–120. [CrossRef]
35. Catarino, M.D.; Silva, A.M.S.; Cardoso, S.M. Fucaceae: A source of bioactive phlorotannins. *Int. J. Mol. Sci.* **2017**, *18*, 1327. [CrossRef]
36. Heffernan, N.; Brunton, N.P.; FitzGerald, R.J.; Smyth, T.J. Profiling of the molecular weight and structural isomer abundance of macroalgae-derived phlorotannins. *Mar. Drugs* **2015**, *13*, 509–528. [CrossRef]
37. Meslet-Cladière, L.; Delage, L.; Leroux, C.J.; Goulitquer, S.; Leblanc, C.; Creis, E.; Gall, E.A.; Stiger-Pouvreau, V.; Czjzek, M.; Potin, P. Structure/function analysis of a type III polyketide synthase in the brown alga *Ectocarpus siliculosus* reveals a biochemical pathway in phlorotannin monomer biosynthesis. *Plant Cell* **2013**, *25*, 3089–3103. [CrossRef]
38. Kasper, P.T.; Rojas-Chertó, M.; Mistrik, R.; Reijmers, T.; Hankemeier, T.; Vreeken, R.J. Fragmentation trees for the structural characterization of metabolites. *Rapid Comms. Mass Spectr.* **2012**, *26*, 2275–2286. [CrossRef]

 *marine drugs*

Article

Microwave-Assisted Extraction of Phlorotannins from *Fucus vesiculosus*

Sónia J. Amarante, Marcelo D. Catarino, Catarina Marçal, Artur M. S. Silva, Rita Ferreira and Susana M. Cardoso *

LAQV-REQUIMTE, Department of Chemistry, University of Aveiro, 3810-193 Aveiro, Portugal; sonia.amarante@ua.pt (S.J.A.); mcatarino@ua.pt (M.D.C.); catarina.marcal@ua.pt (C.M.); artur.silva@ua.pt (A.M.S.S.); ritaferreira@ua.pt (R.F.)
* Correspondence: susanacardoso@ua.pt; Tel.: +351-234-370-360; Fax: +351-234-370-084

Received: 18 October 2020; Accepted: 10 November 2020; Published: 15 November 2020

Abstract: Microwave-assisted extraction (MAE) was carried out to maximize the extraction of phlorotannins from *Fucus vesiculosus* using a hydroethanolic mixture as a solvent, as an alternative to the conventional method with a hydroacetonic mixture. Optimal MAE conditions were set as ethanol concentration of 57% (v/v), temperature of 75 °C, and time of 5 min, which allowed a similar recovery of phlorotannins from the macroalgae compared to the conventional extraction. While the phlorotannins richness of the conventional extract was slightly superior to that of MAE (11.1 ± 1.3 vs. 9.8 ± 1.8 mg PGE/g $DW_{extract}$), both extracts presented identical phlorotannins constituents, which included, among others, tetrafucol, pentafucol, hexafucol, and heptafucol structures. In addition, MAE showed a moderate capacity to scavenge ABTS$^{\bullet+}$ (IC_{50} of 96.0 ± 3.4 µg/mL) and to inhibit the activity of xanthine oxidase (IC_{50} of 23.1 ± 3.4 µg/mL) and a superior ability to control the activity of the key metabolic enzyme α-glucosidase compared to the pharmaceutical drug acarbose.

Keywords: brown seaweeds; phlorotannins; microwave-assisted extraction; response surface methodology; antioxidant; antiradical activity; xanthine oxidase; α-glucosidase

1. Introduction

Seaweeds are claimed to be a sustainable and rich source of bioactive compounds, holding huge application potential in distinct fields. Among seaweeds' bioactive compounds, phlorotannins—i.e., phenolic compounds typical from brown macroalgae—are one of the most promising, since they have been related to numerous beneficial biological properties, including antioxidant [1–3], anti-inflammatory [4,5], antibacterial [6], anticancer [7], and antidiabetic [8] activities. While not fully elucidated, the bioactivity of phlorotannins is accepted as being largely dependent on their structure.

Chemically, these compounds consist of dehydro-oligomers or dehydro-polymers formed through the C-C and/or C-O-C oxidative coupling of phloroglucinol (1,3,5-trihydroxybenzene), which may occur in a wide range of molecular sizes and in different assemblages [9,10]. According to the number of hydroxyl groups and the nature of the structural linkages between phloroglucinol units, they are classified in four groups: phlorethols and fuhalols (containing ether linkages), fucols (containing aryl linkages), fucophlorethols (containing both aryl-aryl and ether linkages), and eckols and carmalols (containing dibenzodioxine linkages) [11].

Fucus vesiculosus is a widespread species, which is naturally found along the coastlines of the North Sea, the western Baltic Sea, and the Atlantic and Pacific oceans [12]. Phlorotannins-rich extracts obtained from *F. vesiculosus* have been described for their promising antioxidant, anti-inflammatory, and antitumor activities among other things, granting them great potential for application in the food, cosmetic, and pharmaceutical industries [12].

As for tannins in general, the extraction of phlorotannins is traditionally performed by the conventional solvent extraction method [8,12,13], using hydroacetonic mixtures, although some authors have also resorted to hydroethanol and hydromethanol mixtures [14–16]. Due to their peculiar characteristics including chemical complexity, susceptibility to oxidation, and interaction with other components of the matrix, the extraction of phlorotannins is a challenging process and the structures found in crude extracts and in purified fractions may depend on the extraction conditions applied [11,17,18].

In addition to the traditional solid–liquid extraction at room temperature, advanced methods such as supercritical fluid extraction (SFE) [19,20], pressurized liquid extraction (PLE) [21], microwave-assisted extraction (MAE) [12,22], and ultrasound-assisted extraction (UAE) [23,24] have been previously used for recovery of phlorotannins from seaweeds. Nowadays, MAE is one of the techniques that allow fast and large extraction of bioactive compounds, including phenolic compounds [12], showing several advantages over other methods. Among others, it allows the rapid heating of aqueous samples with non-ionizing electromagnetic radiation, a lower solvent use, a greater selectivity for the family of compounds of interest, a higher level of automation, a superior efficiency, and lower extraction times [22]. Since several variables influence the extraction of phlorotannins, the optimal operating extraction parameters may be estimated with a statistical optimization method. The response surface methodology (RSM) makes use of the quantitative data of an appropriate experimental design to determine and simultaneously solve the multivariate equation. In order to minimize the number of experiments, this methodology relies on a mathematical model where all the interactions that occur between the test variables are taken into account [25]. This type of approach enables a considerable reduction in the cost and execution time in experimental projects with more than two variables [26]. One of the RSM models most used for experimental planning is the Box–Behnken design (BBD). The main advantage of this experimental design is that the experiments are not carried out under extreme conditions—i.e., the combinations between the different factors are never in their higher or lower levels, since this type of combination usually gives unsatisfactory results [27]. As far as we know, previous studies focusing on the extraction of phlorotannins by MAE have already been applied in seaweeds from the *Saccharina*, *Carpophyllum*, and *Ecklonia* genera, but no study has been performed with *Fucus* genus yet.

In this context, this study aimed to optimize the extraction process of phlorotannins from *F. vesiculosus* using the MAE technique and a green solvent—namely, ethanol. In addition, it was intended to elucidate the potential biological capacity of the resultant extracts, particularly with respect to their ability to act against oxidative events and to control the activity of α-glucosidase (i.e., a key enzyme in diabetes control). All the data were compared with those obtained by the conventional method using hydroacetonic mixtures.

2. Results

2.1. Single-Factor Experiment on MAE

Taking into account the different variables that could mainly affect the phlorotannins extraction, preliminary single-factor experiments were performed to specify the selected factors in the BBD experiment. Different concentrations of ethanol were tested in the range of 0% to 100% (*v/v*). According to Figure 1A, the total phlorotannins content (TPhC) recovered from *F. vesiculosus* increased almost proportionally between 20% and 60% ethanol (1.23 ± 0.03 to 1.59 ± 0.03 mg PGE/g DW$_{algae}$), with the maximum yield obtained for this last concentration. In turn, the use of ethanol above 60% resulted in a decrease in the TPhC to approximately 1.40 mg PGE/g DW$_{algae}$. Based on this, the concentration of ethanol used to study the next variable was 60%. Moreover, considering these results, an ethanol concentration range between 40% and 80% was selected for the BBD experiment.

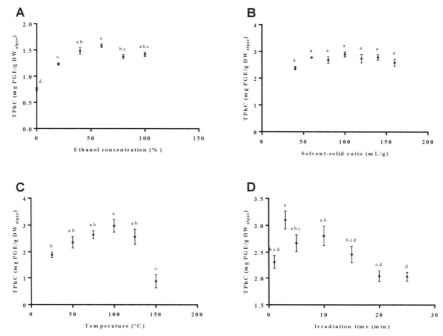

Figure 1. Effect of (**A**) ethanol concentration, (**B**) solvent–solid ratio, (**C**) temperature, and (**D**) irradiation time on the recovery of phlorotannins from *F. vesiculosus* in the single-factor experiments. Data represent the mean ± SEM and the results are expressed in mg of phloroglucinol equivalents/g of dried algae (mg PGE/g DW_{algae}). Different letters represent statistical significance (one-way ANOVA followed by Tukey's post hoc test; $p \leq 0.05$).

The effect of different solvent–solid ratios on the TPhC recovered from *F. vesiculosus* was tested in the range of 40 to 160 mL/g, as for our previous study [8]. As represented in Figure 1B, the variation in this parameter did not significantly influence the TPhC, which accounted for approximately 2.7 mg PGE/g of DW_{algae} from 60 to 160 mL/g. Yet, given that a maximum point was achieved at 100 mL/g (2.90 ± 0.09 mg PGE/g DW_{algae}), this solvent–solid ratio was selected for the following factor study and for the BBD experiment as well.

It is expected that temperature affects the extraction process of thermolabile compounds such as phlorotannins. Taking this into account, different temperatures were selected between 25 and 150 °C. As represented in Figure 1C, a linear increase in the recovery of TPhC was obtained between 25 and 100 °C (1.89 ± 0.10 to 2.98 ± 0.24 mg PGE/g DW_{algae}). However, temperatures of extraction above 100 °C—namely, 125 and 150 °C—caused a decrement of the phlorotannins recovery (TPhC of 2.57 ± 0.28 and 0.878 ± 0.251 mg PGE/g DW_{algae}, respectively). Hence, for the analysis of the next variable, the temperature was set at 100 °C, which presented the maximum TPhC yield. Additionally, for the BBD experiment, the interval chosen was 75–125 °C.

Moreover, based on the literature data for other algae [12,22], the influence of the irradiation time was considered for the interval of 1 to 25 min. As depicted in Figure 1D, while the raising of the irradiation time up to 3 min caused an increase in the amount of TPhC (maximum 3.10 ± 0.17 mg PGE/g DW_{algae}), the opposite tendency was registered for longer extraction periods, reaching levels of 2 mg PGE/g of DW_{algae} for 20 min of extraction time. Based on these results, the irradiation time interval selected for the BBD experiment was 1–5 min.

2.2. Analysis of the Response Surface Methodology

2.2.1. Fitting the Model

The experimental values obtained for TPhC, represented in Table 1, were fitted to a quadratic polynomial model (Equation (1)). This equation allowed the determination of the optimal conditions for the extraction process in order to obtain the maximum phlorotannins recovery and also to determine the different correlations, which are related to the independent variable interactions and respective responses.

Table 1. Experimental TPhC values obtained from the Box–Behnken design matrix.

Extract No.	Independent Variables			Experimental TPhC (mg PGE/g DW$_{algae}$)
	X_1	X_2	X_3	
1	40	125	3	1.17 ± 0.39
2	60	100	3	2.58 ± 0.36
3	80	125	3	1.61 ± 0.20
4	60	75	5	3.09 ± 0.34
5	80	100	5	1.99 ± 0.35
6	60	125	1	2.37 ± 0.25
7	80	75	3	2.16 ± 0.54
8	60	100	3	2.58 ± 0.36
9	40	100	1	2.60 ± 0.23
10	40	100	5	1.95 ± 0.37
11	60	125	5	0.85 ± 0.22
12	60	75	1	2.52 ± 0.21
13	40	75	3	2.42 ± 0.24
14	60	100	3	2.58 ± 0.36
15	80	100	1	2.35 ± 0.33

X_1—ethanol concentration (%); X_2—temperature (°C); X_3—time (min); TPhC—total phlorotannins content. All values are expressed as mean ± SD of mg of phloroglucinol equivalents/g of dried algae (mg PGE/ g DW$_{algae}$).

The experimental data allowed the determination of the coefficients of the model, which were evaluated for statistical significance using a statistical analysis of variance (ANOVA) and are listed in Table 2. Accordingly, the independent variables with a higher impact on TPhC were the temperature (X_2, $p < 0.001$) and time (X_3, $p < 0.01$), while the ethanol concentration revealed no effect. Moreover, significant interactive effects between the ethanol concentration and temperature ($X_1 X_2$, $p < 0.05$) and between the temperature and time ($X_2 X_3$, $p < 0.001$) were observed. The variables ethanol concentration and temperature also showed a significant quadratic effect on TPhC ($p < 0.01$, for both).

$$TPhC = 2.58 - 0.004 X_1 - 0.52 X_2 - 0.24 X_3 - 0.36 X_1{}^2 - 0.38 X_2{}^2 - 0.0001 X_3{}^2 + \\ 0.17 X_1 X_2 + 0.073 X_1 X_3 - 0.36 X_2 X_3. \tag{1}$$

The statistical analysis revealed a high F-value (43.77) and, simultaneously, a low p-value ($p < 0.001$), meaning that the model is significant. Furthermore, the coefficient of multiple determination (R^2) for the response TPhC was 0.99 and the adjusted determination coefficient (R^2Adj) was 0.96. The similarity of these values suggests that there is a good correlation between the observed and predicted values for TPhC. Taking this into account, the fitted model may be assumed as trustworthy and capable of predicting the TPhC response.

Table 2. Regression coefficients and results of the ANOVA analysis of the model.

Parameter	Regression Coefficient
β_0	2.58 ***
X_1	−0.004
X_2	−0.52 ***
X_3	−0.24 **
$X_1 X_2$	0.17 *
$X_1 X_3$	0.073
$X_2 X_3$	−0.52 ***
$X_1 X_1$	−0.36 **
$X_2 X_2$	−0.38 **
$X_3 X_3$	−0.0001
R^2	0.99
R^2_{Adj}	0.96
Model F-value	43.77
Model p-value	<0.001

β_0—constant coefficient; X_1—ethanol concentration (%); X_2—temperature (°C); X_3—time (min). *, **, *** represent statistical significance with $p < 0.05$, 0.01, and 0.001, respectively.

2.2.2. Effect of the Independent Variables on TPhC

The effects of the independent variables and their mutual interactions on TPhC can be visualized on the three-dimensional response surface plots and two-dimensional contour plots shown in Figure 2, respectively. Each plot demonstrates the effects of two independent variables on the target response, while the third variable is maintained at its zero level.

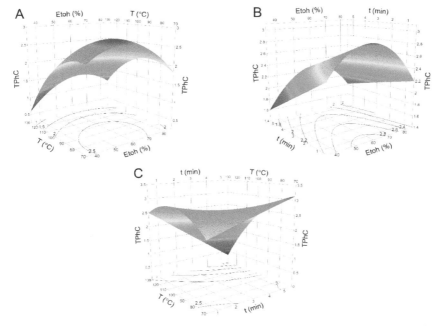

Figure 2. Response surface and contour plots for the total phlorotannins content (TPhC, expressed as mg of phloroglucinol equivalents/g of dried algae ie, mg PGE/g DW$_{algae}$) from *F. vesiculosus* extracts with respect to (**A**) ethanol concentration (%, X_1) and temperature (°C, X_2); (**B**) ethanol concentration (%, X_1) and time (min, X_3); and (**C**) temperature (°C, X_2) and time (min, X_3). The third variable of each graph was kept at its zero level.

According to the results of the regression coefficient shown in Table 2, the interaction between the ethanol concentration and temperature, and temperature and time, revealed a significant effect on the TPhC ($p < 0.05$ and $p < 0.001$, respectively). As observed in Figure 2A, this increased in the range of temperatures between 125 and 75 °C and of ethanol between 50% and 60%. In addition, both independent variables, ethanol concentration and temperature, had a significant quadratic effect ($p < 0.01$, for both). Figure 2B also demonstrated an increase in TPhC for short time extractions (1 to 5 min) and a quadratic effect for ethanol concentrations between 50% and 70%. Moreover, a higher level of TPhC was obtained at lower temperatures (75 °C) and for a longer time interval (5 min) (Figure 2C). In addition, the quadratic effect of temperature was also registered in this figure.

2.2.3. Optimization and Validation of the Models

The optimal MAE conditions for the extraction of phlorotannins from *F. vesiculosus* were estimated according the quadratic polynomial model (Equation (1)). The settled conditions were ethanol concentration at 57% (*v/v*), temperature at 75 °C, and time at 5 min, with a theoretical maximum value of TPhC of 3.01 ± 0.25 mg PGE/g DW_{algae}. These conditions were tested to validate the adequacy of the model prediction, and the experimental value of 3.16 ± 0.06 mg PGE/g DW_{algae} was obtained, thus demonstrating a good correlation between the experimental and predicted values, confirming the appropriateness of this model, which is trustworthy and precise.

2.3. Comparison between MAE and Conventional Solvent Extraction

To conclude on the feasibility of using MAE and ethanol as an alternative extraction method for the extraction of phlorotannins from *F. vesiculosus*, the TPhC (mg PGE/g DW_{algae}) was compared to that of an extract obtained by conventional extraction under optimized conditions, as previously established by our group for this macroalgae species [8]—i.e., a solvent–solid ratio of 70 mL/g with 70% acetone and 1% glacial acetic acid at room temperature for 3 h. Moreover, the extracts' richness in phlorotannins and their potential to hamper oxidative events and the activity of α-glucosidase (a key metabolic enzyme in the control of diabetes) were compared.

2.3.1. Phlorotannins

Using optimal conditions, it was possible to recover similar amounts of phlorotannins with the two methods (3.16 ± 0.06 mg PGE/g DW_{algae} versus 2.94 ± 0.28 mg PGE/g DW_{algae} for MAE and conventional extraction, respectively). On the other hand, when comparing the phlorotannins richness of both extracts, it is possible to conclude that the MAE had a slightly lower concentration of phlorotannins (Table 3), suggesting that, compared to the conventional extraction using hydroacetone, the application of MAE at 75°C and ethanol concentration of 57% can further facilitate the co-extraction of non-phenolic components—most likely, lipophilic compounds such as fatty acids, sterols, and pigments (in particular, fucoxanthin—i.e., the characteristic pigment from brown algae). Nevertheless, the individual phlorotannins detected by UHPLC-DAD-ESI-MS^n were, in general, coincident between the two extracts (Table 3).

2.3.2. Bioactive Potential

The bioactive capacity of the two extracts was evaluated for their capacity to act as antioxidant and antidiabetic agents—namely, through the ability to scavenge radicals ($ABTS^{\bullet+}$ and $O_2^{\bullet-}$), inhibit the activity of xanthine oxidase, and control the activity of α-glucosidase. Consistent with the phlorotannins' superior richness in conventional extract compared to MAE, in general the latter was less active (Table 4). Despite this, it is worth mentioning that this extract had a promising potential to hamper the activity of xanthine oxidase and, in particular, of α-glucosidase, for which the IC_{50} was 115 times lower than that of the commercial drug acarbose.

Table 3. Phlorotannins from *F. vesiculosus* extracts obtained under optimized MAE and conventional extraction conditions.

RT (min)	[M − H]⁻	MS² Main Fragments	Probable Compound	CONV	MAE
1.3	317	225, 165, 207, 125, 249, 153	Phlorotannin derivative	D	D
1.9	497	479, 331, 461, 435, 395, 165	Tetrafucol	D	D
2.5	247	203, 121, 81	Dibenzodioxine-1,3,6,8-tetraol	D	D
2.7	621	603, 455, 585, 331, 529, 559, 577	Pentafucol	D	D
4.4	745	727, 455, 579, 709, 289, 701, 683	Hexafucol	D	D
5.3	623	495, 477, 605, 577, 601, 496	Phlorotannin derivative	D	D
6.2	869	851, 579, 455, 833, 785, 703	Heptafucol	D	D
6.4	479	461, 433, 315, 389, 435, 401	Fucofurodiphlorethol	D	D
10.0	363	345, 257, 319, 138, 182	Phlorotannin derivative	D	D
11.0	637	619, 496, 593, 601, 591, 335	Pentafuhalol	D	D
11.7	497	451, 479, 437, 453, 336, 335, 461	Tetrafucol	D	D
11.8	529	485, 511, 471, 467, 493, 403, 389, 373	Hydroxytetrafuhalol	D	D
12.9	635	575, 617, 557, 335, 466, 273, 531, 229	Phlorotannin derivative	D	D
13.3	587	507, 523, 505, 383, 277, 229	Unidentified	D	D
13.5	723	679, 701, 405, 714, 497, 678, 331	Unidentified	D	D
14.2	635	575, 617, 557,335, 466	Phlorotannin derivative	D	D
14.8	587	507	Unidentified	D	D
14.8	507	277, 461, 439, 489, 479, 382, 229, 275, 231	Phlorotannin derivative	D	D
15.0	950	904	Unidentified	D	D
16.4	603	585, 559, 543, 567, 269, 523, 313	Fucofurotriphlorethol	D	D
18.5	610	225, 538, 299, 592, 226, 486	Unidentified	ND	D
19.9	771	753, 727, 761, 725, 749, 610	Phlorotannin derivative	ND	D
Total Phlorotannins (mg/g₍extract₎) [1]				11.1 ± 1.3	9.8 ± 1.8

RT—Retention time; CONV—Conventional solvent extraction; MAE—Microwave-assisted extraction; D—detected; ND—not detected. [1] Determined by 2,4-dimethoxybenzaldehyde assay (DMBA).

Table 4. Antioxidant activity and inhibition of α-glucosidase of *F. vesiculosus* extracts obtained by optimized MAE and conventional methodologies.

Sample	IC₅₀ (µg/mL)			
	ABTS•⁺	O₂•⁻	Xanthine Oxidase	α-Glucosidase
MAE	95.99 ± 3.40	527.30 ± 47.78	23.07 ± 3.40	6.86 ± 0.70
Conventional	62.55 ± 1.93	457.18 ± 23.97	6.36 ± 2.20	1.73 ± 0.13
Reference compound *	5.07 ± 0.25	5.07 ± 0.77	0.05 ± 0.005	789.93 ± 41.08

MAE—Microwave-assisted extraction. IC₅₀ was determined as the concentration at which ABTS•⁺ and O₂•⁻ were inhibited by 50%. All values are expressed as mean ± SD. * Trolox was used as a reference compound for ABTS•⁺, gallic acid for O₂•⁻, allopurinol for xanthine oxidase, and acarbose for α-glucosidase.

3. Discussion

Since MAE has several advantages compared with regular stirring—namely, allowing the rapid heating of aqueous samples with non-ionizing electromagnetic radiation; a lower extraction time and solvent quantities; and, in turn, a higher level of automation, allied to a superior selectivity and efficiency [22]—one of the main aims of this work was to maximize the recovery of phlorotannins from *F. vesiculosus* using the MAE technique combined with a greener solvent—namely, ethanol.

According to the preliminary single factor experiments, solvent-solid ratio was shown to have neglectable effect on the recovery of phlorotannins, contray to the ethanol concentration in the range of 0–100%, the temperature in the range of 25–150 °C, and the time in the range of 1–25 min. Concerning the effect of ethanol concentration, the maximum TPhC was obtained for an ethanol concentration of 60% (*v/v*), which is slightly higher than that settled in the optimization performed by He and colleagues [22] (55% *v/v*) for the MAE of phlorotannins from *Saccharina japonica*. Moreover, despite Magnusson et al. [12] pointed that water is a better solvent than ethanol for recovering phlorotannin with MAE, our data, combined with those of others, suggest that hydroethanolic mixtures are, in fact, more appropriate.

Afterwards, a maximum TPhC was herein obtained for the extraction temperature of 100 °C, which is higher than that established by He et al. [22] (60 °C) for *Saccharina japonica* but lower than that described by Magnusson et al. [12] (160 °C) for *Carpophyllum flexuosum*, *Carpophyllum plumosum* and *Ecklonia radiata*, using the same technique. Such differences can be explained by the thermolability of these compounds, as well as the differences in the phlorotannins profiles that probably occur between these seaweed species. In turn, the effect of time on MAE revealed that the maximum TPhC could be achieved within 3 min, which is in agreement with the work previously described by Magnusson et al. [12].

According to the BBD model, the predicted optimal conditions of extraction were set as ethanol concentration of 57% (*v/v*), temperature of 75 °C and a time of 5 min. The ethanol concentration was similar to that obtained in the preliminary single factor experiments (60% *v/v*) and to that described by He et al. [22] (55% *v/v*) for the macroalgae *Saccharina japonica*. In turn, the optimal temperature was lower than that established in the preliminary experiments (100 °C), while the extraction time was superior. These differences could be explained by the interactive effects between variables that are not considered when performing the single-factor experiments. Indeed, the results gathered from the BBD experiment allowed to conclude that the main interactions between the different variables were temperature versus time, followed by ethanol concentration versus temperature. This could be because the variables ethanol concentration and temperature revealed a quadratic effect. In the case of ethanol concentration, it is clear that the presence of water and ethanol provides a polar medium more suitable for the phlorotannin extraction than just ethanol [28]. Regarding temperature, the optimal temperature for the extraction process was set at 75 °C. To the best of our knowledge, there is no previous data focusing on the extraction of phlorotannins from *F. vesiculosus* using MAE, thus hampering the comparison of data. Naturally, the optimal conditions obtained for the MAE were quite distinct than those established by Catarino et al. [8] for the conventional solvent extraction using hydroacetonic mixtures (acetone of 70% (*v/v*), solvent-solid ratio of 70 mL/g at temperature 25 °C and time of 3 h). Nonetheless, it must be noted that the herein set conditions for MAE allowed the recovery of identical amounts of phlorotannins than conventional solvent extraction (3.16 ± 0.06 mg PGE/g DW_{algae} versus 2.94 ± 0.28 mg PGE/g DW_{algae}, respectively), hence indicating that the use of MAE associated to green solvents may serve as a good alternative to the extraction of phlorotannins from *F. vesiculosus*. In fact, despite the lower concentration, the phlorotannin constituents from MAE were, in general, concordant with those from the conventional extract (as demonstrated through UHPLC-MS analysis). Regardless several non-identified compounds, it was possible to detect tetrafucol, dibenzodioxine-1,3,6,8-tetraol, pentafucol, hexafucol, heptafucol, fucofurodiphlorethol, pentafuhalol, hydroxytetrafuhalol and other distinct phlorotannin derivatives (*m/z* at 317, 623, 363, 635, 507 and 771) that were previously reported by other authors for this macroalgae species [8,29].

As phenolic compounds, phlorotannins' most characteristic biological effect is their antioxidant activity. Indeed, several authors have proven this capacity using different approaches—namely, through anti-radical systems such as DPPH, ORAC, $O_2^{\bullet-}$, and $NO^\bullet$ [30–33] and biological systems including H_2O_2 and *t*-BHP-induced oxidative stress [15,34] in different cell lines, and even in vivo studies on rats [30]. Overall, our findings revealed a dose-dependent activity observed for both MAE and conventional phlorotannin extracts (data not shown) with the later always showing lower IC_{50} values than the former, which is in agreement with the superior concentration of phlorotannins found in conventional rather than in MAE extract. These observations agree with previous studies supporting the evidence that the phlorotannins content of the extracts is correlated with their antioxidant capacity [32]. Nevertheless, the results gathered in this study suggest that MAE may hold antioxidant potential both through enzymatic and antiradical mechanisms, although further assays may be needed.

In addition to the antioxidant properties of phlorotannins, promising anti-diabetes effects have also been demonstrated for these compounds via the inhibition of key enzymes, prevention of the formation of advanced glycation end products, improving insulin sensitivity and others [34–36]. As responsible for the hydrolysis of carbohydrates into monomers of glucose, α-glucosidase constitutes an important target enzyme for the control and prevention of diabetes. Indeed, our results demonstrate that both MAE and conventional extracts of *F. vesiculosus* strongly inhibited the activity of α-glucosidase, with IC_{50} values remarkably inferior to that of acarbose, which is a pharmaceutical drug currently used to prevent the development of diabetic symptoms. Previous works showed that *F. vesiculosus* phlorotannin extracts or their purified fractions were able to strongly interfere with the activity of this enzyme [8,36,37]. Notably, consistent with the lower phlorotannin concentrations, MAE revealed lower inhibitory potential compared to the conventional extract. Nevertheless, this effect was still approximately 115 times stronger than that of acarbose, meaning that MAE combined with hydroethanolic solvent may represent a fast and safer method to obtain extracts that could help the regulation of diabetes. In any case, further elucidation of phlorotannins' stability through the GI tract and/or bioactivity of metabolites will be required to sustain the in vivo effects.

4. Materials and Methods

4.1. Materials

Ground *F. vesiculosus* from July 2017 was purchased from Algaplus Lda (Aveiro, Portugal). Acetone, ethanol, acetonitrile HPLC grade, hydrochloric acid, and glacial acetic acid were acquired from Fisher Chemical (Pittsburgh, PA, USA). The enzymes α-glucosidase from *Saccharomyces cerevisiae* (EC No.—3.2.1.20) and xanthine oxidase from bovine milk (EC No.—1.17.3.2), together with 2,4-dimethoxybenzaldehyde (DMBA), phloroglucinol, 2,2'-azino-bis(3-ethylbenzothiazoline-6-sulphonic acid)diammonium salt (ABTS-NH_4), nitrotetrazolium blue chloride (NBT), phenazine methosulfate (PMS), 4-nitrophenyl α-D-glucopyranoside (pNPG), allopurinol, ascorbic acid, and formic acid, were purchased from Sigma-Aldrich (St. Louis, MO, USA). Potassium persulfate, potassium di-hydrogen phosphate, potassium hydroxide, sodium di-hydrogen phosphate 1-hydrate, and gallic acid were acquired from Panreac (Barcelona, Spain). β-nicotinamide adenine dinucleotide (β-NADH), trolox, 3,5-dinitrosalicylic acid (DNS), acarbose, and 4-nitrophenol were purchased from Acros Organics (Hampton, NH, USA) and dimethylsulfoxide (DMSO) were acquired from Honeywell Riedel-de Haën (Charlotte, NC, USA), and xanthine were purchased from AlfaAesar (Ward Hill, MA, USA). All reagents were of analytical grade or of the highest available purity.

4.2. Methods

4.2.1. Single-Factor Experiments Using Microwave-Assisted Extraction (MAE)

The method development for the extraction of seaweed phlorotannins by MAE was based on the work of He et al. [22]. The extraction process was performed by systematically varying one condition at a time—namely, the concentration of ethanol (0%, 20%, 40%, 60%, 80%, 100% (*v/v*)), the solvent-solid ratio (40, 60, 80, 100, 120, 140 and 160 (mL/g)), the extraction temperature (25, 50, 75, 100, 125 and 150 °C), and the irradiation time (1, 3, 5, 10, 15, 20 and 25 min). When one variable was not studied, it was kept constant. The constant values for irradiation time, solvent-solid ratio, temperature and microwave power were 20 min, 40 mL/g, 60 °C and 400 W, respectively, and samples were heated to the target temperature within a 2 min ramp. The extract was recovered by filtration through cotton to remove the solid residues, followed by a G4 glass filter, and was maintained at −20 °C until analysis. Experiments were performed with a focused microwave system with an Ethos MicroSYNTH Microwave Labstation (Milestone Inc.) using an 80 mL reactor at atmospheric pressure, and samples were stirred under constant agitation throughout the extraction process.

4.2.2. Experimental Design for the Optimization of Phlorotannins Microwave-Assisted Extraction

An RSM based on a three-level-three-factor Box–Behnken experimental design (BBD) was employed in this study to optimize the phlorotannin extraction process considering the effects of solvent concentration (%, *v/v*, X_1), temperature (°C, X_2), and extraction time (min, X_3). The factor levels of these three variables were coded as −1 (low), 0 (central point or middle), and +1 (high), respectively, according to the single-factor tests outlined bellow (Table 5).

Table 5. Independent variables and their levels used in BBD.

Symbols	Independent Variables	Levels		
		−1	0	+1
X_1	Solvent concentration (%, *v/v*)	40	60	100
X_2	Temperature (°C)	75	100	125
X_3	Time (min)	1	3	5

A total of 15 different experiments, including three replicates at central point (Table 1), were conducted in a randomized order. Using the response surface methodology, the experimental design and analysis of variance (ANOVA) were carried out in the statistical software JMP, version 10.0.0, to generate the following second-order polynomial equation (Equation (2)) that represents the total phlorotannins content (TPhC) as a function of the coded independent variables:

$$Y = \beta_0 + \sum_{i=1}^{k} \beta_1 X_1 + \sum_{i=1}^{k} \beta_{ii} X_i^2 + \sum_{i \neq j=1}^{k} \beta_{ij} X_i X_j, \tag{2}$$

were Y is the predicted response; β_0 is the constant coefficient; β_i, β_{ii}, β_{ij} are the linear, quadratic and interactive coefficients of the model, respectively; and X_i and X_j are the coded independent variables.

The model adequacy was evaluated using the coefficient of determination (R^2) and the lack-of-fit test represented at 5% level of significance, accordingly. Three-dimensional response surface plots and two-dimensional contour plots were used for the visualization of the effects of independent variables and their mutual interactions in the responses. To validate the accuracy of the models, experiments were carried out at the optimal conditions predicted for TPhC, and the obtained experimental data were compared to the values predicted by the corresponding regression model.

4.2.3. Extraction of Phlorotannins under Optimal MAE and Conventional Solvent Extraction

The MAE extract was prepared following the optimal conditions determined through the response surface method. A total of 10.8 g of macroalgae (corresponding to 1.08 L hydroethanolic mixture) were used, with 60 mL in each microwave flask. The conventional solvent extraction of *F. vesiculosus* was performed according to the optimal conditions established by Catarino et al. [8]. Briefly, 11 g of dried algal powder (DW_{algae}) was dispersed in 770 mL of 70% acetone solution with 1% of glacial acetic acid, and incubated for 3 h at room temperature under constant agitation. The combined mixture obtained with MAE, or by conventional solvent extraction method, was filtered through cotton to remove the solid residues and then through a G4 glass filter. Afterwards, the extract solvents were removed by rotary evaporation. The dried extracts were resuspended in DMSO and subsequently stored at −20 °C until further analysis.

4.2.4. Characterization of Phlorotannins

The TPhC was estimated according to the 2,4-dimethoxybenzaldehyde (DMBA) colorimetric method previously described [9]. Briefly, equal volumes of the stock solutions of DMBA (2%, *m/v*) and HCl (6%, *v/v*), both prepared in glacial acetic acid, were mixed prior to use (work solution). Afterwards, 250 µL of this solution was added to 50 µL of each extract in a 96-wells plate and the reaction was incubated in the dark, at room temperature. After 60 min, the absorbance was read at 515 nm in an automated plate reader (Biotek Instrument Inc., Winooski, VT, USA) and the phlorotannins content was determined by using a regression equation of the phloroglucinol linear calibration curve

(0.06–0.1 mg/mL). The results were expressed as mg phloroglucinol equivalents per g of dried algae (mg PGE/g DW$_{algae}$) or per g of dried extract (mg PGE/g DW$_{extract}$).

In addition, the identification of individual phenolic compounds in the extracts was performed by UHPLC-DAD-ESI/MS analysis, after defatting with *n*-hexane, as previously described [13], and filtration through a nylon filter of 0.22 μm (Whatman™, Buckinghamshire, UK). The analysis was carried out in an Ultimate 3000 (Dionex Co., San Jose, CA, USA) apparatus consisting of an autosampler/injector, a binary pump, a column compartment, and an ultimate 3000 Diode Array Detector (Dionex Co., San Jose, CA, USA), coupled to a Thermo LTQ XL (Thermo Scientific, San Jose, CA, USA) ion trap mass spectrometer equipped with an ESI source. The LC separation was conducted with a Hypersil Gold (ThermoScientific, San Jose, CA, USA) C18 column (100 mm length; 2.1 mm i.d.; 1.9 μm particle diameter, end-capped) maintained at 30 °C and a binary solvent system composed of (A) acetonitrile and (B) 0.1% of formic acid (*v/v*). The solvent gradient started with 5–40% of solvent (A) over 14.72 min, from 40–100% over 1.91 min, remaining at 100% for 2.19 more min before returning to the initial conditions. The flow rate was 0.2 mL/min and the UV–Vis spectral data for all peaks were accumulated in the range of 200–700 nm while the chromatographic profiles were recorded at 280 nm. Control and data acquisition of MS were carried out with the Thermo Xcalibur Qual Browser data system (ThermoScientific, San Jose, CA, USA). Nitrogen of above 99% purity was used, and the gas pressure was 520 kPa (75 psi). The instrument was operated in negative mode with the ESI needle voltage set at 5.00 kV and an ESI capillary temperature of 275 °C. The full scan covered the mass range from *m/z* 100 to 2000. CID-MS/MS experiments were performed for precursor ions using helium as the collision gas with a collision energy of 25–35 arbitrary units. All solvents were LC-MS grade.

4.2.5. Antioxidant Properties

ABTS$^{\bullet+}$ Discoloration Assay

The total antioxidant activity of both crude extracts was measured using an adaptation of the ABTS$^{\bullet+}$ discoloration assay based on the procedure described by Catarino et al. [13]. A stock solution of ABTS$^{\bullet+}$ was prepared by reacting the ABTS-NH$_4$ aqueous solution (7 mM) with 2.45 mM of potassium persulfate in the dark at room temperature for 12–16 h to allow the completion of radical cation generation. This solution was then diluted with distilled water until its absorbance reached 0.700 ± 0.05 at 734 nm. Afterwards, 50 μL of each sample were mixed with 250 μL of the diluted ABTS$^{\bullet+}$ solution in a 96-well microplate. The mixture was then allowed to react for 20 min in the dark, at room temperature and the absorbance was then measured at 734 nm in an automated plate reader (Biotek Instrument Inc., Winooski, VT, USA). The percentage of inhibition of ABTS$^{\bullet+}$ was calculated using the Equation (3) described by Yen and Duh [38]:

$$\% \text{ ABTS}^{\bullet+} \text{ scavenging } = \frac{\Delta A_c - \Delta A_e}{\Delta A_c} \times 100 \tag{3}$$

where *Ac* is the absorbance of the control (without extract addition) and *Ae* is the absorbance of the extract. Ascorbic acid was used as the reference compound. The concentration of the extract/standard able to inhibit 50% of ABTS$^{\bullet+}$ (IC$_{50}$) was then calculated by plotting the percentage of inhibition against the plant extract concentrations.

Superoxide Scavenging Assay

In a 96-well plate, 75 μL of nitroblue tetrazolium (NBT) (0.2 mM), 100 μL of β-NADH (0.3 mM), 75 μL of each crude extract, and 75 μL of phenazine methosulfate (PMS) (15 μM) were mixed and incubated for 5 min at room temperature. The absorbance was then measured at 560 nm in an automated plate reader (Biotek Instrument Inc., Winooski, VT, USA). Gallic acid was used as the reference compound. The IC$_{50}$ value for superoxide scavenging activity was determined by plotting the percentage of inhibition of superoxide radical anion generation in the presence of the crude extract and calculated using Equation (3).

4.2.6. Enzymatic Assays

α-Glucosidase Inhibition Assay

The inhibition of α-glucosidase was measured according to the method previously described by Neto et al. [39]. In short, 50 μL of different extract concentrations (0–0.006 mg/mL in 50 mM of phosphate buffer, pH 6.8) were mixed with 50 μL of 6 mM 4-nitrophenyl-D-glucopyranoside (pNPG) dissolved in deionized water. The reaction was started with the addition of 100 μL of α-glucosidase solution and the absorbance was monitored at 405 nm every 60 s for 20 min at 37 °C. Blank readings (no enzyme) were then subtracted from each well and the inhibitory effects towards the α- glucosidase activity was calculated as follows:

$$\% \text{ inhibition } = \frac{\Delta Abs_c - \Delta Abs_e}{\Delta Abs_c} \times 100 \tag{4}$$

where ΔAbs_c is the variation in the absorbance of the negative control and ΔAbs_e is the variation in the absorbance of the extract. Acarbose was used as a positive control of inhibition.

Xanthine Oxidase Assay

The inhibition of xanthine oxidase activity was carried out following the method described by Pereira et al. [40], with slight modifications. Briefly, in a 96-well plate 40 μL of extract (concentrations of 0–2 mg/mL) was mixed with 45 μL of sodium dihydrogen phosphate buffer (100 mM, pH 7.5) and 40 μL of enzyme (5 mU/mL). After 5 min of incubation at 25 °C, the reaction was started with the addition of 125 μL of xanthine (0.1 mM dissolved in buffer), and the absorbance at 295 nm was measured every 45 s over 10 min at 25 °C. The inhibitory effects towards xanthine oxidase activity were calculated using Equation (4). Allopurinol was used as a positive control of inhibition.

4.2.7. Statistical Analysis

All the data were expressed as the mean ± standard deviation (SD) of three similar and independent experiments performed in duplicate. The JMP and Minitab software were used to construct the BBD and to analyze the results. Data from single-factor experiments and BBD were analyzed using ANOVA ($p < 0.05$), followed by Tukey's post hoc test.

5. Conclusions

In this work, a single-factor experimental approach followed by a response surface methodology was carried out for the determination of the optimal conditions that maximize the extraction of phlorotannins from *F. vesiculosus* using microwave-assisted extraction combined with hydroethanolic mixtures as a solvent, as a greener approach to the conventional methods that usually make use of acetone. The optimal conditions were settled on as ethanol concentration at 57% (*v/v*), the temperature at 75 °C, and time at 5 min. When compared to the yield of extraction obtained under optimized conditions for conventional solvent extraction with hydroacetonic solvent, MAE extraction allowed the recovery of similar amounts of phlorotannins (2.94 ± 0.28 mg PGE/g DW$_{algae}$, versus 3.16 ± 0.06 mg PGE/g DW$_{algae}$, respectively). Likewise, the UHPLC-MS analysis revealed that both extracts presented a very similar phenolic profile, allowing the identification of 10 possible phlorotannins and seven other phlorotannin-derivatives. The two extracts were evaluated for their antioxidant properties through ABTS$^{\bullet+}$ and O$_2^{\bullet-}$ scavenging assays and for their ability to inhibit the enzymatic activity of xanthine oxidase and α-glucosidase. In general, the conventional extract revealed better results than MAE extract, most likely due to its higher phlorotannin richness, although they both exhibited exceptional inhibitory activity against α-glucosidase, showing better results than the commercial antidiabetic pharmaceutical drug. In a wider perspective, the investigation of the applicability of seaweeds such as *F. vesiculosus* could lead to the development of nutraceuticals and pharmacological applications to treat a wide spectrum of disorders and/or diseases.

Author Contributions: S.J.A. and C.M. performed the experimental work, the analysis, and data interpretation; S.J.A. wrote the original draft; A.M.S.S. and M.D.C. reviewed and edited the manuscript; R.F., supervising, review and editing of the revised manuscript; S.M.C. project coordination, supervising, review, and editing of the revised manuscript. All authors have read and agreed to the published version of the manuscript.

Funding: This research was funded by project PTDC/BAA-AGR/31015/2017, "Algaphlor—Brown algae phlorotannins: From bioavailability to the development of new functional foods", co-financed by the Operational Programme for Competitiveness and Internationalization (POCI), within the European Regional Development Fund (FEDER), and the Science and Technology Foundation (FCT) through national funds. Science and Technology Foundation/Ministry of Education and Science (FCT/MEC) funded the Associated Laboratory for Green Chemistry (LAQV) of the Network of Chemistry and Technology (REQUIMTE) (UIDB/50006/2020) and the PhD grant of Marcelo D. Catarino (PD/BD/114577/2016) through national funds and, where applicable, co-financed by FEDER, within the Portugal 2020. Project AgroForWealth (CENTRO-01-0145-FEDER-000001), funded by Centro2020, through FEDER and PT2020 financed the research contract of Susana M. Cardoso.

Conflicts of Interest: The authors declare no conflict of interest.

References

1. Shibata, T.; Ishimaru, K.; Kawaguchi, S.; Yoshikawa, H.; Hama, Y. Antioxidant activities of phlorotannins isolated from Japanese Laminariaceae. *J. Appl. Phycol.* **2008**, *20*, 705–711. [CrossRef]

2. Kim, A.-R.; Shin, T.-S.; Lee, M.-S.; Park, J.-Y.; Park, K.-E.; Yoon, N.-Y.; Kim, J.-S.; Choi, J.-S.; Jang, B.-C.; Byun, D.-S.; et al. Isolation and Identification of Phlorotannins from *Ecklonia stolonifera* with Antioxidant and Anti-inflammatory Properties. *J. Agric. Food Chem.* **2009**, *57*, 3483–3489. [CrossRef] [PubMed]

3. Kang, M.-C.; Cha, S.H.; Wijesinghe, W.A.J.P.; Kang, S.-M.; Lee, S.-H.; Kim, E.-A.; Song, C.B.; Jeon, Y.-J. Protective effect of marine algae phlorotannins against AAPH-induced oxidative stress in zebrafish embryo. *Food Chem.* **2013**, *138*, 950–955. [CrossRef] [PubMed]

4. Kim, A.-R.; Lee, M.-S.; Shin, T.-S.; Hua, H.; Jang, B.-C.; Choi, J.-S.; Byun, D.-S.; Utsuki, T.; Ingram, D.; Kim, H.-R. Phlorofucofuroeckol A inhibits the LPS-stimulated iNOS and COX-2 expressions in macrophages via inhibition of NF-κB, Akt, and p38 MAPK. *Toxicol. Vitr.* **2011**, *25*, 1789–1795. [CrossRef]

5. Yayeh, T.; Im, E.J.; Kwon, T.-H.; Roh, S.-S.; Kim, S.; Kim, J.H.; Hong, S.-B.; Cho, J.Y.; Park, N.-H.; Rhee, M.H. Hemeoxygenase 1 partly mediates the anti-inflammatory effect of dieckol in lipopolysaccharide stimulated murine macrophages. *Int. Immunopharmacol.* **2014**, *22*, 51–58. [CrossRef]

6. Eom, S.-H.; Kim, D.-H.; Lee, S.-H.; Yoon, N.-Y.; Kim, J.H.; Kim, T.H.; Chung, Y.-H.; Kim, S.-B.; Kim, Y.-M.; Kim, H.-W.; et al. In Vitro Antibacterial Activity and Synergistic Antibiotic Effects of Phlorotannins Isolated from *Eisenia bicyclis* Against Methicillin-Resistant *Staphylococcus aureus*. *Phytother. Res.* **2013**, *27*, 1260–1264. [CrossRef]

7. Parys, S.; Kehraus, S.; Krick, A.; Glombitza, K.-W.; Carmeli, S.; Klimo, K.; Gerhäuser, C.; König, G.M. In vitro chemopreventive potential of fucophlorethols from the brown alga *Fucus vesiculosus* L. by anti-oxidant activity and inhibition of selected cytochrome P450 enzymes. *Phytochemistry* **2010**, *71*, 221–229. [CrossRef]

8. Catarino, M.; Silva, A.; Mateus, N.; Cardoso, S. Optimization of Phlorotannins Extraction from *Fucus vesiculosus* and Evaluation of Their Potential to Prevent Metabolic Disorders. *Mar. Drugs* **2019**, *17*, 162. [CrossRef]

9. Lopes, G.; Sousa, C.; Silva, L.R.; Pinto, E.; Andrade, P.B.; Bernardo, J.; Mouga, T.; Valentão, P. Can Phlorotannins Purified Extracts Constitute a Novel Pharmacological Alternative for Microbial Infections with Associated Inflammatory Conditions? *PLoS ONE* **2012**, *7*, e31145. [CrossRef]

10. Isaza Martínez, J.H.; Torres Castañeda, H.G. Preparation and Chromatographic Analysis of Phlorotannins. *J. Chromatogr. Sci.* **2013**, *51*, 825–838. [CrossRef]

11. Pal Singh, I.; Bharate, S.B. Phloroglucinol compounds of natural origin. *Nat. Prod. Rep.* **2006**, *23*, 558. [CrossRef] [PubMed]

12. Magnusson, M.; Yuen, A.K.L.; Zhang, R.; Wright, J.T.; Taylor, R.B.; Maschmeyer, T.; de Nys, R. A comparative assessment of microwave assisted (MAE) and conventional solid-liquid (SLE) techniques for the extraction of phloroglucinol from brown seaweed. *Algal Res.* **2017**, *23*, 28–36. [CrossRef]

13. Catarino, M.D.; Silva, A.M.S.; Cruz, M.T.; Cardoso, S.M. Antioxidant and anti-inflammatory activities of *Geranium robertianum* L. decoctions. *Food Funct.* **2017**, *8*, 3355–3365. [CrossRef] [PubMed]

14. Koivikko, R.; Loponen, J.; Pihlaja, K.; Jormalainen, V. High-performance liquid chromatographic analysis of phlorotannins from the brown alga *Fucus Vesiculosus*. *Phytochem. Anal.* **2007**, *18*, 326–332. [CrossRef] [PubMed]

15. O'Sullivan, A.M.; O'Callaghan, Y.C.; O'Grady, M.N.; Queguineur, B.; Hanniffy, D.; Troy, D.J.; Kerry, J.P.; O'Brien, N.M. In vitro and cellular antioxidant activities of seaweed extracts prepared from five brown seaweeds harvested in spring from the west coast of Ireland. *Food Chem.* **2011**, *126*, 1064–1070. [CrossRef]

16. Bahar, B.; O'Doherty, J.V.; Smyth, T.J.; Sweeney, T. A comparison of the effects of an *Ascophyllum nodosum* ethanol extract and its molecular weight fractions on the inflammatory immune gene expression in-vitro and ex-vivo. *Innov. Food Sci. Emerg. Technol.* **2016**, *37*, 276–285. [CrossRef]

17. Arnold, T.M.; Targett, N.M. To grow and defend: Lack of tradeoffs for brown algal phlorotannins. *Oikos* **2003**, *100*, 406–408. [CrossRef]

18. Schoenwaelder, M.E.A. The occurrence and cellular significance of physodes in brown algae. *Phycologia* **2002**, *41*, 125–139. [CrossRef]

19. Subra, P.; Boissinot, P. Supercritical fluid extraction from a brown alga by stagewise pressure increase. *J. Chromatogr. A* **1991**, *543*, 413–424. [CrossRef]

20. Saravana, P.S.; Getachew, A.T.; Cho, Y.-J.; Choi, J.H.; Park, Y.B.; Woo, H.C.; Chun, B.S. Influence of co-solvents on fucoxanthin and phlorotannin recovery from brown seaweed using supercritical CO_2. *J. Supercrit. Fluids* **2017**, *120*, 295–303. [CrossRef]

21. Boisvert, C.; Beaulieu, L.; Bonnet, C.; Pelletier, É. Assessment of the Antioxidant and Antibacterial Activities of Three Species of Edible Seaweeds. *J. Food Biochem.* **2015**, *39*, 377–387. [CrossRef]

22. He, Z.; Chen, Y.; Chen, Y.; Liu, H.; Yuan, G.; Fan, Y.; Chen, K. Optimization of the microwave-assisted extraction of phlorotannins from *Saccharina japonica* Aresch and evaluation of the inhibitory effects of phlorotannin-containing extracts on HepG2 cancer cells. *Chin. J. Ocean. Limnol.* **2013**, *31*, 1045–1054. [CrossRef]

23. Dang, T.T.; Van Vuong, Q.; Schreider, M.J.; Bowyer, M.C.; Van Altena, I.A.; Scarlett, C.J. Optimisation of ultrasound-assisted extraction conditions for phenolic content and antioxidant activities of the alga *Hormosira banksii* using response surface methodology. *J. Appl. Phycol.* **2017**, *29*, 3161–3173. [CrossRef]

24. Kadam, S.U.; Tiwari, B.K.; Smyth, T.J.; O'Donnell, C.P. Optimization of ultrasound assisted extraction of bioactive components from brown seaweed *Ascophyllum nodosum* using response surface methodology. *Ultrason. Sonochemistry* **2015**, *23*, 308–316. [CrossRef]

25. Song, J.; Li, D.; Liu, C.; Zhang, Y. Optimized microwave-assisted extraction of total phenolics (TP) from *Ipomoea batatas* leaves and its antioxidant activity. *Innov. Food Sci. Emerg. Technol.* **2011**, *12*, 282–287. [CrossRef]

26. Cheok, C.Y.; Chin, N.L.; Yusof, Y.A.; Talib, R.A.; Law, C.L. Optimization of total phenolic content extracted from *Garcinia mangostana* Linn. hull using response surface methodology versus artificial neural network. *Ind. Crop. Prod.* **2012**, *40*, 247–253. [CrossRef]

27. Dahmoune, F.; Spigno, G.; Moussi, K.; Remini, H.; Cherbal, A.; Madani, K. *Pistacia lentiscus* leaves as a source of phenolic compounds: Microwave-assisted extraction optimized and compared with ultrasound-assisted and conventional solvent extraction. *Ind. Crop. Prod.* **2014**, *61*, 31–40. [CrossRef]

28. Alothman, M.; Bhat, R.; Karim, A.A. Antioxidant capacity and phenolic content of selected tropical fruits from Malaysia, extracted with different solvents. *Food Chem.* **2009**, *115*, 785–788. [CrossRef]

29. Heffernan, N.; Brunton, N.; FitzGerald, R.; Smyth, T. Profiling of the Molecular Weight and Structural Isomer Abundance of Macroalgae-Derived Phlorotannins. *Mar. Drugs* **2015**, *13*, 509–528. [CrossRef]

30. Zaragozá, M.C.; López, D.; Sáiz, M.P.; Poquet, M.; Pérez, J.; Puig-Parellada, P.; Màrmol, F.; Simonetti, P.; Gardana, C.; Lerat, Y.; et al. Toxicity and Antioxidant Activity in Vitro and in Vivo of Two *Fucus vesiculosus* Extracts. *J. Agric. Food Chem.* **2008**, *56*, 7773–7780. [CrossRef]

31. Catarino, M.D.; Silva, A.; Cruz, M.T.; Mateus, N.; Silva, A.M.S.; Cardoso, S.M. Phlorotannins from *Fucus vesiculosus*: Modulation of Inflammatory Response by Blocking NF-κB Signaling Pathway. *Int. J. Mol. Sci.* **2020**, *21*, 6897. [CrossRef] [PubMed]

32. Wang, T.; Jónsdóttir, R.; Ólafsdóttir, G. Total phenolic compounds, radical scavenging and metal chelation of extracts from Icelandic seaweeds. *Food Chem.* **2009**, *116*, 240–248. [CrossRef]

33. Barbosa, M.; Lopes, G.; Ferreres, F.; Andrade, P.B.; Pereira, D.M.; Gil-Izquierdo, Á.; Valentão, P. Phlorotannin extracts from Fucales: Marine polyphenols as bioregulators engaged in inflammation-related mediators and enzymes. *Algal Res.* **2017**, *28*, 1–8. [CrossRef]

34. O'Sullivan, A.M.; O'Callaghan, Y.C.; O'Grady, M.N.; Hayes, M.; Kerry, J.P.; O'Brien, N.M. The effect of solvents on the antioxidant activity in Caco-2 cells of Irish brown seaweed extracts prepared using accelerated solvent extraction (ASE®). *J. Funct. Foods* **2013**, *5*, 940–948. [CrossRef]

35. Paradis, M.-E.; Couture, P.; Lamarche, B. A randomised crossover placebo-controlled trial investigating the effect of brown seaweed (*Ascophyllum nodosum* and *Fucus vesiculosus*) on postchallenge plasma glucose and insulin levels in men and women. *Appl. Physiol. Nutr. Metab.* **2011**, *36*, 913–919. [CrossRef] [PubMed]

36. Lordan, S.; Smyth, T.J.; Soler-Vila, A.; Stanton, C.; Ross, R.P. The α-amylase and α-glucosidase inhibitory effects of Irish seaweed extracts. *Food Chem.* **2013**, *141*, 2170–2176. [CrossRef] [PubMed]

37. Liu, B.; Kongstad, K.T.; Wiese, S.; Jäger, A.K.; Staerk, D. Edible seaweed as future functional food: Identification of α-glucosidase inhibitors by combined use of high-resolution α-glucosidase inhibition profiling and HPLC–HRMS–SPE–NMR. *Food Chem.* **2016**, *203*, 16–22. [CrossRef] [PubMed]

38. Yen, G.C.; Duh, P.D. Scavenging Effect of Methanolic Extracts of Peanut Hulls on Free-Radical and Active-Oxygen Species. *J. Agric. Food Chem.* **1994**, *42*, 629–632. [CrossRef]

39. Neto, R.; Marçal, C.; Queirós, A.; Abreu, H.; Silva, A.; Cardoso, S. Screening of *Ulva rigida*, *Gracilaria* sp., *Fucus vesiculosus* and *Saccharina latissima* as Functional Ingredients. *Int. J. Mol. Sci.* **2018**, *19*, 2987. [CrossRef]

40. Pereira, O.; Catarino, M.; Afonso, A.; Silva, A.; Cardoso, S. *Salvia elegans*, *Salvia greggii* and *Salvia officinalis* Decoctions: Antioxidant Activities and Inhibition of Carbohydrate and Lipid Metabolic Enzymes. *Molecules* **2018**, *23*, 3169. [CrossRef]

Publisher's Note: MDPI stays neutral with regard to jurisdictional claims in published maps and institutional affiliations.

 *marine drugs*

Article

Optimisation of Ultrasound Frequency, Extraction Time and Solvent for the Recovery of Polyphenols, Phlorotannins and Associated Antioxidant Activity from Brown Seaweeds

Viruja Ummat [1,2], Brijesh K Tiwari [2], Amit K Jaiswal [3], Kevin Condon [3], Marco Garcia-Vaquero [4], John O'Doherty [4], Colm O'Donnell [1] and Gaurav Rajauria [4,*]

[1] School of Biosystems and Food Engineering, University College Dublin, Belfield, D04 V1W8 Dublin, Ireland; viruja.ummat@ucdconnect.ie (V.U.); colm.odonnell@ucd.ie (C.O.)
[2] Teagasc Food Research Centre, Ashtown, D15 DY05 Dublin, Ireland; brijesh.tiwari@teagasc.ie
[3] School of Food Science and Environmental Health, College of Sciences and Health, Technological University Dublin—City Campus, Grangegorman, D07 EWV4 Dublin, Ireland; amit.jaiswal@tudublin.ie (A.K.J.); C15467412@mytudublin.ie (K.C.)
[4] School of Agriculture and Food Science, University College Dublin, Belfield, D04 V1W8 Dublin, Ireland; marco.garciavaquero@ucd.ie (M.G.-V.); john.vodoherty@ucd.ie (J.O.)
*   Correspondence: gaurav.rajauria@ucd.ie; Tel.: +353-1-601-2167

Received: 14 April 2020; Accepted: 6 May 2020; Published: 11 May 2020

Abstract: This study investigates ultrasound assisted extraction (UAE) process parameters (time, frequency and solvent) to obtain high yields of phlorotannins, flavonoids, total phenolics and associated antioxidant activities from 11 brown seaweed species. Optimised UAE conditions (35 kHz, 30 min and 50% ethanol) significantly improved the extraction yield from 1.5-fold to 2.2-fold in all seaweeds investigated compared to solvent extraction. Using ultrasound, the highest recovery of total phenolics (TPC: 572.3 ± 3.2 mg gallic acid equivalent/g), total phlorotannins (TPhC: 476.3 ± 2.2 mg phloroglucinol equivalent/g) and total flavonoids (TFC: 281.0 ± 1.7 mg quercetin equivalent/g) was obtained from *Fucus vesiculosus* seaweed. While the lowest recovery of TPC (72.6 ± 2.9 mg GAE/g), TPhC (50.3 ± 2.0 mg PGE/g) and TFC (15.2 ± 3.3 mg QE/g) was obtained from *Laminaria digitata* seaweed. However, extracts from *Fucus serratus* obtained by UAE exhibited the strongest 1,1-diphenyl-2-picryl-hydrazyl (DPPH) scavenging activity (29.1 ± 0.25 mg trolox equivalent/g) and ferric reducing antioxidant power (FRAP) value (63.9 ± 0.74 mg trolox equivalent/g). UAE under optimised conditions was an effective, low-cost and eco-friendly technique to recover biologically active polyphenols from 11 brown seaweed species.

Keywords: ultrasound assisted extraction; conventional extraction; polyphenols; phlorotannin; macroalgae; antioxidant capacity

1. Introduction

The consumption of seaweeds is a long tradition in many Asian countries and recently has also increased in Europe and North America [1]. Many seaweed species contain significant quantities of compounds such as polyphenols, polysaccharides, carotenoids, fibres, minerals, trace elements, proteins and amino acids [2–4]. Moreover, brown seaweeds are good sources of polyphenolic compounds such as phlorotannins, flavanols and catechins [5]. Phlorotannins are a specific group of polyphenols produced by brown seaweeds that have gained recognition for their broad range of potential biological properties which are beneficial to humans [5–7]. The potential beneficial biological properties of polyphenols include antioxidant, antimicrobial, antiviral, anticancer, anti-inflammatory and antidiabetic activities [8]. These phenolic compounds are not only being explored for their biological activities but also for their potential to prevent several chronic diseases such as cancer, cardiovascular diseases, obesity and diabetes [9]. Due to their nutritional and health benefits, algal polyphenols are increasingly being investigated for their possible use in nutraceuticals, functional foods, cosmetic, and pharmaceutical applications [10].

The increasing interest in utilising brown seaweeds as a sustainable biosource material for bioactive recovery has fuelled the development of new extraction/pre-treatment technologies. *Fucus vesiculosus* or bladderwrack is rich in sulphated polysaccharides (such as fucoidan) as well as polyphenols (such as phlorotannins). The species is reported to have the highest phlorotannin (6% DW) and fucoidan (up to 20% DW) contents among *Fucus* species [2]. However, it has been mainly investigated to produce fucoidan, while its high phenolic content including phlorotannins has not been exploited to date. The recovery of polysaccharide is mainly carried out using acidic water as a solvent, while its phenolic content is either unextracted and stays in the residue or removed during purification, if co-extracted. Extraction of polyphenolic compounds is a huge challenge as they are embedded deeply within the seaweed matrix. Traditional polyphenol extraction from seaweeds has relied on methods that require high energy consumption, long extraction periods, low yields and the use of potentially toxic chemical agents and solvents [11]. A range of solvents both chlorinated (chloroform, chlorobenzene, carbon tetrachloride, tetrachloroethylene) and non-chlorinated (methanol, ethanol, acetone) are used for extraction of these compounds. However, toxicity, low yield of extracts and prevalence of residues in the target compound have always been a concern. Extraction of these compounds at a commercial scale requires high extraction yields as well as intact biological activities, which are difficult to achieve using conventional extraction methods [4]. Therefore, to address these shortcomings, and to facilitate the transition towards more environmentally sustainable extraction technologies, there is a need to develop new, safe, effective and affordable extraction technologies, which give maximum product yield, minimum presence of residues and enable attainment of clean label status. Recently, several greener and cleaner extraction technologies were investigated which have greater extraction efficiency and lower carbon footprints [12].

Novel extraction technologies also referred to as cold extraction techniques such as ultrasound, due to the comparatively low temperatures employed during the process, have minimal impact on the stability of the target compounds, have high potential to reduce or eliminate the use of toxic chemicals, increase process efficiency and enhance yield and quality of the target products [13,14]. Ultrasound technology can be used either as a pre-treatment or in combination with other safe organic solvents to disrupt cell membranes to enable better extractability [13]. Thus, the motive to develop a sustainable extraction process has led to more research on ultrasound assisted extraction (UAE), owing to decreased levels of solvent consumption, shorter duration of extraction and reduced operational costs [15]. Ultrasound involves various physical and chemical phenomena including compression rarefaction, vibration, pressure, shear forces, microjets, agitation, cavitation and radical formation. However, the main driving force for extraction is acoustic cavitation, which involves creation, expansion and implosive collapse of micro bubbles, formed due to a series of compression and rarefactions in molecules, generated by ultrasound waves [13]. The ability of ultrasound to cause cavitation, for extraction and processing applications, depends on UAE process parameters such as frequency and sonication intensity within the range of 10–1000 W/cm^2. Other process parameters which influence extraction include time, temperature, pressure applied during the process and properties of the extraction solvent (e.g., viscosity and surface tension) [13]. Several methods for extraction of phenolic content from seaweed have been investigated, however, limited data are available on the effect of UAE conditions to extract phenolic compounds from a range of seaweeds and the resultant biological properties of the macroalgal extracts. Therefore, this study aims to (1) optimise UAE process parameters (time, frequency, solvent) to obtain high extraction yields and recovery of phlorotannins and total phenols from *F. vesiculosus* (2) assess the damage caused by ultrasound on algal cell surfaces using scanning electron microscopic (SEM) analysis and (3) investigate the extraction efficiency of optimised UAE conditions and conventional solvent extraction techniques to obtain phenolic constituents (phlorotannins, flavonoids and total phenolic compounds) from 11 brown macroalgae species widely available in Ireland.

2. Results and Discussion

2.1. Effect of Ultrasound on Extraction Yield, Total Polyphenols and Phlorotannin Content

F. vesiculosus brown seaweed was used as the raw material for optimisation of UAE conditions (time, frequency and solvent) to obtain high recovery of total polyphenol and phlorotannin content. Figure 1 shows the effects of ethanol concentration (30%, 50% and 70% v/v), ultrasound frequency (0 kHz or control, 35 kHz and 130 kHz) and UAE treatment time (10 and 30 min) on extraction yield (%). There was a statistically significant interaction observed between ultrasonic frequency and solvent type ($p < 0.001$). No statistical differences were observed within each extraction treatment by applying these conditions for 10 or 30 min. Further statistical analyses were performed to elucidate the effects of different solvents at each US frequency on the yield of extracts as shown in Figure 1. Irrespective of extraction solvents, application of ultrasound significantly improved the extraction yield from 78.7% to 201.8% with respect to the controls. Considerable variations in extraction yield were observed between controls and treatments. Among the controls, the highest extraction yield (16%) was recorded for 30% ethanol, while the lowest yield (8%) was recorded from 70% ethanol. Interestingly, the yield started to decrease as the concentration of ethanol increased which is in agreement with previously reported findings [16,17]. Among the UAE treatments investigated, both the highest (33.8%) and the lowest (20.5%) extraction yields were observed at 130 kHz US frequency and 30 min extraction time but at different ethanol concentrations (30% and 70%, respectively) (Figure 1). Therefore, ultrasonic frequency of 35 kHz which is less energy intensive and provides similar extraction yields for the same treatment time, was considered as the best frequency to recover high yields of extract from *F. vesiculosus*.

Similar findings were reported in other studies using other innovative extraction technologies. He et al. [18] observed that phlorotannin yields from *Saccharina japonica* increased with ethanol concentration over the range of 40–50% for microwave-assisted extraction. The same authors reported a reduction in the yields of compounds extracted when using ethanol concentrations above 50% due to an increased extraction of other less polar components. Another study on optimisation of the extraction variables on the yields of crude extracts from *Ascophyllum nodosum*, also observed that ethanol concentration had a significant effect on the yield of extract obtained, and that lower ethanol concentrations enhanced the extraction yield, suggesting that most of the extractable compounds were high in polarity [19].

Table 1 summarises the recovery of total polyphenols and phlorotannins extracted as well as retained (unextracted) in the residue of *F. vesiculosus* seaweed after ultrasound treatment. There was a statistically significant ($p < 0.001$) influence of the interaction solvent × ultrasonic frequency × extraction time on the recovery of bioactive compounds. The influence of the extraction time (10 and 30 min) on the recovery of TPC and TPhC for each extraction condition is shown in Table 1. Extraction using 50% ethanol yielded more polyphenols and phlorotannins in control samples while 30% ethanol generated more in treated samples, irrespective of US treatments frequencies. As shown in Table 1, values of TPC and TPhC obtained in treated samples were in the range of 422.7–579.7 mg GAE/g and 327.2–471.5 mg PGE/g, respectively, of dried extract, while the control samples exhibited values in the range of 306.8–358.5 mg GAE/g and 222–286.2 mg PGE/g, respectively. Among the treatments, samples treated with 35 kHz US frequency in 30% ethanol for 10 min yielded the lowest amount of TPC (422.7 ± 4.4 mg/g) and TPhC (327.2 ± 7.2 mg/g). Samples treated with 130 kHz US frequency in 30% ethanol for 30 min yielded the highest amount of TPC (579.7 ± 9.2 mg/g) and TPhC (471.5 ± 7.5 mg/g), however these values were statistically similar ($p > 0.05$) with the values obtained from the samples treated with 35 kHz US frequency in 50% ethanol for 30 min (TPC: 571.1 ± 10.0 mg/g and TPhC: 462.6 ± 2.1 mg/g) (Table 1). As 35 kHz US frequency utilises less energy compared to 130 kHz in the same extraction time (30 min) and yields were statistically similar for TPC and TPhC values, 35 kHz US frequency, 50% ethanol and 30 min were considered the optimum conditions for recovery. Interestingly, TPhC contributed 70.6–87.9% to both the lowest and highest TPC amounts. Moreover, control samples exhibited a strong solvent and time effect. All the control samples showed lower TPC values at 10 min treatment time compared to 30 min treatment time. Control samples extracted with 50% ethanol contained the highest TPC at 30 min, while the lowest TPC content was observed for 70% ethanol at 10 min. Compared to the control, ultrasound treatment significantly improved the extraction efficiency of TPC and TPhC, irrespective of treatment time and solvent used. The extraction efficiency calculated by using Equation (1) (Equation (1)) revealed that the highest extraction efficiency (70.3%) of TPC was achieved at 130 kHz for 30 min in 30% ethanol, while the lowest efficiency (32.2%) was observed at 130 kHz for 30 min in 50% ethanol (Table 1). The highest extraction efficiency (89.6%) of TPhC was achieved at 130 kHz in 30% ethanol for 30 min, while the lowest efficiency (28.1%) was observed at 130 kHz for 30 min in 50% ethanol. Though the extraction efficiencies for TPC and TPhC were higher at 130 kHz for 30 min in 30% ethanol compared to 35 kHz for 30 min in 50% ethanol (selected optimum conditions), extraction at 35 kHz is less energy intensive and yielded statistically similar ($p > 0.05$) TPC and TPhC to 130 kHz (Table 1), and was thus considered the best condition for extraction. Similar results were obtained in a study by Kadam et al. [20], in which the effects of ultrasound amplitude (22.8–114 μm), extraction time (5–25 min) and acid concentration (0–0.06 M HCl) on total phenolics, fucose and uronic acids from *A. nodosum* were investigated. The authors reported that the highest recovery of phenolics and fucose were observed at 114 μm ultrasound frequency, 0.03 M HCl solvent concentration and 25 min extraction time.

$$\text{Extraction efficiency } (\%) = \frac{(V_t - V_c)}{V_c} \times 100 \tag{1}$$

where V_t = values of TPC or TPhC obtained after US treatment and V_c = values of corresponding controls of TPC or TPhC.

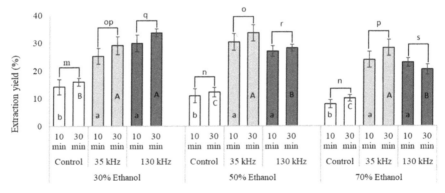

Figure 1. Effects of ultrasound assisted extraction (UAE) conditions (solvent concentration (30%, 50% and 70% ethanol), ultrasonic frequency (control, 35 kHz and 130 kHz) and UAE treatment time (10 and 30 min)) on the extract yield obtained from *F. vesiculosus*. Different letters indicate statistical differences ($p < 0.05$) on the yields of seaweed extract obtained using different solvents at each US frequency: control (m-n), 35 kHz (o-p) and 130 kHz (q-s). [abc] columns with similar letters are not significantly different ($p < 0.05$) treated for 10 min; ABC columns with similar letters are not significantly different ($p < 0.05$) treated for 30 min. The extraction yield is calculated by using following formula: Extraction yield (%) = (weight of dry extract / weight of dry sample) × 100.

Table 1. Effect of ultrasound assisted extraction process parameters (time, ultrasound frequencies and solvents) on total phenolic content (TPC) and total phlorotannin content (TPhC) of *F. vesiculosus* seaweed.

Extraction Solvent	US Frequencies	Extraction Time (min)	TPC (Extract) (mg GAE/g)	TPC (Residue) (mg GAE/g)	TPC (Total) (mg GAE/g)	TPhC (Extract) (mg PGE/g)	TPhC (Residue) (mg PGE/g)	TPhC (Total) (mg PGE/g)
30% ethanol	Control *	10	314.5 ± 5.9 c	262.9 ± 2.8 a	577.5 ± 8.7 c	222.0 ± 4.8 c	179.9 ± 2.8 a	401.9 ± 7.6 c
		30	340.3 ± 9.4 C	247.8 ± 1.4 A	588.1 ± 4.9 C	248.7 ± 5.9 C	178.9 ± 1.2 A	427.7 ± 7.1 C
	35 kHz	10	422.7 ± 4.4 b	218.1 ± 2.8 b	640.9 ± 7.2 b	327.2 ± 7.2 b	144.7 ± 2.3 b	471.9 ± 9.5 a
		30	463.7 ± 5.3 B	201.2 ± 3.2 B	664.9 ± 8.5 B	392.3 ± 5.5 B	122.1 ± 2.6 B	514.4 ± 8.2 B
	130 kHz	10	453.9 ± 6.7 a	190.9 ± 2.9 c	644.9 ± 9.7 a	347.8 ± 4.4 a	116.1 ± 2.4 c	463.9 ± 6.8 b
		30	579.7 ± 9.2 A	125.2 ± 0.4 C	704.9 ± 9.6 A	471.5 ± 7.5 A	97.4 ± 2.9 C	568.9 ± 9.9 A
50% ethanol	Control *	10	338.0 ± 4.9 b	235.6 ± 1.8 a	573.6 ± 6.8 b	262.2 ± 4.0 c	144.1 ± 1.5 a	406.2 ± 5.6 c
		30	358.5 ± 5.3 C	232.7 ± 3.9 A	591.3 ± 9.2 C	286.2 ± 4.4 C	125.4 ± 3.2 A	411.6 ± 7.6 C
	35 kHz	10	464.7 ± 9.8 a	179.6 ± 3.5 c	644.3 ± 13.4 a	408.7 ± 4.1 a	121.7 ± 2.9 b	530.4 ± 7.0 a
		30	571.1 ± 10.0 A	125.6 ± 1.1 C	696.7 ± 11.7 A	462.6 ± 2.1 A	99.8 ± 2.9 B	562.4 ± 5.1 A
	130 kHz	10	458.6 ± 10.8 a	190.0 ± 1.9 b	648.7 ± 12.7 a	350.1 ± 6.9 b	144.6 ± 1.6 a	494.8 ± 8.5 b
		30	474.1 ± 12.7 B	180.6 ± 1.1 B	654.7 ± 13.9 B	366.5 ± 4.0 B	119.8 ± 1.9 A	486.3 ± 6.0 B
70% ethanol	Control *	10	306.8 ± 7.3 c	272.8 ± 3.2 a	579.6 ± 10.5 c	234.8 ± 5.0 c	147.3 ± 2.5 a	382.1 ± 7.5 b
		30	316.4 ± 6.1 C	271.8 ± 0.9 A	588.2 ± 6.9 C	253.6 ± 5.0 C	153.4 ± 1.7 A	406.9 ± 6.7 C
	35 kHz	10	436.9 ± 7.3 b	195.6 ± 3.7 b	632.5 ± 12.9 a	362.4 ± 5.3 a	119.5 ± 3.0 b	481.9 ± 8.3 a
		30	503.2 ± 9.2 A	158.8 ± 3.3 B	662.0 ± 12.5 A	389.9 ± 3.5 A	106.4 ± 2.9 C	496.3 ± 6.4 A
	130 kHz	10	453.2 ± 6.8 a	154.2 ± 2.2 c	607.4 ± 9.1 b	347.2 ± 2.2 b	122.7 ± 6.7 b	469.9 ± 8.9 a
		30	468.1 ± 10.4 B	149.2 ± 2.6 B	617.3 ± 13.0 B	334.8 ± 2.2 B	121.7 ± 2.1 B	456.4 ± 4.4 B

* Control: no ultrasound. Results are expressed as average ± standard deviation (n = 3); different letters indicate statistical differences ($p < 0.05$) on the recovery of compounds by applying either 10 (a–c) or 30 min (A–C) within the same extraction treatment conditions; TPC (total phenolic content) and TPhC (total phlorotannin content) are expressed as mg gallic acid equivalents (GAE)/g dried weight extract and mg phloroglucinol equivalents (PGE)/g dried weight extract, respectively.

Ultrasound treatment was found to be more effective in extracting phlorotannin as the acoustic cavitation generated by ultrasound enhanced the release of these compounds from the matrix. It was noted that the duration of treatment affected TPhC in ultrasound treated samples; 30 min resulted in extraction of higher TPhC values in all treated and control samples compared to 10 min, except for UAE conditions (130 kHz, 30 min and 70% ethanol) which resulted in lower TPhC values. Ultrasound treatments showed better extraction yields compared to control in all cases. It was also observed that the extraction time influenced the amount of TPC and TPhC recovered.

While testing the residues of the control and ultrasound treated samples, the opposite trend was observed. The TPC values in the dried treated residue samples ranged from 125.2 to 218.1 mg GAE/g while TPhC values ranged from 97.4 to 144.6 mg PGE/g. Likewise, TPC and TPhC values in the dried control residue samples ranged from 232.7 to 272.8 mg GAE /g and 125.4 to 179.9 mg PGE/g, respectively. TPC and TPhC values were higher in all extracts compared to residues, indicating that the solvents enabled extraction of most of the phenolics and phlorotannins from the seaweed samples. Samples that showed the highest (579.7 ± 9.2 mg/g) and the lowest (422.7 ± 4.4 mg/g) TPC in the extracts retained the lowest (125.2 ± 1.1 mg/g) and the highest (218.1 ± 2.8 mg/g) phenolic content in their respective residues. Similarly, samples that showed the highest (471.5 ± 7.5 mg/g) and the lowest (327.2 ± 7.2) TPhC in the extracts retained the lowest (TPhC: 97.4 ± 2.9 mg/g) and the highest (TPhC: 144.7 ± 2.3 mg/g) phlorotannins in their respective residues. While summing the TPC and TPhC values of respective extracts and residues, total polyphenols and total phlorotannins (total in seaweed) varied significantly ($p < 0.05$) in control as well as treated samples. Total polyphenols (extract + residue) ranged from 573.6 ± 6.8 to 591.3 ± 9.2 mg/g in control samples, and ranged from 607.4 ± 9.1 to 704.9 ± 9.6 mg/g of dried extract in treated samples. Similarly, total phlorotannins (extract + residue) in control samples ranged from 382.1 ± 7.5 to 427.7 ± 7.1 mg/g, and ranged from 456.4 ± 4.4 to 568.9 ± 9.9 mg/g of dried extract (Table 1) in treated samples. Variations in total polyphenols and phlorotannins (extract + residue) content may be caused by the non-specific nature of extraction solvents, ultrasound frequencies and spectrophotometric tests for TPC and TPhC. During extraction, other polar compounds (such as sugars, proteins) may be released along with polyphenols [21] which may be detected and quantified along with TPC and TPhC.

2.2. Scanning Electron Microscopic Analysis

Scanning electron microscopy (SEM) was used to investigate the impact of the extraction treatments on the structure of the treated macroalgal cells. Figure 2 shows the microscopic structure of *F. vesiculosus* biomass prior to extraction (Figure 2a), seaweed residue of control sample (50% ethanol, 30 min, no ultrasound (Figure 2b)) and treated sample after UAE under optimum conditions (35 kHz, 50% ethanol, 30 min (Figure 2c)). It can be observed in the SEM images that the cell surface of the initial seaweed biomass is intact and surrounded by other impurities and residual materials, while the cell surfaces of the control samples (without ultrasound treatment) appear to be smooth with an increased number of pores that allowed diffusion of bioactive compounds to the media. The impact of the optimised UAE conditions (35 kHz, 50% ethanol, 30 min) on the macroalgal biomass are more evident in Figure 2c. The cell surfaces of treated samples exhibit an increased porosity that facilitated the extraction of higher yields of TPC and TPhC compared to control samples (Table 1). Previously, Rodriguez-Jasso et al. [22] and Garcia-Vaquero et al. [23] reported similar findings when evaluating the efficiency of MAE, UAE and UMAE (ultrasound-microwave assisted extraction) to generate extracts from *F. vesiculosus* and *A. nodosum* seaweed.

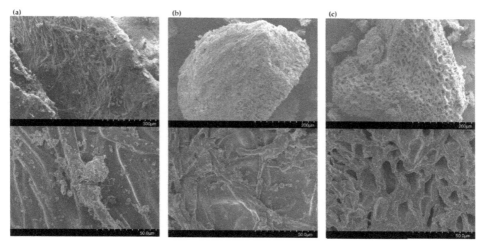

Figure 2. SEM images of *F. vesiculosus* (**a**) untreated samples (intact and dried macroalgae), (**b**) control samples (50% ethanol, 30 min, no ultrasound) and (**c**) UAE treated samples (35 kHz, 50% ethanol, 30 min). Scale bars: 300 μm (magnification 150×), 200 μm (magnification 250×) and 50 μm (magnification: 1000×).

2.3. Comparison of Optimised UAE Conditions and Conventional Solvent Extraction

2.3.1. Extraction Yield and Phenolic Constituents

UAE efficiency for the extraction of phenolic compounds was further evaluated by applying the optimised UAE conditions (35 kHz, 50% ethanol, 30 min) on 11 seaweed species and comparing the yields obtained to conventional solvent extraction. Extracts recovered from all 11 seaweeds using both UAE and solvent extraction techniques were analysed for extraction yield, TPC, TPhC, total flavonoid content (TFC) and antioxidant capacity (Table 2 and Figure 3).

Table 2. Extraction yield (%) obtained from selected seaweed species using UAE and conventional solvent extraction techniques.

Seaweed Species	Extraction Yield (%)	
	UAE	Conventional
Pelvetia caniculata	20.5 ± 0.35 **h**	14.0 ± 0.28 *e*
Fucus vesiculosus	35.1 ± 0.33 **b**	11.2 ± 0.29 *g*
Laminaria saccharina	30.9 ± 0.41 **d**	17.0 ± 0.47 *c*
Laminaria hyperborea	36.9 ± 0.11 **a**	19.3 ± 0.21 *a*
Fucus spiralis	25.2 ± 0.50 **f**	14.7 ± 0.45 *de*
Ascophyllum nodosum	24.4 ± 0.31 **f**	12.7 ± 0.15 *f*
Fucus serratus	20.4 ± 0.19 **h**	10.5 ± 0.12 *g*
Himanthalia elongata	23.4 ± 0.30 **g**	15.0 ± 0.26 *d*
Halidrys siliquosa	29.0 ± 0.25 **e**	14.5 ± 0.31 *de*
Laminaria digitata	29.4 ± 0.16 **e**	18.4 ± 0.21 *b*
Alaria esculenta	33.0 ± 0.06 **c**	17.2 ± 0.09 *c*

Results are expressed as average ± standard deviation (n = 6). Different letters indicate statistical differences (*p* < 0.05) between the yield of extracts obtained from multiple seaweed species by using UAE (lowercase letters) and conventional extraction conditions (italic letters). The extraction yield is calculated by using following formula: extraction yield (%) = (weight of dry extract / weight of dry sample) × 100.

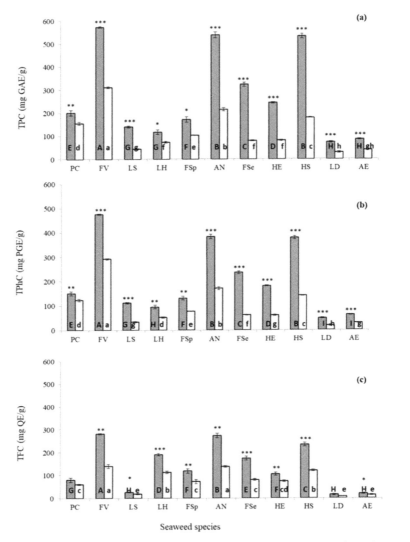

Figure 3. Total polyphenol (**a**), total phlorotannin (**b**) and total flavonoid (**c**) content from 11 seaweed species obtained using UAE (grey bars) and conventional solvent extraction (white bars) technologies. The statistical differences in bioactive compounds extracted using UAE or conventional solvent extraction technologies for each seaweed are represented as * $p < 0.05$, ** $p < 0.01$ and *** $p < 0.001$. Different letters indicate statistical differences ($p < 0.05$) in the yields of bioactive compounds between seaweed obtained by UAE (uppercase letters) or conventional solvent extraction (lowercase letters). TPC (total phenolic content), TPhC (total phlorotannin content) and TFC (total flavonoid content) are expressed as mg gallic acid equivalents (GAE)/g dried weight extract, mg phloroglucinol equivalents (PGE)/g dried weight extract and mg quercetin equivalents (QE)/g dried weight extract, respectively. Abbreviation of seaweed species are as follows: PC (*Pelvetia caniculata*), FV (*Fucus vesiculosus*), LS (*Laminaria saccharina*), LH (*Laminaria hyperborea*), FSp (*Fucus spiralis*), AN (*Ascophyllum nodosum*), FSe (*Fucus serratus*), HE (*Himanthalia elongata*), HS (*Halidrys siliquosa*), LD (*Laminaria digitata*) and AE (*Alaria esculenta*).

UAE extraction yields were statistically ($p < 0.05$) higher than the yields obtained from conventional solvent extraction for all seaweed studied. The yields obtained from conventional extraction were in the range of 10.5%–19.3%, while yields obtained using UAE were in the range of 20.4–36.9% (Table 2). Ultrasound improved the extraction yield 1.5–2.2 fold in all tested seaweeds. Statistical differences were also observed for the yields obtained between different seaweed species even for the same extraction conditions (Table 2). With UAE, the highest extraction yield was obtained from *Laminaria hyperborea* (36.9%), while the lowest yield (20.4%) was obtained from *F. serratus*. Compared to solvent extraction, the highest increase in yield (3.1-fold) using UAE was obtained from *F. vesiculosus*, while the lowest increase in yield (1.5-fold) was obtained from *Pelvetia caniculata*, indicating an extraction yield variation between seaweed species. Extraction yield varies with composition, type and quantity of compounds present in seaweed. Previously, Farvin et al. [24] conducted an extraction study involving 16 seaweeds species using ethanol and water. They observed variation in extraction yields which was attributed to the polarities of compounds present. They also reported that the *F. vesiculosus* extracts obtained using water were more viscous and difficult to extract by passing through a filter, and thus had the lowest yield. Similar findings were reported in a study by Bixler et al. [25] which investigated ultrasound for extraction of phenolic compounds from *Laminaria japonica* using an ionic liquid as solvent. They observed that ultrasound enhanced the extraction yield, by acting as a driving force for dispersing the solvent into the solid samples.

As shown in Figure 3, the levels of TPC, TPhC and TFC extracted using UAE and conventional solvent extraction technologies varied significantly for all 11 seaweeds investigated. UAE enhanced the recovery of bioactive in all 11 seaweeds investigated compared to conventional solvent extraction. The highest recovery of TPC (572.3 ± 3.19 mg GAE/g), TPhC (476.3 ± 2.19 mg PGE/g) and TFC (281.0 ± 1.65 mg QE/g) was recorded from *F. vesiculosus*, while the lowest recovery (TPC: 72.6 ± 2.92 mg GAE/g; TPhC: 50.3 ± 2.01 mg PGE/g; and TFC: 15.2 ± 3.30 mg QE/g) was obtained from *Laminaria digitata* seaweed using optimised UAE conditions. Solvent-led extraction yielded values ranging from 28.7 to 310.1 mg GAE/g for TPC, from 19.0 to 292.0 mg PGE/g for TPhC and from 8.1 to 138.4 mg QE/g for TFC in the 11 seaweeds investigated. The values of TPC (1.3–4.1-fold), TPhC (1.2–3.8-fold) and TFC (1.3–2.1-fold) obtained from UAE treated samples were higher than the values obtained using convention solvent extraction. The highest values of TPC, TPhC and TFC were observed in *F. vesiculosus* while the highest increase of TPC (4.1-fold), TPhC (3.8-fold) and TFC (2.1-fold) was obtained from *Fucus serratus* seaweed. The highest and the lowest values of TPC, TPhC and TFC were recorded in *F. vesiculosus* and *L. digitata* while values of extraction yield recorded the highest and the lowest in *L. hyperborea* and *F. serratus* seaweeds, respectively. Due to the high value of phenolic constituents and extraction yields obtained from *Fucus* species, it can be considered a good source of phenolics compared to other seaweed species investigated. Holdt et al. [2] observed that genus *Fucus* accumulates the highest amount of phlorotannins (up to 12% dry weight). The quantity accumulated depends on the geographical location, season, solar exposure and salinity.

2.3.2. Antioxidant Capacity Determination

The antioxidant capacity of seaweed extracts recovered from UAE and conventional solvent extraction were analysed using 1,1-diphenyl-2-picryl-hydrazyl (DPPH) and ferric reducing antioxidant power (FRAP) assays (Figure 4). The extract from *F. serratus* obtained by UAE had the strongest DPPH free radical scavenging activity (29.1 ± 0.25 mg TE/g) and the highest FRAP value (63.9 ± 0.74 mg TE/g), while the extract obtained from *L. digitata* had the weakest free radical scavenging activity (5.2 ± 0.15 mg TE/g) and the lowest FRAP value (7.8 ± 0.30 mg TE/g) (Figure 4a). However, for conventional solvent extraction, the extract from *H. elongata* had the strongest DPPH free radical scavenging activity (20.7 ± 0.10 mg TE/g) while the extract from *Pelvetia caniculata* exhibited the highest FRAP value (42.0 ± 1.20 mg TE/g). However, similar to the results observed for UAE, the extract from *L. digitata* obtained by conventional solvent extraction showed the weakest DPPH free radical scavenging activity (2.5 ± 0.09 mg TE/g) and the lowest FRAP value (4.4 ± 0.11 mg TE/g) (Figure 4b).

It is likely that that ultrasound facilitated the release of phenolic compounds from the seaweed which exhibited a strong antioxidant capacity. Compared to conventional extraction, UAE increased the DPPH scavenging capacity from 23.6% to 146.4% and reduced the power (FRAP) from 16.6% to 86.7% in the 11 seaweeds investigated. The highest enhancement in DPPH scavenging capacity and FRAP reducing power was observed for *F. serratus* seaweed. However, the lowest increase in DPPH scavenging capacity was recorded for *F. vesiculosus*, while the lowest enhancement in FRAP reducing power was recorded in *Alaria esculenta* seaweed.

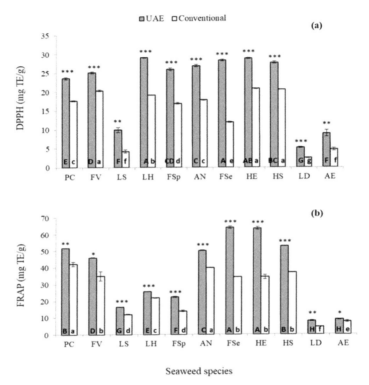

Figure 4. Antioxidant capacity measured as 1,1-diphenyl-2-picryl-hydrazyl (DPPH) activity (**a**) and ferric reducing antioxidant power (FRAP) (**b**) of 11 seaweed extracts obtained from UAE (grey bars) and conventional solvent extraction (white bars) techniques. The statistical differences in antioxidant activity extracted by using UAE or conventional solvent extraction for each seaweed are represented as * $p < 0.05$, ** $p < 0.01$ and *** $p < 0.001$. Different letters indicate statistical differences in the antioxidant activity among seaweed species obtained by UAE (uppercase letters) or conventional solvent extraction (lowercase letters). DPPH and FRAP: expressed as mg trolox equivalent (TE)/g of dry weight extract. Abbreviation of seaweed species are as follow: PC (*Pelvetia caniculata*), FV (*Fucus vesiculosus*), LS (*Laminaria saccharina*), LH (*Laminaria hyperborea*), FSp (*Fucus spiralis*), AN (*Ascophyllum nodosum*), FSe (*Fucus serratus*), HE (*Himanthalia elongata*), HS (*Halidrys siliquosa*), LD (*Laminaria digitata*) and AE (*Alaria esculenta*).

Overall, the extracts recovered from UAE treated seaweeds showed higher antioxidant activity compared to extracts from conventional solvent extraction. Similar results were obtained by Kadam et al. [26], who studied extraction of laminarin from *A. nodosum* and *L. hyperborea*, with conventional solvent extraction and UAE. They reported that the antioxidant activity and total phenolic content were higher in samples treated with ultrasound. The combined ferric reducing/

antioxidant power (FRAP) value of antioxidants in the sample is proportional to the antioxidant potential [27]. A study by Dang et al. [28] reported that when UAE and conventional extraction was performed for extraction of phenolic compounds from *Hormosira banksii* algae, UAE was more effective than conventional solvent extraction to TPC and antioxidant capacity in terms of ABTS, DPPH and FRAP. They reported that ABTS, DPPH and FRAP values using UAE were higher (166.8%, 154.6% and 150.6%, respectively) compared to conventional extraction techniques employed. They also reported that the TPC levels in UAE samples were 142.6% higher than conventionally extracted samples, indicating that ultrasound is effective for extraction of bioactive compounds, as well as giving a good quality extract.

3. Materials and Methods

3.1. Chemicals

Vanillin, ethanol, methanol, ferric chloride, aluminum chloride, sodium nitrite, sodium carbonate and sodium hydroxide chemicals were purchased from Fisher Scientific (Loughborough, UK). Reagents and standards including 6-hydroxy-2,5,7,8-tetramethylchromane-2-carboxylic acid (trolox), 1,1-diphenyl-2-picryl-hydrazyl (DPPH), 2,4,6-tripyridyl-s-triazine (TPTZ), Folin-Ciocalteau reagent, gallic acid, phloroglucinol, quercetin and catechin were purchased from Sigma-Aldrich Chemical Co. (Steinheim, Germany). All other chemicals used were of analytical grade and were purchased from Sigma-Aldrich.

3.2. Seaweed Biomass

The 11 seaweeds used in this study were Fucus serratus, Fucus vesiculosus, Fucus spiralis, Himanthalia elongata, Halidrys siliquosa, Laminaria digitata, Laminaria saccharina, Laminaria hyperborea, Ascophyllum nodosum, Alaria esculenta and Pelvetia caniculata. All seaweeds were harvested (Quality Sea Veg., Co. Donegal, Ireland) and washed to remove any debris and epiphytes attached. Seaweed samples were dried using an air circulating oven following industry practices (50 °C, 9 days), and milled to 1 mm particle size using a hammer mill (Christy and Norris, Chelmsford, UK). The samples were stored at room temperature in dark conditions prior to use.

3.3. Ultrasound assisted Extraction (UAE) Procedures

F. vesiculosus seaweed was selected for optimisation of UAE conditions to recover high yield of phenolic compounds with antioxidant properties. Dried and milled seaweed samples were mixed with aqueous ethanolic solutions (30%, 50% and 70% v/v) at a ratio of 1:10 (w/v) for phlorotannins and phenolics extraction. The UAE treatments were performed using an ultrasonic water bath (Fisher Bioblock Scientific, Pittsburgh, PA, USA) at 35 and 130 kHz for 10 and 30 min. Control samples were also treated following the same procedures while omitting the use of ultrasound. All the extraction procedures were performed in duplicate. After the treatment, control and ultrasound treated samples were centrifuged at 3000× g for 15 min at 20 °C. The supernatants and residues (pellet) were collected separately and ethanol was evaporated under vacuum. The remaining aqueous fraction of supernatants and residues were freeze-dried and stored at −20°C prior to subsequent analyses.

The most effective optimised UAE conditions (35 kHz, 30 min, 50% ethanol) were further tested on the other 10 seaweed species and the efficiency of UAE was compared to conventional solvent extraction.

3.4. Conventional Solvent Extraction

Solvent extraction was carried out on all 11 seaweed species to compare the efficiency of conventional solvent extraction with the optimised UAE conditions determined in this study. Seaweed samples were extracted following the method described by Li et al. [29] with minor modifications. Briefly, samples were mixed with 50% ethanol (1:15, w/v) and extracted in a shaking water bath (20 °C, 200 rpm and 4 h). The supernatant was filtered through Whatman #1 filter paper (Whatman

International Limited, Maidstone, UK). The macroalgal pellets were re-extracted following the same procedure, and both supernatants were pooled together. Ethanol was evaporated under vacuum and the remaining aqueous fraction was freeze-dried and stored at −20 °C for further analyses.

3.5. Scanning Electron Microscopy

Scanning Electron Microscopy (SEM) was used to investigate the effect of ultrasound on *F. vesiculosus* surface characteristics. Dried seaweed samples and pellets from the extraction conditions achieving high extraction yields (35 kHz, 50% ethanol, 30 min) and control (no ultrasound, 50% ethanol, 30 min) were collected and prepared as described by Garcia-Vaquero et al. [23]. The images were recorded using a SEM Regulus 8230 (Hitachi Ltd., Tokyo, Japan).

3.6. Phenolic Composition and Antioxidant Capacity Analysis

All the extracts were analysed for total phenolic content (TPC), total phlorotannin content (TPhC), total flavonoid content (TFC) and antioxidant capacity using 1,1-diphenyl-2-picryl-hydrazyl (DPPH) activity and ferric reducing antioxidant power (FRAP) assays.

3.6.1. Total Phenolic Content (TPC) and Total Phlorotannin Content (TPhC)

The amount of TPC and TPhC in the extracts was determined as outlined by Rajauria et al. [7]. Briefly, 100 μL of sample or standards (gallic acid and phloroglucinol for TPC and TPhC, respectively) was mixed with 2 mL of 2% Na_2CO_3, and left to stand for 2 min before adding 100 μL of Folin-Ciocalteau reagent (1:1, v/v). The solutions were mixed and incubated for 30 mins at room temperature in dark conditions. The absorbance of the reaction was read at 720 nm using a spectrophotometer (UVmini-1240, Shimadzu, Kyoto, Japan). All the measurements were done in triplicate. The TPC was expressed as mg gallic acid equivalent per gram (mg GAE/g) dried extract and the TPhC was expressed as mg phloroglucinol equivalents per gram (mg PGE/g) dried extract.

3.6.2. Total Flavonoid Content

The total flavonoid content (TFC) was determined using the method described by [30] with slight modifications. Briefly, 250 μL of each extract or standard solution was mixed with 1.475 mL distilled water and 75 μL sodium nitrite (5%) solution followed by the addition of 150 μL aluminum chloride hexahydrate (10%) after 6 min, and then mixed. After 5 min, 0.5 mL of sodium hydroxide (1 M) solution was added to the reaction mixture and the absorbance against blank was determined at 510 nm. Quercetin was used as a reference compound and the results were expressed as mg quercetin equivalents per gram (mg QE/g) dried extract.

3.6.3. DPPH Radical-scavenging Assay

The DPPH free radical scavenging activity was conducted as per the method reported by Sridhar and Charles [31]. Briefly, 700 μL of sample or standard was mixed well with 700 μL of 100 μM DPPH methanolic solution in a test tube. The reaction mixture was incubated at room temperature in the dark for 20 min and read against a blank of methanol (without DPPH solution) at 515 nm using a UV-Vis spectrophotometer. Samples were prepared in triplicate. Trolox standard was used to generate a standard curve and results were expressed as mg trolox equivalents (TE)/g dry weight extract. The inhibition percentage of scavenging of DPPH was calculated using Equation (2).

$$\text{DPPH radical scavenging capacity } (\%) = \frac{\left(A_{control} - A_{sample}\right)}{A_{control}} \times 100 \qquad (2)$$

where "$A_{control}$" is the absorbance of the control (DPPH solution without sample/standard), "A_{sample}" is the absorbance of the test sample (DPPH solution plus test sample/standard).

3.6.4. Ferric Reducing Antioxidant Power (FRAP) Assay

The total antioxidant reducing power of seaweed extracts and standard was measured using the ferric reducing antioxidant power (FRAP) assay as reported by Benzie and Strain [32]. Preheated 2.5 mL FRAP reagent (300 mM acetate buffer, pH 3.6; 10 mM 2,4,6-Tri(2-pyridyl)-s-triazine in 40 mM hydrochloric acid and 20 mM Iron(III) chloride hexahydrate in the ratio of 10:1:1, *v/v/v*) at 37 °C was mixed with 83 µL of samples or standard and incubated in dark at room temperature for 10 min. Trolox was used as a standard and the absorbance of the standard or samples was recorded at 593 nm against a reagent blank containing FRAP reagent only. The results were expressed as mg trolox equivalents (TE)/g dry weight extract.

3.7. Statistical Analysis

All the experiments were carried out in triplicate. Results are expressed as mean ± standard deviation. Statistical analysis was performed using SPSS version 24.0 (IBM, Armonk, NY, USA). The effects of ultrasound frequency, extraction time and solvents on the recovery of total polyphenols, total phlorotannins and associated antioxidant activities were analysed using ANOVA and the differences analysed further by Student's *t*-tests and Tukey's HSD post-hoc tests. In all cases, differences were considered statistically significant at $p < 0.05$.

4. Conclusions

UAE was found to be more effective than conventional solvent extraction for extraction of bioactive compounds compared. It was observed that ethanol concentration, ultrasound frequency and duration of extraction influenced extraction yield and the phenolic compounds obtained. The optimised UAE treatment based on extraction yield, total phenol and total phlorotannin content recovered was found to be 35 kHz for 30 min with 50% ethanol. UAE resulted in higher extraction yield of extracts and higher values of TPC, TPhC and TFC and antioxidant capacities for all 11 seaweeds studied compared to conventional solvent extraction. It was also noted that the response to the extraction technique was species specific. The TPC obtained using UAE was highest for *F. vesiculosus* but the largest enhancement of extraction (4.1-fold) was achieved for *F. serratus* seaweed. Likewise, despite the highest extract yield and polyphenolic concentration observed for *F. vesiculosus*, the highest antioxidant capacity was recorded for *F. serratus*, using UAE, which indicates that there is no direct relationship between phenolic content and antioxidant activities. It is recommended that a full profiling of phenolic compounds in the extract should be carried out prior to its utilisation for commercial applications.

Author Contributions: V.U. and K.C. performed the experiments and analysed the data. G.R. designed and supervised the work and wrote and revised the manuscript. M.G.-V. performed the statistical treatment of the data. A.K.J. collaborated in the design of the work and analysed the data. B.K.T., C.O. and J.O. provided funds and revised the manuscript. All authors have read and agreed to the published version of the manuscript.

Funding: This research work was supported by Science Foundation Ireland (SFI). Grant numbers: [16/RC/3889] and [14/IA/2548].

Conflicts of Interest: The authors declare no conflict of interest.

References

1. Mouritsen, O.G.; Dawczynski, C.; Duelund, L.; Jahreis, G.; Vetter, W.; Schröder, M. On the human consumption of the red seaweed dulse (*Palmaria palmata* (L.) Weber & Mohr). *J. Appl. Phycol.* **2013**, *25*, 1777–1791.
2. Holdt, S.L.; Kraan, S. Bioactive compounds in seaweed: Functional food applications and legislation. *J. Appl. Phycol.* **2011**, *23*, 543–597. [CrossRef]
3. Wang, L.; Park, Y.J.; Jeon, Y.J.; Ryu, B. Bioactivities of the edible brown seaweed, *Undaria pinnatifida*: A review. *Aquaculture* **2018**, *495*, 873–880. [CrossRef]
4. Bordoloi, A.; Goosen, N. Green and integrated processing approaches for the recovery of high-value compounds from brown seaweeds. *Adv. Bot. Res.* **2019**. [CrossRef]

5. Gómez-Guzmán, M.; Rodríguez-Nogales, A.; Algieri, F.; Gálvez, J. Potential role of seaweed polyphenols in cardiovascular-associated disorders. *Mar. Drugs* **2018**, *16*, 250. [CrossRef]
6. Namvar, F.; Mohamad, R.; Baharara, J.; Zafar-Balanejad, S.; Fargahi, F.; Rahman, H.S. Antioxidant, antiproliferative, and antiangiogenesis effects of polyphenol-rich seaweed (*Sargassum muticum*). *BioMed. Res. Int.* **2013**, *2013*. [CrossRef]
7. Rajauria, G.; Foley, B.; Abu-Ghannam, N. Identification and characterization of phenolic antioxidant compounds from brown Irish seaweed *Himanthalia elongata* using LC-DAD–ESI-MS/MS. *Innov. Food Sci. Emerg. Technol.* **2016**, *37*, 261–268. [CrossRef]
8. Audibert, L.; Fauchon, M.; Blanc, N.; Hauchard, D.; Ar Gall, E. Phenolic compounds in the brown seaweed *Ascophyllum nodosum*: Distribution and radical-scavenging activities. *Phytochem. Anal.* **2010**, *21*, 399–405. [CrossRef]
9. Déléris, P.; Nazih, H.; Bard, J.M. Seaweeds in human health. In *Seaweed in Health and Disease Prevention*, 1st ed.; Fleurence, J., Levine, I., Eds.; Elsevier: London, UK, 2016; Chapter 10; pp. 319–367.
10. Fernando, I.S.; Kim, M.; Son, K.T.; Jeong, Y.; Jeon, Y.J. Antioxidant activity of marine algal polyphenolic compounds: A mechanistic approach. *J. Med. Food* **2016**, *19*, 615–628. [CrossRef]
11. Soria, A.C.; Villamiel, M. Effect of ultrasound on the technological properties and bioactivity of food: A review. *Trends Food Sci. Technol.* **2010**, *21*, 323–331. [CrossRef]
12. Tierney, M.S.; Smyth, T.J.; Hayes, M.; Soler-Vila, A.; Croft, A.K.; Brunton, N. Influence of pressurised liquid extraction and solid–liquid extraction methods on the phenolic content and antioxidant activities of Irish macroalgae. *Int. J. Food Sci. Technol.* **2013**, *48*, 860–869. [CrossRef]
13. Tiwari, B.K. Ultrasound: A clean, green extraction technology. *TrAC Trends. Anal. Chem.* **2015**, *71*, 100–109. [CrossRef]
14. Kadam, S.U.; Tiwari, B.K.; Álvarez, C.; O'Donnell, C.P. Ultrasound applications for the extraction, identification and delivery of food proteins and bioactive peptides. *Trends. Food Sci. Technol.* **2015**, *46*, 60–67. [CrossRef]
15. Chemat, F.; Rombaut, N.; Sicaire, A.G.; Meullemiestre, A.; Fabiano-Tixier, A.S.; Abert-Vian, M. Ultrasound assisted extraction of food and natural products. Mechanisms, techniques, combinations, protocols and applications. A review. *Ultrason. Sonochem.* **2017**, *34*, 540–560. [CrossRef]
16. Rajauria, G.; Jaiswal, A.K.; Abu-Ghannam, N.; Gupta, S. Antimicrobial, antioxidant and free radical-scavenging capacity of brown seaweed *Himanthalia elongata* from western coast of Ireland. *J. Food. Biochem.* **2013**, *37*, 322–335. [CrossRef]
17. Wang, T.; Jonsdottir, R.; Ólafsdóttir, G. Total phenolic compounds, radical scavenging and metal chelation of extracts from Icelandic seaweeds. *Food Chem.* **2009**, *116*, 240–248. [CrossRef]
18. He, Z.; Chen, Y.; Chen, Y.; Liu, H.; Yuan, G.; Fan, Y.; Chen, K. Optimization of the microwave-assisted extraction of phlorotannins from *Saccharina japonica* Aresch and evaluation of the inhibitory effects of phlorotannin-containing extracts on HepG2 cancer cells. *Chin. J. Oceanol. Limnol.* **2013**, *31*, 1045–1054. [CrossRef]
19. Liu, X.; Luo, G.; Wang, L.; Yuan, W. Optimization of antioxidant extraction from edible brown algae *Ascophyllum nodosum* using response surface methodology. *Food Bioprod. Proc.* **2019**, *114*, 205–215. [CrossRef]
20. Kadam, S.U.; Tiwari, B.K.; Smyth, T.J.; O'Donnell, C.P. Optimization of ultrasound assisted extraction of bioactive components from brown seaweed *Ascophyllum nodosum* using response surface methodology. *Ultrason. Sonochem.* **2015**, *23*, 308–316. [CrossRef]
21. Brglez Mojzer, E.; Knez Hrnčič, M.; Škerget, M.; Knez, Ž.; Bren, U. Polyphenols: Extraction methods, antioxidative action, bioavailability and anticarcinogenic effects. *Molecules* **2016**, *21*, 901. [CrossRef]
22. Rodriguez-Jasso, R.M.; Mussatto, S.I.; Pastrana, L.; Aguilar, C.N.; Teixeira, J.A. Microwave-assisted extraction of sulfated polysaccharides (fucoidan) from brown seaweed. *Carbohydr. Polym.* **2011**, *86*, 1137–1144. [CrossRef]
23. Garcia-Vaquero, M.; Ummat, V.; Tiwari, B.; Rajauria, G. Exploring Ultrasound, Microwave and Ultrasound–Microwave Assisted Extraction Technologies to Increase the Extraction of Bioactive Compounds and Antioxidants from Brown Macroalgae. *Mar. Drugs* **2020**, *18*, 172–186. [CrossRef] [PubMed]
24. Farvin, K.S.; Jacobsen, C. Phenolic compounds and antioxidant activities of selected species of seaweeds from Danish coast. *Food Chem.* **2013**, *138*, 1670–1681. [CrossRef] [PubMed]
25. Bixler, H.J.; Porse, H. A decade of change in the seaweed hydrocolloids industry. *J. Appl.Phycol.* **2011**, *23*, 321–335. [CrossRef]

26. Kadam, S.; O'Donnell, C.; Rai, D.; Hossain, M.; Burgess, C.; Walsh, D.; Tiwari, B. Laminarin from Irish brown seaweeds *Ascophyllum nodosum* and *Laminaria hyperborea*: Ultrasound assisted extraction, characterization and bioactivity. *Mar. Drugs* **2015**, *13*, 4270–4280. [CrossRef]

27. Benzie, I.F.; Szeto, Y. Total antioxidant capacity of teas by the ferric reducing/antioxidant power assay. *J. Agric. Food Chem.* **1999**, *47*, 633–636. [CrossRef]

28. Dang, T.T.; Van Vuong, Q.; Schreider, M.J.; Bowyer, M.C.; Van Altena, I.A.; Scarlett, C.J. Optimisation of ultrasound-assisted extraction conditions for phenolic content and antioxidant activities of the alga *Hormosira banksii* using response surface methodology. *J. Appl. Phycol.* **2017**, *29*, 3161–3173. [CrossRef]

29. Li, Y.; Fu, X.; Duan, D.; Liu, X.; Xu, J.; Gao, X. Extraction and identification of phlorotannins from the brown alga, *Sargassum fusiforme* (Harvey) Setchell. *Mar. Drugs* **2017**, *15*, 49. [CrossRef]

30. Liu, S.C.; Lin, J.T.; Wang, C.K.; Chen, H.-Y.; Yang, D.J. Antioxidant properties of various solvent extracts from lychee (*Litchi chinenesis* Sonn.) flowers. *Food Chem.* **2009**, *114*, 577–581. [CrossRef]

31. Sridhar, K.; Charles, A.L. In vitro antioxidant activity of Kyoho grape extracts in DPPH and ABTS assays: Estimation methods for EC50 using advanced statistical programs. *Food Chem.* **2019**, *275*, 41–49. [CrossRef]

32. Benzie, I.F.; Strain, J.J. The ferric reducing ability of plasma (FRAP) as a measure of "antioxidant power": The FRAP assay. *Anal. Biochem.* **1996**, *239*, 70–76. [CrossRef] [PubMed]

MDPI

St. Alban-Anlage 66

4052 Basel

Switzerland

Tel. +41 61 683 77 34

Fax +41 61 302 89 18

www.mdpi.com

Marine Drugs Editorial Office

E-mail: marinedrugs@mdpi.com

www.mdpi.com/journal/marinedrugs

9 783036 502649